PSYCHIATRY IN AN ANTHROPOLOGICAL
AND BIOMEDICAL CONTEXT

STUDIES IN THE HISTORY
OF MODERN SCIENCE

Editors:

ROBERT S. COHEN, *Boston University*

ERWIN N. HIEBERT, *Harvard University*

EVERETT I. MENDELSOHN, *Harvard University*

VOLUME 15

GERLOF VERWEY

Faculty of Philosophy, University of Nijmegen, The Netherlands

PSYCHIATRY IN AN ANTHROPOLOGICAL AND BIOMEDICAL CONTEXT

Philosophical Presuppositions and Implications of German Psychiatry, 1820-1870

D. REIDEL PUBLISHING COMPANY

A MEMBER OF THE KLUWER ACADEMIC PUBLISHERS GROUP

DORDRECHT / BOSTON / LANCASTER

Library of Congress Cataloging in Publication Data

CIP

Verwey, Gerlof, 1939–
    Psychiatry in an anthropological and biomedical context.

    (Studies in the history of modern science; v. 15) Translation of:
    Psychiatrie tussen antropologie en natuurwetenschap.
    Bibliography: p.
    Includes indexes.
        1. Psychiatry–Germany–Philosophy–History–19th century.
2. Philosophy, German–19th century. 3. Science–Germany–Philosophy–
History–19th century. 4. Anthropology–Germany–Philosophy–History–
19th century. I. Title. II. Series. [DNLM: 1. Psychiatry–
history–Germany. 2. Philosophy, Medical. 3. Anthropology–
history–Germany.   WM 11 GG4 V5p]
    RC450.G3V4713   1984   616.89′00943   84-9951
    ISBN 90–277–1713–3

Published by D. Reidel Publishing Company,
P.O. Box 17, 3300 AA Dordrecht, Holland.

Sold and distributed in the U.S.A. and Canada
by Kluwer Academic Publishers,
190 Old Derby Street, Hingham, MA 02043, U.S.A.

In all other countries, sold and distributed
by Kluwer Academic Publishers Group,
P.O. Box 322, 3300 AH Dordrecht, Holland.

Translated by Lynne Richards with Philip Hyams and Johan Staargaard.

Printed in The Netherlands

TABLE OF CONTENTS

v

PREFACE

In the period between about 1820 and about 1870 German psychiatry was born and reborn: first as anthropologically orientated psychiatry and then as biomedical psychiatry. There has, to date, been virtually no systematic examination of the philosophical motives which determined these two conceptions of psychiatry. The aim of our study is to make up for this omission to the best of our ability.

The work is aimed at a very diverse readership: in the first place historians of science (psychiatry, medicine, psychology, physiology) and psychiatrists (psychologists, physicians) with an interest in the philosophical and historical aspects of their discipline, and in the second place philosophers working in the fields of the history of philosophy, philosophy of science, philosophical anthropology and philosophy of medicine.

The structure and content of our study have been determined by an attempt to balance two different approaches to the historical material. One approach emphasises the philosophical literature and looks at the question of the way in which official philosophy determined the self-conception (*Selbstverständnis*) of the science of the day (Chapters 2 and 4). The other stresses the scientific literature and is concerned with throwing light on its philosophical implications (Chapters 1 and 3).

It is our claim that, having proceeded in this way, we have avoided over-simplification and have laid the foundation of a balanced account of the relationship between philosophy and science, doing justice to the historical complexity of the ways in which philosophical and scientific thought have contributed to the rise of the anthropological and biomedical conceptions of psychiatry.

The content of the four chapters can be summarised as follows.

Chapter 1 deals with the origins of the anthropological conception of psychiatry. Within anthropological psychiatry (from about 1820 to about 1845) two dominant trends can be distinguished: that of the so-called psychicists (*Psychiker*) and that of the so-called somaticists (*Somatiker*). The analysis of the work of the two most typical representatives of these trends, the psychicist Heinroth and the somaticist Jacobi, makes it clear, that, contrary to current opinion, the great controversy between psychicists and somaticists must in the first place be conceived as a controversy bearing on anthropological presuppositions and not as a controversy about method. It appears that Heinroth is committed to a Platonic tradition and Jacobi to an Aristotelian tradition of anthropological thought. In the concluding section we discuss the clinical psychopathological phenomenology of the institutional psychiatrist Leo Snell. We show the way in which — beyond differences in psychiatric conceptions, as represented by somaticism and psychicism on the one side and anthropological and scientific psychiatry on the other — the continuity of the clinical psychiatric tradition was preserved.

Chapter 2 provides an analysis of the scientific and philosophical factors which were responsible for the spread of the belief in mechanism in science (especially in physiology and psychology). The aim of this analysis is to contribute to an adequate understanding of materialism as conceived and advocated by leading scientists since the eighteen-forties.

It is argued that materialism was here essentially a methodological position, and that it should not be confused with the metaphysical materialism of scientists like Vogt, Moleschott and Büchner (which is usually — but erroneously — taken to be the typical form of scientific materialism in nineteenth-century Germany). Of special heuristic significance in this connection is Schopenhauer's criticism of naturalism (or materialism), as is Lotze's criticism of materialistic methodology, because in both cases materialism appears to be conceived as a position that primarily implies the advocacy of 'materialistic' (i.e., natural scientific) *method*.

Chapter 3 deals with Wilhelm Griesinger and attempts to assess his role in the emergence of the conception of psychiatry as a mechanical science in the period from about 1845 to about 1868. In the attempt to give a precise definition of Griesinger's own materialistic *Selbstverständnis*, his relationship to the philosopher Hermann Lotze is examined, as is his relationship to Johann Friedrich Herbart and his relationship to the anthropological psychiatry of his teacher Albert Zeller. In particular, the revision of the common view of Griesinger's relation to Zeller tends to cast doubt on Binswanger's contention that Griesinger's psychiatry and anthropological institutional psychiatry were divided by a sharp break. It is shown how Zeller, Griesinger, and Binswanger can be said to belong to one and the same tradition of clinical psychiatry.

In Chapter 4 we return to the philosopher's perspective: the names of Schopenhauer, Rokitansky and Lange mark the successive stages in a process of progressive clarification of the position of methodological materialism. This position, fairly widespread since the middle of the nineteenth century, found acceptance among scientists of idealistic as well as of pragmatic—agnostic ('positivist') leaning. The transcendental naturalism of F. A. Lange, a synthesis of idealistic and 'materialistic' motives that gave a philosophical justification of the claim of both scientific rationality and of 'the standpoint of the ideal', is presented as the horizon that encompasses the different attempts of scientists from about 1840 onwards to attain an articulate 'philosophical' view of science (cf. Chapter 2).

I should like at this point to make an observation about the translation. In many cases, instead of current, generally-used expressions such as 'mental illness, disease, disorder, disturbance', we have used the terms 'psychic disease, illness, disorder', etc., occasionally varying this, for stylistic reasons, by using 'disease (illness, disorder, etc.) of the soul'. This is not simply terminological arbitrariness. The expression 'mental illness' assumes the framework of Cartesian ontology (the differentiation of body and mind), and it can therefore not be used without conceptual ambiguity and historical distortion in the description of psychiatric theories which rest on a differentiation between body, soul, and mind — in which the mind is conceived as the higher part of the soul.

In preparing this English edition I have been greatly assisted by the critical and often stimulating remarks — both verbal and written — of friends and colleagues inside and outside the Faculty of Philosophy of the University of Nijmegen. I hope they will not be too annoyed at discovering that their observations have not always left those traces in the text which they might have expected or approved of. My thanks are due to Dr. G. J. Renes, Professor H. A. G. Braakhuis, Mr. M. L. J. Karskens, Dr. W. Oudemans, Professor C. Sanders and Mr D. Tiemersma. I should also like to take this opportunity to express my gratitude to the persons who contributed most to the completion of this study during the preparation of the Dutch version (1980), whether through advice and encouragement, critical comment or the creation of favourable working conditions: Professor T. de Boer, Professor A. A. Derksen and Professor C. E. M. Struyker Boudier. My special thanks go to Dr. T. Baumeister for much time and energy devoted to discussing points of philosophical relevance and for the patience with which he talked over the many problems of interpretation and translation of the German texts quoted in the study. In the preparation of the bibliography and indexes the help I received from Mr. R. Corbey proved invaluable.

I consider myself fortunate to have found in Mrs. Lynne Richards a translator whose courage in undertaking the laborious and difficult task of translating my book was equalled by her skill in translation. Thanks are also due to Mr. Philip Hyams and Dr. J. A. Staargaard, whose well-considered translation of German texts often delighted me.

Finally, I should like to thank the Netherlands Organisation for the Advancement of Pure Research (Nederlandse organisatie voor zuiver-wetenschappelijk onderzoek) for the grant which made the English translation of this work possible.

Department of Philosophical Anthropology   GERLOF VERWEY
   and Philosophy of Medicine
University of Nijmegen, Netherlands.
December 1983

INTRODUCTION

In the period from about 1820 to about 1870 German psychiatry was born twice: first as an anthropologically oriented discipline and then as a natural science. There has so far been no systematic examination of the philosophical motives which determined the self-conceptions of these two forms of psychiatry. A whole complex of questions has been denied even the chance of an answer because these problems did not fall directly, or in their totality, within the scope of the alternative approaches available — the strict history of philosophy or history of science approach. In this study our aim is to make up for this omission to the best of our ability.

In order that the intentions and pretensions of this study may be clearly understood, it is important that the type of examination we are undertaking be accurately distinguished from other current conceptions of the history of science and the history of philosophy. Two points are relevant here. The first is that our study is essentially limited to the consideration of a subclass of the motives which may possibly have determined the two conceptions of psychiatry — as an anthropological discipline and as a natural science — in Germany, namely the philosophical motives (particularly the philosophical anthropological motives and those concerned with the philosophy of science). This limitation has been chosen deliberately; an examination adopting this approach does not imply the superfluousness of history of science studies aiming at the analysis of other subsets of motivational factors (social, economic, political, psychological, etc.); it can at best supplement them. However, our approach also means (and this is the second point) that if, in what follows, we talk about the *self-conception* (self-understanding, self-interpretation) of psychiatry, it may not be concluded that our examination is meant to make a purely philosophical point. To be precise, where we take as our theme the problem of the self-conception of nineteenth-century German psychiatry, the question is not whether, philosophically speaking, psychiatry did or did not have an adequate understanding of itself, but rather what this self-understanding or self-conception implies, how it can be descriptively *identified* as historical datum and studied as a factor that (to some degree) determined the course of nineteenth-century psychiatric thinking in Germany.

It is obvious that anyone who studies the history of science from a philosophical viewpoint will tend in his description of this history to emphasise
the *discontinuity* in the philosophical presuppositions which played a part in
determining the self-conception of this science. It was therefore not surprising
that in the early stages of our research the outline of the history of German
psychiatry led to the supposition that there had been little continuity of
psychiatric tradition. However, as our analysis of the historical material progressed, it became clear that, and in what sense, there existed (and still exists)
beyond the difference in philosophical self-conception, that is beyond the level
of philosophical discontinuities, a continuity of clinical psychiatric tradition.

A single observation should be made concerning the method used. It
was determined primarily by the sort of problem which occupies such a large
part of our study: the problem of the historical identification of psychiatry's
self-conception. It is obvious that the way in which a science of a particular
period understands itself is influenced by philosophy, contemporary or
otherwise. In the history of twentieth-century psychiatry we can find good
examples of this: Minkowski was influenced by Bergson, Binswanger by
Husserl and Heidegger, Von Gebsattel by Scheler, Sullivan by Mead, etc.
This is not to say that there were not great difficulties in arriving at a, to
some degree, reliable and convincing reconstruction of such a self-conception.
as a rule scientists do not philosophise readily or at length. The historian of
science therefore often has little more to go on than the usually very scanty
philosophical observations in the introduction or footnotes, and what he can
discover about the philosophical views of important scientists from a pains-
taking textual study of their scientific works. In such a situation he will
make grateful use of the assistance which he can obtain by studying the
relevant philosophical literature, which is often, but by no means always,
contemporary. In fact, to put it more strongly, in research aimed at the
identification of the self-conception(s) of science, method demands that the
philosophical literature be systematically included in the research.

The value of this approach became particularly clear to me when studying
Griesinger's work. In Griesinger's case the prospects of achieving a reconstruction of his philosophical self-understanding which would be, to some degree,
convincing were extremely poor. Not only was there a dearth of explicit
philosophical reflection, but Griesinger's scientific work proved to have
undergone such varied treatment in its *Wirkungsgeschichte* that any attempt
at reconstruction of its internal unity seemed to be doomed to failure from
the start. The opinion of the historian of science, Ackerknecht, that Griesinger
was an eclectic[1] seemed to be justified. In addition to this, the lack of any

comprehensive interpretation in the Griesinger literature[2] certainly did not make it easy to gain an insight into the fundamental motives of his scientific thought. In this discouraging situation, the clarification of his relationships with contemporary philosophers such as Herbart and Lotze contributed to a better understanding of Griesinger's philosophical position and thus opened the door to a better-founded interpretation of his conception of psychiatry.

Although to a lesser degree than in Griesinger's case, it was nevertheless essentially the same sort of problem which was encountered in researching the anthropological–psychiatric literature. Here, we even found it necessary to look way back in the history of philosophical anthropological thinking in order to get a precise understanding of the philosophical presuppositions inherent in the anthropological conception of psychiatry which we were seeking, because the contemporary philosophical literature gave us no clues to it. In short, in all cases the relating of expressions of psychiatric self-reflection to explicitly formulated philosophical standpoints in philosophy itself proved to be important, if not indispensable, in furthering the interpretative reconstruction of the different self-conceptions of psychiatry.

Indeed, the significance of studying the philosophical literature in a history of science study such as this one cannot easily be overestimated (although the *danger* of overestimating it is constantly present; more on this point later). This significance lies in the fact that, after the event, philosophy frequently expresses clearly that *Selbstverständnis* of a science which seems only to be dimly conceived and often poorly expressed by earlier and/or contemporary science itself. In a study of this kind, such a philosophy serves to define and frame the horizons within which the existing expressions of scientific self-reflection, incomplete and fragmentary as they are, can be placed. In our study, such a 'horizon-function' is most clearly to be seen in Chapters 2 and 4. In my view, this is best pointed out by the interpretation of F. A. Lange's philosophy in the final chapter. Thematically, this no longer belongs to the history of (natural science) psychiatry in the chapter preceding it, but nevertheless it is indirectly relevant to it because it makes it possible to gain insight from above, as it were, into the fundamental philosophical motives which, since the eighteen-forties, have played a part in self-reflection in natural science in general (and thus, *a fortiori*, also in the self-reflection of the natural science school of psychiatry at the time). In short, what is thus revealed by a 'bird's eye view' can sometimes put the historian of science on a track which had hitherto eluded him for no other reason than that he was attempting to reach his goal solely from the 'grass-roots perspective' of the strictly scientific literature. By this we mean no more than that in some fortuitous cases the study of philosophical texts can provide an increase in clarity for the science historian's analysis which could not have been achieved in any other way.

Experience compels us, however, to acknowledge the validity of another point of view, and to guard against self-deception: history of science practised by philosophers is always in danger of succumbing to an inherent tendency towards idealisation and simplification which, if it is not self-critically supervised, leads to the acceptance as historical truth of that which is neither historical nor true.

It must be admitted that in our own history of science study, too, the danger of idealisation and simplification was never far away: the tendency to absolutise the bird's eye view of the philosopher and the historian of philosophy at the expense of the grass-roots perspective of the historian of science, the inclination uncritically to give greater weight to the testimony of the philosophical literature than to that of the scientific literature — these were constantly present and demanded vigilant self-control. It was no less important to guard continually against that comfortable self-deception that leads the philosopher to assume that scientific self-reflection follows, both chronologically and logically, the conceptual trail blazed by prominent philosophers. On this point, however, the research carried out from the grass-roots perspective (see particularly Chapters 1 and 3) revealed a very different picture. My experience is that the philosophical viewpoints which emerge within the sphere of scientific self-reflection, however incompletely developed, are often more varied and, above all, different from what one might in the first instance be led to expect by the philosophical literature of the time. The relative independence and philosophical originality of some expressions of scientific *Selbstverständnis* cannot be denied. Certainly as far as the period between about 1840 and about 1870 is concerned, there is the additional fact that in Germany this unofficial philosophy of the scientists (I am thinking particularly of the natural sciences) becomes a significant factor in determining the official philosophical reflection of the philosophers, so that instead of science following the lead of philosophy on this point, philosophy finds itself, to a certain extent, dependent on science. This fact is not new in itself. What is new is the idea (prompted by history of science work from the grass-roots perspective) that outside the chronological limits of 1840 and 1870, both before 1840 and after 1870, one must take into account a philosophical 'surplus', or at least a philosophical potential, of science which is exploited only after the event and in a selective fashion by official philosophy. New, too, is the emphasis thus placed on the methodic consequences of such an idea for a history of science study like ours, which I should like to formulate as the requirement that the scientific literature itself be taken seriously as a potential source of philosophical motives.

Thus, in conclusion, I should like to describe the work on *Psychiatry in an Anthropological and Biomedical Context*, in terms of its form and content, as a search for the right balance between the two views outlined above — the bird's eye view of the philosopher (or historian of philosophy) and the grass-roots perspective of the historian of science. Whether that balance has been achieved remains for the reader to decide. The author can only testify that this balance was the ideal for which he strove.

ANTHROPOLOGICAL PSYCHIATRY IN GERMANY DURING THE FIRST HALF OF THE NINETEENTH CENTURY

1.1. INTRODUCTION

1.1.1. The Terms 'Anthropology' and 'Anthropological' in Medical and Psychiatric Literature (First Half of the Nineteenth Century)

When, in 1844, H. Damerow, the moving force behind organised institutional psychiatry'[3] in Germany, wrote in the introduction to the first issue of the *Allgemeine Zeitschrift für Psychiatrie*, of which he was the editor, about "the anthropological factor which is the hidden root of all the theories and systems on the tree of psychiatry",[4] he was pointing to a philosophical presupposition which had, until then, determined nineteenth-century German psychiatry's conception of itself. It is precisely this (philosophical anthropological) presupposition which is meant when we refer to this psychiatric and medical literature as *anthropological* psychiatry or medicine.

The word 'anthropological' when used in this connection means "relating to (man as) the unity of body and soul (or mind)". Anthropological psychiatry is thus, to put it briefly, psychiatry which revolves around the philosophical idea of man as a psychophysical unity. As such it can, with good reason, be described as *philosophically oriented* psychiatry.

Immediately underlying this conception of psychiatry there was, among other things, the more 'philosophical' variant of empirical human studies which had become familiar, particularly in German medicine, in the last quarter of the eighteenth century, chiefly under titles like *Anthropologie für Aerzte und Weltweise* ('Anthropology for Physicians and Philosophers') and *medizinische Anthropologie* ('Medical Anthropology').[5]

An outstanding example of the anthropological conception we are referring to, in which the idea of the connection between body and mind was central and which was therefore a determining factor in what was subsequently meant in medical circles by the word 'anthropology', was E. Platner's influential work *Anthropologie für Aerzte und Weltweise* (1772).[6] Besides the term 'Anthropologie für Aerzte', the expression 'medizinische Anthropologie' also became fashionable towards the end of the eighteenth century.[7]

1

Also of interest in this connection is the almost simultaneous appearance of the term 'medizinische Psychologie'. The concept of medical psychology probably originated with the Berlin doctor and philosopher (and pupil of Kant!) M. Herz, specifically in his review of Platner's anthropology (of 1772) in 1773, in which he writes about the necessity for medical psychology and describes Platner's anthropology as a work that "indeed partly fulfils the demands that came to mind".[8] Herz's appreciation of Platner's work must be viewed against the background of the dissatisfaction with the existing explanations of the mind-body problem which was then widespread in medical circles. The fact that Platner's *Anthropologie* went some way towards resolving this dissatisfaction helps to explain why this concept of medical psychology had become so established in medical circles by the end of the eighteenth century that the conception and content of Platner's anthropology had come to be identified with that of medical psychology.[9]

A nice, and at the same time highly significant, detail is that it was Herz's *philosophical* correspondent, Kant, who in reference to Herz's Platner review gave his opinion, at the end of 1773, that "subtle ... research concerning the way in which physical organs are associated with mental processes would (in his opinion) always be a futile exercise".[10] Kant's own anthropology, the college lectures which he gave in the winter term of 1772–1773 and which appeared a good quarter of a century later in 1798 under the title *Anthropologie in pragmatischer Hinsicht*, did indeed take a different direction.

Kant's example made a great impression: the influence of his 'pragmatic' anthropology can be seen in the works of almost all anthropologists at the beginning of the nineteenth century (in chapter headings, concepts, taxonomy, and terminology) so that this era has been described, not without justification, as the 'pragmatic epoch' in the history of anthropology.[11]

The success of Kant's work must not, however, cause us to lose sight of the fact that, as well as the school of anthropology which addressed itself to practical psychology and which took Kant's *Anthropologie in pragmatischer Hinsicht* as its model, there was in German anthropological literature at the end of the eighteenth and beginning of the nineteenth centuries an equally important group of physician-anthropologists who, following the example of Platner's *Anthropologie* of 1772 (1790), took the connection between the body and the mind as their central idea.

Although in the work of Platner (whose thinking, generally speaking, displays a certain closeness to that of Leibniz)[12] the Cartesian doctrine of two substances is preserved[13] and the conception of anthropology which he made fashionable can therefore not be considered as *identical* to the conception of anthropology in anthropologically-oriented institutional psychiatry,[14] it can reasonably be said that the conception of anthropology which, through Platner, became so influential, that is, in which the emphasis was placed on the problem of the mind–body relationship, effectively sparked off and stimulated the emergence and spread of anthropological medicine (psychiatry) in the period that we are studying.

It is plausible, too, that when the perspective of man as a psychophysical unity came to be acknowledged as 'the' specifically anthropological viewpoint from the first quarter of the nineteenth century onwards, this development was not detached from a shift in interest and emphasis in the anthropological literature away from psychology and towards medicine. On this last point it may safely be assumed that the natural philosophical medicine inspired by Schelling's natural philosophy, which became so popular after about 1800, contributed to this shift.

If, finally, Hegel crops up in our terminological excursus, it is for no other reason than that his use of the word anthropology (namely as a definition of that part of the doctrine of the 'subjective spirit' which deals with the spirit at the lowest level of its development, the spirit as 'Naturgeist' ('the natural spirit'), i.e. the spirit in its immediacy and *an sich*)[15] entirely apart from the special meaning which this term has in the context of his philosophical system, is a true reflection of the dominating influence of medicine (and psychology) on the anthropological literature of his period, as a result of which anthropology was, by definition, concerned with that aspect of man relating to the union of body and soul (or mind).[16]

After Hegel's death, the term 'anthropology' continued to retain the meaning of a doctrine of man as a psychophysical individual, both in institutional psychiatry (as evidenced by, among other things, the extract from Damerow's writings dating from 1844, quoted above) and among the ranks of those disciples of Hegel who set out (but ultimately failed) to complete the work on psychology which Hegel had planned but never written.[17] Thus, in 1837, J. E. Erdmann wrote a booklet called *Leib und Seele* (Body and Soul), subtitled *Ein Beitrag zur Begründung der philosophischen Anthropologie* (A contribution to the foundation of philosophical anthropology) in which (philosophical) anthropology is defined as "that science ... which has to represent the necessary evolution of the still nature-determined mind or, in other words, to show the dialectic evolution of the mind to be a natural one".[18] According to Erdmann, if everyday linguistic usage takes the word anthropology to be completely analogous to words like ornithology, ichthyology, etc., that is as a branch of "Naturgeschichte" (natural history) or more

precisely as the "Naturbeschreibung des Menschen" (i.e. a description of
man, of all his functions, etc. in as far as he is a natural being), and if 'philosophische Geisteslehre' (philosophical theory of the mind) also regards man
as a natural being, that is as a natural individual, then one can, without
transgressing accepted usage, describe that part of the philosophical theory of
the mind which considers man in his natural individuality as philosophical
anthropology.[19] In Erdmann's *Grundriss der Psychologie*, published in
1840 (1873[5]), the first section ('Der Geist als Individuum') is specified as
'Anthropology'.[20]

Such definitions are reminiscent of conceptions of the body—soul relationship from the Aristotelian (-scholastic) tradition and, I would add, as such
they also point to a possible source of the conception of anthropology as it
was taken for granted in the *Selbstverständnis* of the institutional psychiatry
of the period.[21]

1.1.2. *Anthropology – Philosophy or Empiricism?*

The definition of the *subject* of anthropology implied in the above explanation of the use of the term anthropology — the theory of man as a union
(or interconnection) of soul (or mind) and body — leaves unanswered the
question of whether we are dealing with a philosophical or an empirical
discipline. If we go by what is said on this subject in the existing text books
in the fields of the history of philosophy and the history of science (psychology, medicine, etc.) — which is little or nothing [22] — the only possible
conclusion seems to be that that anthropology cannot be considered either as
a science or as a part of philosophy. To what extent this negative conclusion
is justified must now be examined in more detail.

In this connection it is important to note at the outset that what was
understood as anthropology, particularly from the eighteenth century onwards in Germany, implied an emancipation (express or otherwise) from the
metaphysical tradition, which was in general theologically oriented. This
emancipation was not, however, so far-reaching that anthropology became
wholly detached from philosophy; it was a dissociation which took place
within the framework of the *Schulphilosophie* in force at the time[23] and
resulted in the formation of a discipline which at least, through its *holistic*[24]
intention, preserved an unmistakable 'philosophical' stamp.

This anthropological literature was also 'philosophical' for another reason
that is connected with its (philosophically speaking) 'practical' aim: insofar as

anthropology was concerned with the systematic arrangement of practical psychological and/or medical experience according to general principles of philosophy it understood itself as the repository of wordly wisdom for the benefit of life in general, or medical practice in particular.

But anthropology was of old not only 'philosophical' in the sense just outlined, but also 'empirical', be it in the broad sense of the word that permits the inclusion of such things as (practical) wisdom, character insight, etc.. We therefore speak of anthropology as a member of the family of the so-called empirical studies of man.

We find both characteristics − the 'philosophical' and the 'empirical' − in the literature of the period under discussion. Indeed, the last phase in the history of German 'Schulphilosophie im Zeitalter der Aufklärung' ('School philosophy in the Age of Enlightenment'), i.e. 1750−80,[25] has, not without reason, been described in the history of philosophy as the generation of 'Die Lehre vom Menschen'[26] ('the Theory of Man') or, more specifically, as the 'Zeitalter der Menschenkenntnis und Erfahrungsseelenkunde' ('the age of the knowledge of human nature and of empirical psychology'). The first description expresses the greater involvement of the philosophy of the period in the human subject; the second bears witness to the clear shift of emphasis among philosophers, away from *a priori* deductive reasoning, towards (psychological) empiricism. The same characteristics ('philosophical', 'empirical') apply to the anthropological literature of the time that belonged to the medical tradition.

1.1.3. *A Note on the Traditional Situation*

While the broad description of anthropology given above may appear straight-forward, the *actual* traditional situation which confronts the historian of anthropology is far from clear. This is not only because no more than a part of what can be considered as contributing to empirical studies of man since the Italian Renaissance was called anthropology, but also because various works published during the next two centuries which were called, or called themselves, anthropology were still so subservient to the dictates of authority, so eclectic and encyclopedic, that it is almost impossible to talk of a discipline which has any degree of independence.

Even if we restrict ourselves to the period between about 1750 and about 1820 in Germany (the era in which German anthropology, which had become independent, reached its peak), the situation remains at first sight extremely unclear: on the one hand because the empirical human science which was spawned by the 'Schulphilosophie' of the so-called *Aufklärungs*-era and stamped it as the era of 'die Lehre vom Menschen', was published under *different* discipline titles (psychology and anthropology); on the other hand because a malignant fate has decreed that the very scarce, earlier history of philosophy literature and the equally scarce, even earlier history of science literature about this period [27] give the impression that the interest in (empirical) studies of man at that time was exclusively devoted to (*practical*) *psychology*.[28] In addition, the history of science literature on medicine has, until recently, contributed nothing which would correct this one-sidedness.

Only recently has there been any change in this situation: in the last few years a modest start has been made on retrieving and taking stock of that literature which, within the thinking in empirical studies of man since about 1750 (to about 1820), made up the corpus of this German medical–anthropological tradition. Although it is still too early in this research for us to be able to point to any specific results, it has at any rate already become sufficiently clear that the (German) *Aufklärungsphilosophie* of the second half of the eighteenth century was not only the soil in which practical psychology-oriented anthropology — Kant's idea of pragmatic anthropology — germinated and bore fruit, but was at the same time, and no less importantly, the source from which, with renewed emphasis on the view of man as a psychophysical unity, the so-called physicians' school of anthropology drew its inspiration.[29]

1.2. THE RISE AND SPREAD OF THE ANTHROPOLOGICAL VIEWPOINT IN GERMAN PSYCHIATRY FROM ABOUT 1820 TO ABOUT 1845

1.2.1. *Introduction*

If the rise and spread of the anthropological conception of psychiatry in Germany, as far as we can establish it for the period from about 1820 to about 1845, is still unclear in many respects, this is not only because there has been to date remarkably little systematic research carried out into it, but no less because, in the scanty research that has been done, the philosophical motives which played a part in this conception of psychiatry have not been sufficiently taken into account. And yet the fact that we are dealing here with *philosophically-oriented* psychiatry makes it likely that a certain familiarity with philosophical tradition is a *sine qua non* for a good understanding of the fundamental motives behind this form of psychiatry. It is, therefore, no coincidence if the historian of philosophy can come to the aid of the historian of science on precisely this point. Thus, the aim of this chapter can be described as the elucidation of the philosophical presuppositions of this sort of psychiatry in order to arrive at an adequate identification, in history of science terms, of the two representative variants of it, and of the distinctions between them.

We have already suggested that the (medical) anthropological tradition which reached its peak between about 1770 and about 1820 favoured the development of an anthropologically-oriented form of psychiatry. On the other hand, one might equally well assume that the movement in medicine inspired by Schelling's speculative natural philosophy,[30] which also came to the fore after about 1800, helped to shape the intellectual environment from which this anthropologically-oriented form of psychiatry eventually came. Anyhow, this much is clear: anthropological psychiatry, judged by the work of its most typical representatives, can be regarded neither as a simple and straightforward continuation of the medical anthropology practised by Platner *et al.* nor as the 'natural' descendant of the speculative medicine of the romantic school, although it had – unmistakably – certain traits in common with both.

As far as its relationship with the romantic school of medicine is concerned, it is hard to overlook the fact that the specific Christian anthropological orientation which was pioneered in German psychiatry after about 1815 was to put a strain on the relationship with 'romantic' medicine inspired by natural philosophy.[31]

This fact cannot, however, stop us from assuming that the interaction between philosophy (i.e. natural philosophy) and medicine,[32] which characterised the years between 1800 and 1815 and is unparalleled in history, was an important stimulus to the growth and success of the form of psychiatry which was subsequently developed. The fact that Schelling had called medicine the zenith of the (speculative) natural sciences and had said that the doctor was a high priest in the service of Nature [33] is bound to have increased the self-confidence and prestige of the medical profession at that time, while the typical romantic interest in the pathological, especially the psychopathological, must have promoted the development of the philosophy-oriented form of psychiatry which we will be studying.

Viewed against this background, it comes as little surprise when we see how prominent psychiatry was in early nineteenth-century medical literature.

This was largely due to the efforts of C. Friedrich Nasse (1778–1854). In the history of medicine (psychiatry) Nasse is notorious as the indefatigable founder and editor of a number of journals whose existence, although always short-lived, was a clear indication of the interest in the psychiatric disciplines between about 1818 and 1830. I am referring here to the *Zeitschrift für psychische Ärzte* (1818–1822), the *Zeitschrift für die Anthropologie* which appeared between 1823 and 1826, and the 1830 *Jahrbücher für Anthropologie und zur Pathologie und Therapie des Irreseyns* which Nasse edited together with M. Jacobi.[34]

Although the psychosomatic theories advocated by Nasse himself had important points of contact with the medical views of his colleagues in the field of psychiatry, they can hardly be called representative of either of the two anthropological-psychiatric schools of thought which were to dominate the psychiatric scene from the eighteen-twenties onwards. Far more important in this respect are two other writers on psychiatry who, although very different from each other, both left a mark on anthropological psychiatry between about 1820 and about 1845. One was J. C. A. Heinroth (1773–1843), 'Professor der psychischen Heilkunde' (Professor of Psychical Medicine)[35] at Leipzig; the other was his great opponent, the influential institutional psychiatrist Maximilian Jacobi (1775–1858). Together they belong in the context of the controversy between the so-called psychicists and somaticists, which split German psychiatrists into two camps for many years from about 1820 onwards.

This controversy between psychicists and somaticists, I will maintain, was a controversy rooted in two different standpoints in the anthropological debate, that is in two different interpretations of the soul–body relationship. The linking factors in this dispute were – on the negative side – the common opposition to the one-sided, physically-oriented, 'mind-less' medicine of the *Aufklärungs* era, and – on the positive side – a Christian humanitarian ethos with a corresponding (philosophical) view of man. If the psychicists and somaticists were at variance in the ways in which they understood the relationship between soul and body and the conclusions which they drew from this in relation to the theory and practice of 'psychic medicine', on the question of the ultimate *meaning* of the psychiatric enterprise there was, certainly among the most important of them (Heinroth, Zeller, Jacobi, etc.), no essential difference of opinion. It was always self-evident to the great representatives of German institutional psychiatry that the doctor who tries to alleviate the suffering of a fellow-being who is asking for help is fulfilling a Christian duty, and that the purpose of his thinking and treatment is, and can be nothing other than, this way of discharging one's duty.

1.2.2. *J. C. A. Heinroth as an Exponent of 'Psychicism'*

1.2.2.1. *Anthropology.* In the foreword to his *Lehrbuch der Anthropologie*, published in 1820, Heinroth wrote: "The author of this handbook agrees with others that man can only be understood as a moral being. This has been the basic assumption of his treatment of anthropology". This means that there is no real possible explanation of the nature and destiny of human existence which does not start from the fact of human freedom. (I shall discuss later the consequences for medical theory and practice which Heinroth links with this.)

According to Heinroth, anthropology thus conceived forms the pinnacle of science, nature, and spirit in all respects. It is the alpha and the omega of both.[36] He bases this idea on the following: "The key to nature is given by the senses; that to the spirit is given by consciousness or by reason. The consideration of both, i.e. of the senses and of consciousness, is the real business of anthropology. The highest object of nature known to us is man; in the same way, the highest subject of the spirit has its origin only in human consciousness. To put it briefly, the doctrine of nature has its roots as well as its summit in anthropology, and so has philosophy". But not only that; man is also the object of history and "history only derives its meaning and its elucidation from the knowledge of the essence of man, of his talents, and of his destination, developed from intrinsic laws". Heinroth therefore concludes that: "Anthropology is the focal point of the highest scientific ambitions of mankind, as well as the centre of them all".[37]

Generally speaking, the point of anthropology (Heinroth speaks of the "dignity of anthropology"), according to Heinroth, is that it reminds man of his ultimate goal and holds it up before him (a point which it shares with art, science and religion); more particularly in that it enables man to know himself, teaches him about his organic and mental make-up, and by, as it were, exposing his inner workings leads him to the recognition of the creative wisdom manifested in them. It also teaches him, by drawing his attention to the seed of an infinite development contained in his inward self, to respect himself and to guard against injuring or dissipating his physical and spiritual powers; it prompts him to further develop his innate abilities, to ennoble his existence; in short to strive for his highest goal.[38]

What this in fact amounts to is that anthropology is essentially, and by its very meaning, an instrument of self-realisation in the sense of the Christian human ideal. In the final analysis, anthropological research thus serves a practical purpose in life: the fulfilment of the religious, ethical duty to be a true Christian.

The originality which Heinroth claims for his anthropological work is, nevertheless, an originality of method.[39] His views on the various 'anthropological methods'[40] are thus essential for the precise placing of his work in relation to that of his (philosophical and scientific) contemporaries. They are also particularly interesting because they tell us something about the yardstick against which he wanted his work measured and judged as a scientific achievement.

Summarised briefly, this amounts to the following. In (existing) anthropological research there is a demonstrable multiplicity of methodological perspectives ('anthropologische Standpunkte') which all have a right — albeit relative — to exist. Heinroth distinguishes here four points of view: (1) purely empirical, (2) analytic, (3) synthetic, and (4) conciliatory ('*ausgleichend*').

In the first case, the purely empirical standpoint, man is regarded as a natural being in the narrowest sense of the term. It is the standpoint of *physiological* anthropology in the old, traditional (i.e. pre-Schelling) sense of the word. In this school of thought man ultimately appears as a 'mechanical-chemical-galvanic construction or machine' which, in any event, is distinguished in a real sense from machines in that he is conceived as having a physical life force.[41] The views resulting from this 'purely empirical' standpoint can only be defended in so far as serious anthropology cannot function without an empirical foundation. One must, however, guard against turning this empirical basis into a principle for explaining the phenomena of life or fathoming out mankind, that is to say, constructing man from the outside.[42] It is, in short, the standpoint of *external observation*, which was thought at that time to have been realised in exemplary fashion in the so-called old empirical physics.

In the second case, the analytical approach, anthropological investigation is directed towards the unfolding of 'empirical life, conditioned by psychical principles' or in other words 'the analysis of the inner man, in respect of the external man'.[43] The anthropological view of the physiological and empirical standpoint is here supplemented by 'psychic anthropology'. This 'psychic' anthropology, according to Heinroth, is rooted in critical philosophy.[44] The advocates of this school are careful to restrict themselves to accounting for the psychic phenomena and law-like regularities, which they tried to order by means of a classification of the underlying fundamental powers or basic psychic abilities (emotions, thoughts, practical abilities, etc.). The positive significance of this analytical approach, which, as to its object, can be characterised as psychological, lies in the fact that it fills a gap left by physiological anthropology; from a 'higher' point of view (which we will discuss later), however, this analytical view reveals itself as being of only relative validity, because life in its totality eludes analysis.[45]

The third anthropological standpoint is that of synthesis. The advocates of this approach aim at comprehending human life in its totality (or rather the life of which human life is only a '(small) part'). By the very nature of such an approach, anthropological study, as it were, overreaches itself in the direction of an all-embracing theory of nature, and man is thus only of interest as a part of the more comprehensive whole. By adopting the standpoint of synthesis, the researcher does, it is true, avoid the pitfalls of the analytical approach, but that does not mean that this method does not have its drawbacks; the danger that threatens here is that of a sort of philosophical megalomania, the delusion of omniscience. Considering the immense whole of which we make up a part and the sublime unity which this totality contains, it is better to confess our ignorance, rather than expose ourselves to the dangers of attempting to penetrate the heights and depths of the Immeasurable. These attempts are as vain as they are foolhardy, doomed to failure and bound to humiliate us. In fact the only value of such over-exertion lies in this humiliation itself, because the humiliation makes us understand the futility of the pride which inspired us to undertake this arrogant endeavour.[46] This ceases to be criticism of method and becomes a religiously motivated protest: the philosophical pretension to omniscience is condemned as an expression of unchristian pride.[47]

The fourth and last standpoint in anthropological research which Heinroth discusses is that of harmonisation, *Ausgleichung*. Strictly speaking, the standpoint of synthesis is also a standpoint of *Ausgleichung* because it is concerned with the fusion of (theoretically) disparate extremes (nature—spirit) into a 'proper construction of knowledge', but the 'truly harmonising researcher' is speaking on the fourth level. He is concerned with uniting the extremes of empirical and analytical investigation, or conversely with uniting 'the lucidity and unity of thought with the reality of observation, between which there is, however, no direct relationship', by applying a way of thinking which we owe (according to Heinroth) to a genius who is usually regarded only as a poet and not as a thinker — Goethe.[48] It is the method of *gegenständliches Denken* (objective thinking).

The particular superiority of Goethe's objective thinking, as opposed to normal philosophical abstract thinking, is seen by Heinroth as lying in the fact that this thinking "is not separated from its objects. The elements of the objects, the phenomena, are incorporated in this thinking as well as intensively impregnated by it. In this way his perception (*Anschauung*) becomes thinking, his thinking perception: a process which we have to admit is the most perfect one".[49] It is the method which Heinroth says he himself used in his *Anthropology*.[50]

The four anthropological standpoints or methodological perspectives correspond with categories of human cognitive faculties: the empirical-physiological standpoint lays the emphasis on the senses, the analytical standpoint rests primarily on *Urteilskraft* (judgement) and the intellect, the standpoint of synthesis on *Einbildungskraft, Phantasie* (imagination), and the standpoint of true *Ausgleichung* asserts the significance of *Vernunft* (reason).[51]

The distinction between realists and idealists, in terms of this classification of anthropological standpoints, can be defined thus: the former restrict themselves to the empirical-physiological and analytical standpoints, while the latter adopt the position of synthesis or of *Ausgleichung*. It must be noted here that the eventual reconciliation between realists and idealists is considered to be possible only through the intervention of researchers on the fourth level: if one can say of the advocates of *Ausgleichung* (reason) that they too are idealists (Heinroth does not say this explicitly, but the text suggests it),[52] then *Ausgleichung*, as Heinroth means it, would be something between realism and a wrongly-conceived idealism (in this case Schelling's) in the name of the only true idealism (that of Heinroth inspired by Goethe). When we speak, in the last case, of idealism, we are using the word in a very special sense, in which it refers to a conception, according to which the reality of the ideas (or in the sense of the eternal archetypes (*Urbilder*) of nature) can *only* be grasped by way of a special sort of cognition – Goethe's *anschauendes Denken* or *denkendes Anschauen* (perceptual thinking or thinking perception).

It is known that Goethe was pleased with Heinroth's characterisation of his way of thinking[53] although, in the final analysis, he was not satisfied with Heinroth's *Anthropologie*.[54] The reverse is also true: Heinroth, in spite of that initial, surprising appeal to Goethe, can hardly be described as a disciple of Goethe. The theological and personalistic orientation of Heinroth's anthropology and Goethe's pantheistic *Weltanschauung* (conception of life) are ultimately too far apart to allow us to speak of a real relatedness. *Gegenständliches Denken* thus clearly had a different meaning for Heinroth than it had for Goethe. To Heinroth, the *gegenständliches Denken* practised by the anthropologist is not a goal in itself, and the ultimate purpose of anthropology is not fulfilled with the knowledge about man acquired by means of *gegenständliches Denken*, but is, so to speak, religious and pedagogic.

According to Heinroth, what anthropology is about (as we have seen earlier [55]) is making man familiar with himself (his organic and mental organisation) in order to make him aware of the *schöpferische Wahrheit* (creative truth) which is revealed in it. The individual, living man in his physical and spiritual totality is, as it were, the best proof of God one could wish for, and the anthropology which brings an awareness of this miracle of God has thus the special purpose of reminding man of his highest goal: to develop that which is highest in him – *Vernunft* (reason).

To Heinroth, reason is essentially a religious cognitive faculty; its proper object is God.[56] The relation to God (the Deity) which is thus achieved through the medium of human reason (which – most characteristically – Heinroth conceives in this context as *Gewissen* (conscience)) is essential for the full realisation of humanness.[57] Reason in man is what makes him one, both in himself and for himself, and this reason must therefore be of central significance in anthropology because "the cardinal rule in understanding man is to start from unity, to take this as the base, and to relate to it, if only indirectly, all phenomena, even those that apparently are most remote".[58] Since reason without its object, God, is without content and meaningless, man's relation with God (through the medium of his reason) must be the ultimate point of reference of an adequate anthropological view.[59]

The central significance for anthropology of this religious relationship of man with God is – repeatedly and emphatically – underlined by Heinroth: the introduction of religion and religious principles into anthropology has nothing to do with blending something heterogeneous, a μετάβασις εἰς ἄλλο γένος. On the contrary (and on this point Goethe certainly could not agree): "It is the rounding off and finishing, the organic completion of anthropology, which without this essential part has so far only represented a fragment of the depiction of man. The highest relation of man, as adopted and developed here, is indeed more than a rounding off; it is the ray of light illuminating the entire picture of man, without which the essence and meaning of man cannot be known".[60] And indeed, as Heinroth remarked elsewhere about the religious inspiration of his anthropological views: "Thus, this entire anthropological insight is based in fact on revelation and religion, and stands or falls with these supports".[61]

At this point we shall leave the discussion of Heinroth's *Anthropologie*. I do not propose to go into more detail on the subject, because the preceding pages were concerned solely with placing Heinroth's anthropological work in relation to (contemporary) philosophical and scientific positions, and describing the anthropological model he created. This can serve as an introduction to his famous – and notorious – theory of mental derangement as he developed it in his *Lehrbuch der Störungen des Seelenlebens oder der Seelenstörungen und ihrer Behandlung* (Manual of the Disorders of Inner Life or the Disorders of the Soul, and their Treatment) published in 1818 (i.e. four years before the appearance of his *Anthropologie*).[62]

1.2.2.2. *Psychiatry*. What distinguishes man from the animals, according to Heinroth in his *Lehrbuch der Störungen des Seelenlebens* (1818), is his consciousness. In man's conscious life, however, various levels must be distinguished. The goal of human life is to reach the highest rung in the development of that consciousness. In the first instance, this development takes place, individually and collectively (that is to say, in man and in mankind), from *Weltbewusstsein* (world-consciousness) to *Selbstbewusstsein* (consciousness of self). On the level of world-consciousness, man is "all senses and a sensory creature. His sensations, feelings and impulses belong to the outside world . . . pleasure is his aim, and chance his God".[63] In contrast to world-consciousness, consciousness of the self develops at a later stage. At this level, man comprehends body and soul together in the individual totality of the self (I): man becomes conscious of himself as "a unique self or I (*Individuum*), consisting of soul and body, of an inward and an outward self, impossible to conceive one without the other; not as two different entities that were joined together, but as one and the same (life), developing into two contrary directions, manifest for the external perception (in space) as body, for the internal one (in time) as soul".[64] The great majority of educated mankind develops no further than this level of the life of the consciousness: "Everything for the sake of the 'I', for the sake of its self-being, is the law of this phase of consciousness".[65] But the meaning of being human is not fully comprised in having (the world) and being (oneself). Although very few people can get this far, it should be everybody's ultimate concern to reach the third and highest rung in the development of the consciousness. Just as self-awareness springs from an internal antithesis between the internal and the external, so the highest consciousness comes from an internal antithesis (*Entgegensetzung*) within self-awareness itself, because natural egoism and conscience come into conflict.[66] With the awakening of the conscience, the seed of the highest consciousness and life is sown, the (negative) requirement of being not-self makes itself felt over the requirement of self-awareness and, considered from a 'higher' point of view, reveals itself as the "positive commandment of the surrender of self, i.e. of love".[67] The receptivity to the 'above us', which asserts itself on the level of our emotions, i.e. is only vaguely made conscious, is called, where there is great clarity of consciousness: reason. Reason (*Vernunft*) is that which receives (*das Vernehmende*), through it we receive that which is higher than ourselves and the world; it is the sense of the infinite, the unbounded, the eternal, that is, of God.[68] Reason is the only instrument through which God reveals himself to us; only through reason can man come to God. And because, for Heinroth, only the religious relationship between man and God can bring out the full meaning of humanness, he can say that the "development of reason, i.e. the education of the conscience into consciousness, which pervades our being, is the condition of the really human, i.e. free and blessed, life".[69] This condition is, however, not fulfilled of its own accord. The freedom given to man by God as manifested in the conscience has two sides: it is on the one hand the freedom to lose oneself "in earthly having and being" and on the other the freedom, through renunciation of the world and the self, to share in the world of eternal being.[70]

In the light of these general anthropological ideas a conception (anthropological, of course) of psychic health and illness was developed. First, let us look at Heinroth's conception of psychic health.

Where it is believed that the ultimate meaning of human existence can only be perfected through a constant directedness towards the higher life, towards God, it is understandable that 'health' is conceived as the predicate of a condition of well-being, characteristic of the total man, which stems from the unimpeded turning towards the higher life and is expressed in a feeling of life which pervades the whole man and fills him with happiness. "The feeling of this harmonious, untroubled life, the delight of which cannot be compared with any deeper, more pleasant feeling, is that of the really healthy human condition".[71] The crucial point in this conception of psychic health is that of freedom: only the individual who fulfils the religious purpose of his existence can be said to be truly free, only the truly free individual can be truly healthy.

The reverse of this argument is that as "the principle of health lies in freedom, in the same way that of illness [lies] in the restriction of life . . . In the same way as complete freedom is the highest life, complete restriction of all activities of life, imposed from all sides, which cannot be relieved is death".[72] Corresponding to the anthropological conception of health there is an anthropological conception of illness. "A state of diseased humanity is one in which man finds himself more or less restricted in his consciousness and consequently every consciousness that has not entered the domain of conscience or reason is a consciousness in a state of disease".[73]

Once the conscience has been awakened in the (individual) man, a life which is lived in a completely wordly way or is dominated by self (I) becomes sinful; that is to say, a life which, going against the nature and destiny of man and, as a consequence, hindering the free development of that which is highest in man, becomes "a state of life of diseased humanity".[74] Here we come up against the nucleus of Heinroth's sin-theory of psychic derangement.[75]

In general, one is dealing with a disorder of the soul (*Seelenstörung*) when man himself (in the freedom given him by God) upsets the divine creative plan in him, that is to say, when he is guilty of allowing himself to be enslaved, of relinquishing his own God-given freedom (through passions, delusions, and sinful living).[76] Taken in this general sense, disorder of the soul (or 'psychic disorder') is the term for "the inner life, which is impeded as it were in its straight, natural growth".[77] Or, rather more narrowly defined, disorder of the soul should be understood as "complete stagnation, pure standstill, even as *an inner urge of the creative force, predestined for the highest development, towards its complete antithesis, towards self-destruction*"[78] (my italics). This definition makes clear an essential element of Heinroth's conception: 'disorder of the soul' is not the name for a process (of illness) which overtakes an individual in spite of himself, but describes an *attitude* (in the 'deranged' individual) which bears the stamp of being not free.

According to Heinroth, the terms 'psychic disorder' or 'disorder of the life of the soul' are far preferable as generic terms to the fashionable, vogue words like *Wahnsinn* (insanity), *Verrücktheit* (madness), *Narrheit* (idiocy), *Manie* (mania), *Gemüthskrankheit* (melancholia), and *Geisteszerrüttung* (alienation).[79] Even the term *Seelenkrankheit* (psychic disease) is unsatisfactory because (1) there are psychic disorders that are not psychic *disease*, and (2) there are psychic disorders that are not *psychic* diseases. Every psychic disorder is a 'diseased state' but not necessarily a disease, because in the case of disease there must be a *process* of disease (with a good or bad outcome) and signs of *living reaction*. Chronic *Verrücktheit* is not a *disease* of the soul although it is a diseased state. It is thus also true that there are psychic disorders which are due to organic defects, particularly defects of the brain and the nervous system. These include mania, madness and imbecility which occur as a result of brain damage. In such cases one cannot speak of *psychic* disease. (By this Heinroth means that if one wants to talk about disease in these cases, one can only do so in reference to the *organic substratum* of the psychic disorders, and not in reference to the psychic disorders themselves.) The expression 'diseases of the organ of the soul' must therefore also be rejected (as a generic term for all psychic disorders) because, although it is true that the whole body as an organ of the soul can cause psychic disorders, in the great majority of cases, says Heinroth, it is not the body but the soul itself through which, directly and in the first instance (indeed exclusively), psychic disorders are generated and through which the physical organs are only indirectly affected.[80]

As a consequence of this conception of the aetiology of psychic disorders, Heinroth makes — almost unnoticed — a transition to a narrower conception of psychic disorder ('true *Seelenstörung*'), which in fact excludes all somatically defined psychic disorders.[81] Heinroth's argument in support of this exclusion is that in all cases the psychic disorder does not belong to the essence, the true character of the diseased state (the *independent* state of disease), but is only a symptom attendant upon it; or, to put it even more strongly, where the life of the soul has been suppressed as a result of deficient organic, physical conditions and is no longer existent, one can no longer, strictly speaking, talk of a *disorder* of psychic life.

The conception of psychic disorder in the real sense in which Heinroth uses it in this argument implied the primary psychic causation of psychic disorder. On the basis of this postulate, one could at most only refer to the disease or diseased state of the soul in an *analogous* sense. Since Heinroth neglected to emphasise the fact that the terms disease and diseased state had this analogous meaning in his psychiatric theory, his ultimate definition of the concept of psychic disorder[82] was bound to give rise to confusion and obscure the originality of his approach. If, however, one reads his definition against the background of everything he has previously said about psychic disorders it becomes clear that the point of Heinroth's psychiatric theory lies in precisely the fact that it defended a meaning of 'disease' and 'health' that could not be asserted unless the somatic prejudice, which was always strongly represented in medical thinking, was abandoned. This must have been a challenge to many people and Heinroth certainly did not lack opponents.[83]

The criticism voiced by the so-called somaticists, however, was levelled not so much at the anthropological presuppositions of Heinroth's theory as at the consequences they had for the problem of the *aetiology* of psychic disorders. We must therefore conclude our discussion of Heinroth by looking at his *Theorie der Störungen des Seelenlebens* (Theory of the Disturbances of Psychic Life).

For Heinroth the question concerning the cause of something is the question concerning the totality of the conditions or the essence of that

thing[85] and the aetiology of psychic disorder (we may conclude) is in fact
nothing other than the theory which makes explicit the causes[86] of psychic
disorder. This is precisely what the 'elemental theory' is concerned with. The
general question here is which 'elements', under which circumstances (and in
what sort of combination), can 'explain' the possibility of a psychic disorder
(of one sort or another).

In general terms, every psychic disorder is the product of a sort of interaction of internal (i.e. of the soul) and external forces, and an adequate understanding of the nature of such an interaction is not possible without prior
elucidation of the essence of the soul which is considered to become disordered as the result of an interaction of this kind. "The soul appears to be
a free force, which can be *stirred* by stimuli but which, however, is not
necessarily determinable by them. The soul has the power, the vocation, *to
determine itself, self-determination* is its innate activity, its character, its
essence".[87] Man's body is related to his soul or (inward) self as its outward
manifestation, its form; that the two are inextricably bound together, that
the human individual is the indivisible unity of body and soul, is evidenced by
feelings, more particularly, by our feeling of self.[88] With all that, it is as well
to bear in mind that where the soul-body relationship is concerned, the main
point is not so much the difference between them as the relationship of
subordination of the one to the other. The body is not something independent, something destined for independence, but merely the organ of the
soul – as it were, the soul-made organ – "which appears to itself in bodily
form as a being that, estranged from itself, serves itself unconsciously. Therefore it [the body] is not to be thought of as separate from the soul, but only
in relation to it".[89]

The ontological precedence of the soul over the body lies at the root of
the conviction – already validated by the evidence of our feeling of self –
that man, irrespective of how low he is on the ladder of his development, "is
to be considered and honoured only as *a free being, gifted with reason*, and
consequently as a being which is *capable of being moved in a morally relevant
way*".[90] (With this, Heinroth is opposing the 'customary' conception, in
which the soul, and more particularly the soul as moral force, is regarded as
"attendant upon corporeal life".) There is, therefore, nothing about man that
is *purely physical*, says Heinroth. His whole being is, so to speak, immersed
in this moral predisposition and takes part in it from the moment he becomes
human, or more precisely from the moment he becomes conscious to the
moment when his consciousness is extinguished.[91]

In this moral predisposition is expressed the supermundane destination of man: "Man is, without knowing it, dedicated to the Deity as soon as he comes into the world; consciousness, reason want to guide him to the Deity. That this happens so seldom is his *guilt*, and *from guilt all his faults originate, as do the disturbances of the life of the soul*".[92]

In fact, although the law of freedom is the fundamental law of the soul, and of the life of the (human) soul, strictly speaking man is by nature no more than *potentially* free and is not (necessarily) free *in reality*. This is bound up with a tendency, inherent in man and in human nature, to deviate from reason: the so-called tendency to evil or — the term Heinroth suggests — the tendency to inertia (because the essence of reason is pure activity).[93] The point at which the two principles — that of reason and that of evil — come into conflict is "the free-floating life of man itself, man set free", who is left with the responsibility of deciding which of these two principles he will allow to govern his conduct; whether he will incline towards freedom or towards a lack of freedom.[94]

The existence of disease or psychic disorder presupposes a fundamental lack of freedom in man, and since, in the final analysis, man is reponsible for his own lack of freedom, he is essentially at least partly responsible for the existence of his own diseases and psychic disorders. In order to understand how psychic disorders or diseases in general come about, it is necessary to rid oneself of the habit, ingrained in many people, of explaining what occurs by going from the outside inward, because this leads to an incorrect conception and a false judgement of life and mind.[95] The only model that can help us towards an adequate conception of the genesis of psychic disorders is 'the model of procreation'[96] and Heinroth does his utmost to make it clear to his readers that he is not just making a comparison, but is concerned with the objective rendering of a state of affairs.[97]

The explanation Heinroth gives deserves to be quoted at length; it contains the quintessence of his theory and is the part of his work which drew the sharpest criticism from his contemporaries and from posterity. "Who are the parents of this family?" he asks. And he answers:

The mother obviously is the soul itself, because in it and from it these pseudo-products of life (i.e. psychic disturbances) originate. The begetter too is not difficult to discover; it is always *evil* with which the soul mates, always the same though approaching it in different forms. It might be more difficult to discover the nature of the union itself, but here too the analogy helps us out. The soul and evil are united in the same way the sexes are united everywhere: by love. The love of the soul for evil is called the inclination towards evil, a very appropriate word, as the soul can unite with evil only by *bending downwards*, by *descending*. The association of the soul with evil is always a *downfall*, precisely because of that inclination. By this the soul is *pulled down*, as evil lives in the *abyss of darkness*. Therefore the soul of every *disturbed person* is *darkened*. The act, the moment in which the soul becomes the possession of evil is that in which the psychic disturbance is *conceived* and *begotten*. The product differs according to the diversity of psychic mood and the form in which evil is absorted. From these arise the elements of all psychic disturbances, i.e. *psychic mood* and *determinative stimulus*. Obviously the first one has to be considered as the inner, the other as the external element of psychic disturbances.[98]

Heinroth discusses at length[99] what the nature of man's psycho-organic make-up and the 'external' stimulus[100] must be in order to make an individual disposed to psychic disorder, and what the nature of the *relationship* between these two must be actually to generate a psychic disorder. I do not propose, however, to pursue this, as it adds nothing material to the elucidation of the fundamental principles of his psychiatry.

1.2.2.3. *Heinroth's Platonism*. Heinroth's 'sin-theory' of psychic derangement suffered fierce criticism, not only from his contemporaries (as we shall see shortly), but also from all later schools of psychiatry which were built on the foundations of Griesinger's natural science psychiatry. Even a historian like Ackerknecht could see nothing but nonsense in Heinroth's conceptions. Such cirticism is, however, too indiscriminate to be convincing. It overlooks the fact that the despised sin-theory is a consequence of a certain conception of man — in this case a conception which defines man's mode of being as a religious attitude — and that it can therefore only be understood and judged in the context of this conception. My impression is that in many cases the criticism of the sin-theory amounts to little more than an unspoken reluctance to consider its possible (philosophical anthropological) meaning. In the field of psychiatric self-reflection it is not until Jaspers that we find anything which links an idea of the 'existential' sense, transcending science, of mental derangement with a full recognition of the methodological demands of strict scientific research.[101]

However, it is not only among later generations of psychiatrists with a natural science orientation[102] that we find criticism of the 'speculative' (meaning 'unscientific') psychiatry practised by Heinroth, or the psychicists in general, but also among his contemporaries in the 'somaticist' camp, particularly Maximilian Jacobi, son of the poet and philosopher F. H. Jacobi.[103]

The memory of the controversy between psychicists and somaticists was preserved in later psychiatric thinking in the popular conception that for the psychicists the body was of no significance in the genesis of mental derangement, while the somaticists considered that the psyche was irrelevant in this respect. In this form, this view is incorrect because it presupposes a separation of soul and body (of psychogenesis and organogenesis of derangement) which was never made in so radical a form either by the psychicists or the somaticists. The dispute between psychicists and somaticists, as it was fought out in the field of the aetiology of mental disorder, was, in the final analysis, rooted in a contrast between two *philosophical anthropological* positions, a contrast between two different interpretations of the soul–body relationship[104] which can perhaps best be described as the 'Platonic' (Heinroth) and the 'Aristotelian' (Jacobi and his followers) conceptions of the soul–body relationship. (I shall be dealing with the position of the somaticists elsewhere and will, for the moment, therefore confine my explanation of these descriptions to Heinroth.)

If we call Heinroth's position 'Platonic' it is because his anthropology, in defining the soul–body relationship, repeats a familiar feature of Platonism: according to Heinroth, the soul, to Plato, is the real man, the 'Idea of Man', and the body is only a 'shadow' of it;[105] that is to say, the body's relationship to the soul (or the soul's substance) is an accidental one. (The fact that Plato himself considered the soul not to be an idea, but only to be 'idea-like', is not important in this context.)

An interesting detail here is that Heinroth's Plato seems to be closer to the results of modern work on Plato than traditional spiritualistic–dualistic Platonism, which saw Plato (in contrast to Aristotle) only as the arch-dualist. Thus, not so long ago, C. J. de Vogel pointed out that Plato, contrary to one-sided spiritualistic interpretations which arose in later (Christian) Platonism, saw man as a *unity* of body and soul. The recognition of the unity of Plato's image of man, she argued,[106] runs through the whole of his work, from *The Apology* to *Laws*; Plato's term for this (in *Timaeus* and *Phaedo*) is τὸ συναμφότερον, i.e. complex, compound. The issue here is the unity of two heterogeneous 'elements' where one, the soul, is clearly higher in the onto-logical order than the other, the body. The mutual relationship of the two is thus established as one of subordination: the soul takes precedence both in value and temporally (i.e. in the sequence of genesis) over the body. This must be understood in the specific sense that the material (the 'corporeal' as Plato calls it) cannot be explained by the material, but only by something of a higher order, the mind. Heinroth uses an analogous argument when he states that, in explaining the genesis of psychic disorders, we must go from 'the inside out' and not from 'the outside in'.[107] According to Plato's conception (and, likewise, according to Heinroth's) the body is ὄργανον, instrument of the soul. This body must be maintained and developed because physical health is a prerequisite for meaningful existence.[108]

The agreement between Heinroth and Plato is obvious. It will be clear that when I describe Heinroth's anthropology and psychiatry as 'Platonic' I do not mean by this that he wants to *reduce* man to his psyche in a biased, spiritualistic fashion (as current criticism of the psychicists seems to assume), but that − in the true Platonic sense − he adheres to the view of the *connection* of soul and body as the connection of a 'higher' and a 'lower' part of man in his entirety.[109]

This is borne out by the fact that the criticism which Heinroth levelled at the psychiatric theories of the period was that on the issue of the aetiology of psychic disorders they were relative and not "holistic", i.e. concerned with the totality of man.[110] By this he means that an aetiology which only takes into consideration one aspect of man − the soul *or* the body − is one-sided, since it is relative either to the psychic or to the somatic. The only truly adequate aetiology is the holistic one. In this context, 'holistic' is synonymous with 'anthropological'.

The following examination of the standpoint of the somaticists will make it clear that in this standpoint, too, there was a very definite anthropological conception in the background, which determined the fundamental principles of somatic psychopathology and psychotherapy.

1.2.3. *The Standpoint of the 'Somaticists' (Nasse, Jacobi, Friedreich, etc.)*

According to J. B. Friedreich, who himself defended a variant of the so-called somatic theory of psychic disease or psychic disorder, [111] the somatic theory is characterised by the following fundamental principles. (1) All psychic diseases are the result of somatic dysfunctions; only the physical can become ill and not the soul as such. (2) The soul only appears to be disordered (alienated) in the expression of its various functions because the somatic, to which its activity is bound or through which it expresses itself, has become diseased or is so pathologically altered that it no longer functions as a normal intermediary for the expressions of psychic activity.[112]

In its abbreviated, less well-defined form − "underlying all psychic diseases there is a somatic dysfunction" − the somatic conception has roots going back into antiquity,[113] but in the form which interests us, and to which Friedreich's definition applies, the somatic theory dates from the early nineteenth century.[114]

The great trend-setter and moving spirit of the somatic movement was "the brilliant Nasse"[115] who, in his article *Ueber die Abhängigkeit oder Unabhängigkeit des Irreseyns von einem vorausgegangen körperlichen Krankheitszustande* (1818),[116] heralded the counter-movement to Heinroth's psychistic psychiatry.[117] With Maximilian Jacobi's monumental work on frenzy published in 1844 (*Die Hauptformen der Seelenstörungen, in ihren Beziehungen zur Heilkunde, nach der Beobachtung geschildert*) − the first and only volume published of a planned trilogy − this movement reached, if not its immediate end, at least its undisputed zenith.[118]

L. Snell, who calls himself a pupil of Jacobi,[119] was to say in 1871 that "the eternal value of Jacobi's work is to be found in the method of his investigation, rather than in its direct results". It is the unremitting concern for "meticulous observation" which has made his work exemplary and through which he "firmly set out the only direction which can lead to progress in all fields of natural science".[120] However, Snell (although himself an institutional psychiatrist) was writing at a time when methodological consciousness in psychiatry had received significant impetus from the 'university psychiatry' promoted by Griesinger. It is therefore understandable that where a younger generation was concerned to throw light on the continuing significance of Jacobi and others' somatic psychiatry, they would accentuate precisely that aspect of it which, in their eyes, was the basis of the superiority of the somatic school over the psychicist school: the methodological rigour of natural science research. Thus I get the impression that when J. Bodamer states that "the contradiction between psychicists and somaticists is only one of method, not a fundamental one",[121] he is allowing himself to be too greatly influenced in this judgement by the bias of this later psychiatry, and that he fails really to appreciate the fact that the differences of method between the psychicists and the somaticists are of secondary importance when compared with the fundamental difference in the philosophical anthropological presuppositions (i.e. those concerned with the soul−body relationship).

The following considerations reinforce this point. When psychiatry after about 1860 places the somatic psychiatry of the first half of the nineteenth century above that of the psychicists, specifically concerning the question of method, this implies that the conflict between somaticists and psychicists was perceived as a conflict between (natural) *scientific* and *non-scientific* ('philosophical') method. That this version of the case does not agree with the self-conception of either the somaticists or the psychicists may be deduced, in my view, from the fact that (1) Heinroth (as the most important exponent

of psychicism) *opposed* the method of 'pure and complete observation' which
he championed to the 'synthetic' method of speculative philosophy, [222] and
that (2) nowhere in the extremely lengthy confrontation between the two
points of view does Friedreich (a convinced adherent of the somatic standpoint) mention a conflict of methods in the sense described above. Friedreich,
unlike Snell, Griesinger and others, thus sees Jacobi's greatest merit primarily
in the fact that he "has created a mighty opposition to Heinroth's false
doctrine and established the only correct point of view, i.e. of the somatic
origin of psychic diseases, more firmly",[123] by which he means that Jacobi
demonstrated, as nobody else did, the possibility and meaning of a psychiatry
which was *not* tied to the presuppositions of Heinroth-style psychicist psychiatry.

Indeed, what was at issue was more than, and something other than, a
conflict of methods; it was, as we have already observed, a conflict of fundamental anthropological conceptions. In the following discussion of the
somaticists in general, and Jacobi in particular, I shall try to develop these
ideas further.

If Heinroth's anthropological conception can be described as 'Platonic',
then, as far as their anthropology is concerned, the somaticists can be described as 'Aristotelian'. It is to J. Wyrsch's great credit that he was the first
to draw attention to the Aristotelian–Thomistic background of somaticist
psychiatry.[124] If I disagree with Wyrsch, it is only because I am convinced
that he (like J. Bodamer) occasionally allows his vision and judgement to be
coloured by the assumptions of the later, so-called 'scientific' psychiatry, and
that this weakens his defence of his own historical thesis.

At first sight it is certainly difficult to accept the paradox we present
when we maintain, contrary to the established conceptions of 'Platonic'
and 'Aristotelian' anthropology, that Heinroth must be considered as a
'Platonic' writer because he emphasises the *connection* of soul and body,
and that the so-called somaticists must be described as psychiatrists inspired
by the 'Aristotelian' tradition because they defend a *dualism* of soul and
body. Anyone who goes more deeply into the background of the controversy
between psychicists and somaticists cannot, however, avoid acknowledging
the fact that the *fundamental* difficulty which divided the somaticists and
the psychicists was indeed that the somaticists assumed a rigorous *separation*
of soul and body which was not accepted in this form by the psychicists.
It is particularly striking that in the somaticist literature (from about 1800
onwards) referred to and quoted by Friedreich we find cropping up time
and again, in an almost endless series of repetitions and variations, the

fundamental somaticist thesis which states that the soul is not corporeal and cannot therefore become diseased.[125] It is therefore certainly no coincidence when Griesinger, in his review of Jacobi's *Die Hauptformen der Seelenstörungen* (1844), characterises the "general standpoint also adopted by the author [Jacobi]" as that of the dualism of soul and body, and regards it as the opposite of, in his view, the empirically "much more probable hypothesis of a direct unity of corporeal and mental phenomena".[126]

Nevertheless, the notion of a dualistic anthropology inspired by Aristotelian thinking is not as preposterous as it may seem on superficial examination. The text of Aristotle's *"Psychology"* proved to be capable of opposing interpretations. Thus, it was possible in the course of the history of anthropological ('psychological') thinking, that the qualification 'Aristotelian' was sometimes used to emphasise the *unity* of soul and body, and at other times to emphasise their *dualism*. In this connection, I recall that in the Christian reception of Aristotle in St. Thomas Aquinas, the existence of an immaterial, i.e. not incarnate, soul was assumed (soul as *forma subsistens*) in order that the doctrine of the immortality of the (individual) soul could be based on it.[127] It was obvious that later, too, this 'Platonic' element would be emphasised, where the concern to maintain the Christian personalistic view of man forced scholars to adopt a strong position against naturalistic and materialistic opponents.

The fact that this element featured so largely in the writings of the somaticist psychiatrists can be explained not so much by the opposition to the materialistic medicine of the Age of Enlightenment (particularly in France: Cabanis and others) as by the conflict with psychicist medicine (psychiatry), which was a much more burning issue in Germany at that time. In the case of psychicist and somaticist psychiatry there were two opposing parties, each of which cast doubt on the other's claims to Christian religious orthodoxy and used this as a basis for fundamental (scientific) psychiatric criticism. The criticism levelled by the psychicists, who were of an older generation, at the younger somaticists' standpoint on the incarnate soul (bodily soul) would probably have sounded something like this : "an unchristian view of the issue which leaves no room for the immortality of the soul!". It is, therefore, understandable that the somaticist theoreticians believed in underlining their view that the somatic theory, besides being in their opinion superior on clinical psychiatric grounds, was not tainted by the reprehensible mixing of psychiatric viewpoints with moral—religious ones, which they felt was characteristic of the work of someone like Heinroth, and that it was therefore also more sutiable to do justice to the Christian religious truth of the immortality of the individual soul.

It will be clear that the methodological limitation of the somatic theory to the '(psycho) somatic' cannot be regarded in abstraction from this religious anthropological setting, for was it not precisely this dualism which provided the justification for this methodological limitation, since the soul was only susceptible to scientific investigation in as far as it was incarnate (or bodily)? A glance at some of the statements made by St. Thomas Aquinas can throw some light on the historical philosophical background in which this view of early nineteenth-century somaticism — whether consciously or not — had its roots.[128]

Thus, we find in a passage quoted by J. Wyrsch[129] from St. Thomas's commentary on Aristotle's *De Anima* a discussion of the question of how psychic or mental disorder is possible. (In support of his own opinion, Aristotle cites the conception of certain philosophers, according to which the soul in its totality [i.e. including the spirit] is imperishable.)

This conception — Thomas's commentary on Aristotle begins here, if my reading is correct — rests on the fact that we see that all defects in the area of our thinking and perception do not affect the soul itself, but are the result of an organic defect. From this it follows that the mind (intellect) and the soul are actually imperishable, and that a defect that manifests itself in the workings of the mind is attributable not to the fact that the mind itself is sick, but to the fact that the organs (namely those of the mind) are defective. For, supposing that the soul can perish, this would have to be caused in the first place by the infirmities of old age; and yet this is not the case, for if you give an old man the eye of a young man, he will be able to see like a young man. The infirmities of old age do not show that the soul itself or the perceptive faculty suffer, but are related to that in which the soul is. Thus, for example, in the case of illness or drunkenness it is not the soul which changes or displays defects, but the body.[130]

Translated into modern psychiatric language, what this says, as Wyrsch observes,[131] is that there exist only organic psychoses. In particular, present-day psychiatrists will have no difficulty in recognising the organic or intoxication psychoses in St. Thomas's observation on mental disorders resulting from illness or drunkenness.

Diseases of the soul, according to this Thomistic formula, are by definition diseases of the organs of the soul. The phenomena of mental derangement are in fact therefore manifestations of disordered psychic functioning and thereby automatically rank as (psychological) symptoms of the underlying organic defects (*'debilitationes'*).

This is indeed precisely the aetiological figure which we encounter in nineteenth-century somaticist psychiatry. The agreement which is thus revealed between the Aristotelian-scholastic standpoint in respect of the essence and origin of mental disorders, on the one hand, and the views of nineteenth-century somaticist psychiatry, on the other, seems to me to be too striking to be coincidental. I therefore consider Wyrsch's suggestion that Aristotle and St. Thomas should be placed in the ranks of the forefathers of the 'somatic theory' as, at the very least, an extremely attractive hypothesis. This despite the fact that research into the historic connection (transmission of his work, etc.) between St. Thomas Aquinas and the nineteenth-century somaticists is to date virtually non-existent,[132] so that the historical support for this hypothesis remains perforce, for the moment, far from complete.

1.2.4. *M. Jacobi as a Representative of Psychiatric 'Somaticism'*

Jacobi can, without doubt, be regarded as the most important representative of the somaticist school of psychiatry, but it must be added that the conception of psychiatry which he developed was not accepted by all the advocates of the 'somatic theory'.[133]

Jacobi's position is characterised by a radicalism which is not found, in the form in which he defended it, among his somaticist contemporaries and which was not followed up by later generations of psychiatrists. One of the most characteristic features of Jacobi's psychiatric theory is his conviction of the *dependent* nature of psychic disorders. So-called diseases of the soul are regarded merely as symptoms; however, not as symptoms of a brain disease (as they are, for instance, by Combe, whose position was akin to Jacobi's),[134] but as symptoms of other somatic diseases. This means that it is not only diseases of the brain and nervous system that can cause the phenomena of derangement, but equally those of all other tissues and organs, bones, skin, etc.[135] According to Jacobi, the facts made plausible the "principle that *all* morbid psychical phenomena ... can only be considered as symptomatic, as concomitant to states of disease formed and developed elsewhere in the organism: a principle which may be regarded as the most important one for the entire science of psychiatry".[136] In other words, mental derangement must be seen as being attendant upon other somatic diseases. For this reason, Jacobi insists on talking about "diseases associated with madness" rather than diseases of the mind or soul.

This 'symptomatological'[137] interpretation of derangement is based on the hypothesis of an *absolute difference between psychic health and disease.* "The acknowledgement of a total separation of the states of health and disease, in the aforementioned relations, must ... not be too unconditional because it forms the basis of faith in all divine and human justice, without which neither ethics nor legislation could exist. Each of them would become a monstrosity, since human existence in general would entirely lose the foothold from which it is destined to attain morality, religion and blessedness by means of the same power of the implanted ideas of goodness, truth and beauty".[138] We can without difficulty recognise here the underlying somaticist idea, which we pointed out earlier, that "the soul cannot become ill (and therefore all so-called psychic diseases must have a somatic basis)".

In order to see the extent to which the dualism of soul and body assumed by Jacobi in his psychiatry differs from a dualism of Cartesian origin, one must begin by ridding oneself of all 'modern', present-day conceptions of the distinction between the somatic and the psychic. The soul–body dualism which is implicit in the distinction between somatic and psychic factors in the genesis of (forms of) mental disorder, familiar to current psychiatry, is a dualism which the nineteenth-century somaticists found suspect. This is not odd if one accepts that — in accordance with the Aristotelian viewpoint — they conceived the (living) body as the body animated by the soul, and that the viewpoint that 'psychic disorders' had a somatic basis meant no more than that in the explanation of so-called diseases of the soul the stress in the aetiology had to be placed on the somatic side of the animated body. This idea excluded a true ('Cartesian') dualism of somatic and psychic factors as two logically independent categories of causal factors, indeed, it presupposed the *unity* of soul and body. Given this presupposition, all 'diseases of the soul', in the strict sense of the term, are 'psychosomatic', that is (manifestations of) disorders in the soul–body unity. A 'somaticist', then, is someone who in his psychiatric theory adopts a 'psychosomatic' standpoint of this kind.[139]

From this we can conclude that when, as in the case of Jacobi, we are dealing with a dualism of body and soul, the distinction between 'somatic' and 'psychic' means something very different from what present-day scientific psychiatric usage suggests. Moreover, we must point out that while, for somaticists like Jacobi and others, 'somatic' was not dissimilar to 'psychosomatic' (in the sense just outlined), the word 'psychic', as a term of contrast, was more or less equivalent to 'moral'.[140] The passage from Jacobi, quoted above, illustrates this point to some extent: it is a requirement of religious ethics to assume an absolute difference between psychic health and disease, which presupposes an absolute separation of the soul (which cannot become ill) and the (animated) body (which can become ill). The context makes it clear that "soul" must be understood here as the 'higher' soul or non-incarnate soul (that is, mind or spirit) of Aristotelian-scholastic tradition and "psychic", accordingly, as "(only) relating to the non-incarnate mind".

It is also precisely this separation which — as an axiom — lies behind the criticism of the principle of Heinroth's 'psychicism', as Jacobi formulates it when he censures Heinroth's sin-theory of psychic derangement on the grounds that it contains an undesirable confusion of ethical and medical viewpoints. What characterises Heinroth's ideas, says Jacobi, is "that he mixes up all that belongs exclusively to psychic life itself [i.e. the life of the incorporeal soul], especially moral deterioration, degeneration and degradation, and certain states of disease accompanied by abnormal psychical phenomena, originating from disturbances of the somatic life. He wants the psychic disturbances to originate from extreme moral deterioration, that is from a purely moral origin, whereas the accompanying organic disturbances, according to him, should be considered strictly as consecutive to those moral ones".[141] And elsewhere: "The moral and religious spirit of the patient . . . has to be considered as playing no part whatsoever, if one is not to abandon either, on the one hand, the idea of disease or, on the other, the idea of moral freedom".[142]

The absolute separation of 'psychic' (for which read 'moral') and medical matters is one of two essential elements in the self-conception of Jacobi's somaticist psychiatry. The other element emerges when he states that the physician as such . . . is somatologist, physiologist and physicist, who in that capacity has to do with psychic phenomena but who simply and solely observes "the organic phenomenon of nature as natural scientist" and whose area of work is that of the "physiology of psychic phenomena".

Jacobi believes that the job of the psychologist is to consider the spiritual, moral and aesthetic significance of the psychic phenomena as a whole, that is to restrict his attention to the aspects actually concerned with the life of the mind as such. The physician, on the other hand, does not concern himself with these matters and keeps his sphere of work strictly segregated from that of metaphysics and religion: research into what the life of the soul itself is, what is its purpose and meaning, what is good and evil, moral and immoral, holy and unholy, and questions about God, freedom and immortality lie outside the scope of his work.

The *non-identity* of medicine and psychology postulated here must be seen in terms of the previously-mentioned distinction between (animated) body and (incorporeal) soul; the 'psychology' meant is a matter of metaphysics and religion. What, on the other hand, gives medicine, or rather psychiatry as a branch of medicine, its *identity*, in positive terms, is that it is — by virtue of the object of its study — the true science of the animated body. This identity and this non-identity are the constituent elements in the self-conception of Jacobi's somatic psychiatry. Neither of these elements, separately or in combination, appears in its true light if one fails to see the context in which they ultimately belong: the *Wirkungsgeschichte* of Aristotelian-scholastic psychology.

1.2.5. *Reinterpretation of the Conflict between Somaticists and Psychicists*

Against this background it becomes possible to get a clearer picture of precisely where the critical point in the opposing views of the somaticists and psychicists must lie. Heinroth's point of view, as we have seen, excluded a somatic cause of psychic disorder. The soul can be stirred by external factors but it cannot be determined by them; it is self-determining and the body is its outward manifestation, that is, the body is not independent but is an organ of the soul. Only the soul has independence. In the somaticists' view, however, the soul Heinroth refers to can only be regarded as the soul in as far as it was incorporeal (and thus immortal), because only this incorporeal soul could claim to be independent. Heinroth's psychiatry was thus in fact concerned with the area of metaphysics and religion which Jacobi had allotted to 'psychology' and which he had therefore expressly excluded from the domain of medicine and psychiatry as a 'somatological' discipline. From Jacobi's point of view, Heinroth undoubtedly had an inadequate understanding of the nature and business of psychiatry: it is not the (incorporeal) mind, but the (animated) body which is the proper object of psychiatric study. The πρῶτον ψεῦδος of the psychicists or, more specifically, the objectionable element of Heinroth's conception, as far as the somaticists were concerned, must have been that it did not acknowledge the distinction between, and separation of, the bodily and incorporeal soul which was fundamental to the somaticist conception.

It is therefore, in my view, the deep dissatisfaction with this lack of distinction which governs the somaticists's criticism of the psychicists, because where they deal with fundamental issues they use either the argument of the incorruptibility of the soul (by which they mean the incorporeal soul) — "the mind cannot become ill" — or the argument of the organic (physical) determination of psychic disorder which is implicit in the assumption of a bodily soul or an animated body. In both cases they take for granted the validity of a distinction which their opponents would not be able to accept — a point which, needless to say, further underlines the fact that the disagreement between the psychicists and the somaticists was based on a question of principle.

If my interpretative reconstruction is essentially correct, then this rendering of the relationship between psychicists and somaticists introduces an important correction of perspective to the accepted view of this controversy. The anachronistic interpretation of this disagreement as a conflict of methodological ideas is no more than superficial. It disregards the ultimate grounds on which these methodological differences are based: the difference in the philosophical conception of the soul and the soul—body relationship.

At a later stage in the history of psychiatry, with mechanism and materialism gaining ground in Germany in the second half of the nineteenth century, the philosophical—anthropological background of psychicist and somaticist psychiatry sank into oblivion and was no longer recognised, and this, coupled with the rise of the natural science view of psychiatry, which was the reverse side of this oblivion, favoured an interpretation which saw somaticist psychiatry as a sort of pre-scientific forerunner of this natural science psychiatry and perceived the conflict between psychicists and somaticists as an anticipation of the problem, always topical in psychiatry, of the possibility or impossibility of psychic causality.

In the eyes of this later generation of psychiatrists, self-assured and proud of the scientific standing of the new psychiatry, psychicism could find no favour. Somaticist psychiatry, on the other hand — although still not accorded the status of true scientific psychiatry — enjoyed, as the pre-scientific forerunner of natural science psychiatry, the prestige of being grandfather to an illustrious grandson. (In this analogy, the 'son' who comes between the two is the psychiatrist W. Griesinger, who will be the subject of discussion later in this book).

Nowadays one would want to remark that these interpretations and evaluations, however understandable given the prejudice against anthropology felt by so-called scientific psychiatry, contain a distortion and are based on insufficient grounds. Heinroth's psychicist psychiatry was not *per se* really methodologically inferior to Jacobi's psychiatry. If we remember that Heinroth defends his method of "pure and complete observation" inspired by Goethe's reflections on natural science, against the speculative excesses of romantic natural philosophy, while — according to the words of L. Snell, quoted earlier — Jacobi's work derives its great significance from the concern expressed in it for "meticulous observation", it becomes clear that the differences which existed between psychicists and somaticists were not primarily differences of method.

Thus, one would like to reconsider the often oversimplified condemnation of Heinroth's psychicism in the light of the twentieth-century anthropological psychiatric way of thinking, which — unaware of having in Heinroth's theory an early nineteenth-century predecessor and kindred spirit — has made us familiar with a concept like 'existential neurosis'. I am thinking here of V. E. von Gebsattel's personalistic anthropological psychiatry, where we come across lines of thought which sometimes display a striking similarity to ideas we encounter in Heinroth's work.

And as far as the psychiatry of the somaticists is concerned, one would like to emphasise that it *differs* too greatly in its philosophical presuppositions from the later 'university psychiatry' to qualify *without more ado* as the origin of the natural science psychiatry of the second half of the nineteenth century.

In general, little or no attention has been paid up to now to the significant differences between the earlier and the later brand of somatic psychiatry, and a continuity has been suggested which did not actually exist. Even a historian of science like E. H. Ackerknecht, so reliable and careful in most cases, cannot, it seems to me, be absolved from blame on this point for, when he states that the somaticists treated mental illnesses as if they were *purely* physical disorders with more or less important psychological symptoms (my italics),[144] this description seems to exclude the mutual *implication* of soul and body which was such an essential element of the philosophical anthropological model of somaticist psychiatry. 'Organic' infirmity or disease, in the anthropological view of the somaticists, is in a certain sense never a 'purely' physical matter, if one implies by this a 'pure' separation of mind and body in the Cartesian sense. It is always an affliction of the 'animated' body, which implies that it is pointless and inadequate to consider physical disease in total isolation from the psychic symptoms which accompany it, just as, in reverse, it is equally pointless and inadequate to see the psychic symptoms as completely divorced from their somatic side.

Jacobi's conception of psychic disorder as *formal* psychic disorder points the same way. Assuming the soul—body unity of the "animated body" psychic disorders manifesting themselves in psychic symptoms can only be conceived as disorders in (or of) the *form* of the (animated) body.[145]

Indeed, the conception of formal psychic disorder and the emphasis characteristic of the somaticists (in contrast to the psychicists) on the somatic aspect of psychic disorder go hand in hand. For, if psychic disorder is related to the form (viz. of the body which is capable of life), this necessarily implies that on the somatic level there is something to be found corresponding to this formal disorder of the soul, which is manifesting itself in the psychic symptoms.

It is the mutual implication of the psychic and the somatic, as we have remarked, which St. Thomas and others see confirmed in the case of mental illnesses in old age, disease and drunkenness, and these 'psychoses' therefore acquire a paradigmatic significance in Thomistic-inspired psychiatry. From this it is understandable that psychiatry of this kind should strike one as 'somatically oriented' psychiatry, since it sets out to conceive *all* psychoses according to the model of 'organic' psychoses.

By now it will have become clear how, and in what sense, the body could acquire methodological precedence in the somaticists' approach to mental illness, and also why this methodological precedence cannot possibly be construed as evidence of a materialistic ontology. As far as this is concerned, it is perhaps nowhere shown more clearly just how dubious were the claims of a later, self-styled 'materialistic' trend in psychiatry that its origins lay in the psychiatry of the somaticists. The early nineteenth-century somaticists were *not* (in Griesinger's words) "uninspired materialists". On this point the passages we have quoted from Jacobi about man's 'higher soul' and about man's moral and religious destination leave no room for doubt that the somaticists' psychiatry, no less than that of the psychicists, presupposed the moral, 'higher' nature of man.

Looking back on the controversy between the somaticists and the psychicists, one would emphasise not so much the differences between these two forms of psychiatry as what distinguished them, taken together as forms of anthropological psychiatry, from the non-anthropological university psychiatry, with its wholly or semi-materialistic thinking in its philosophy of science assumptions, which, switching to the track of Galilean scientific tradition, allowed psychiatry to share in the prestige of modern natural science.

It is, however, hard to deny that of the two of them — Heinroth and Jacobi — it was Jacobi who stood for a psychiatric concept which, despite differences in philosophical assumptions, was more compatible with the views of the later natural science school of psychiatry than Heinroth's 'copulation theory' of psychic derangement.[146] In that sense, perhaps L. Snell was not so wide of the mark when in 1872, in a memorial to Jacobi, he wrote the prophetic words:

... and when, after hundreds of years, a historian of psychiatric science smilingly rejects

much that our time was proud of, he will linger reverently on Jacobi, and indicate the
turning-point in our science brought about by this faithful worker.[147]

1.2.6. Tradition in Clinical Psychiatry despite Discontinuity of Philosophical Presuppositions. Some Reflections on the Psychiatry of L. Snell

Die Hauptformen der Seelenstörungen, Jacobi's principal work and the high
point in somaticist institutional psychiatry, was published in 1844.[148] In the
same year, on the initiative of three prominent directors of institutions, the
psychiatrists Damerow, Roller and Flemming, the first issue of the *Allgemeine
Zeitschrift für Psychiatrie* appeared. Striking evidence of the French influence
on psychiatry at the time, it was modelled on the French journal *Annales
medico-psychologiques*, which had first seen the light of day in the previous
year. It was not until 1867, with the appearance of the *Archiv für Psychiatrie
und Nervenkrankheiten*, published by Griesinger, that the *Allgemeine Zeitschrift* lost its leading position.

The fact that the psychiatry of this period became a truly European
phenomenon, as the result of a rare mutual exchange between French and
German psychiatry, must not be allowed to blind us to the diversity of
psychiatric trends which can be distinguished within it. On the one hand
there are the total views of anthropological psychiatry, which were characteristic of both the psychicists and the somaticists as well as of those
representatives of institutional psychiatry (Damerow, Groos and others) who
sought to achieve a synthesis of these two positions. On the other hand there
is the up-and-coming natural science school, which was to reach its zenith in
Griesinger's psychiatry.

Between these two, both chronologically and in terms of content, the
work of L. Snell, modest in quantity but striking in content, stands alone.[149]
Snell owes his reputation in the field of psychiatry in the first instance to his
'discovery' of monomania as a primary mental disorder. The independence
of this syndrome had until that time not been recognised; Snell carefully
differentiated it from melancholia on the one hand and mania on the other.
The importance of Snell's conception of primary monomania in the history
of psychiatry lies in the fact that, in this conception, before the major work
of pioneering clinicians like Kahlbaum, Kraepelin and Bleuler, the later
conception of paranoia is anticipated.[150]

Less well-known, but certainly as interesting, is Snell's methodologically
new approach to derangement by way of language. Thus in his first study,

Ueber die veränderte Sprechweise und die Bildung neuer Worte und Ausdrücke im Wahnsinn (1852),[151] (On the Changed Speech Patterns and the Coining of New Words and Expressions in Insanity), he takes as his starting-point the use of language by the deranged. In the twentieth century this approach was to prove remarkably viable, particularly in the attempts to understand the world of the schizophrenic. In this context there springs automatically to mind the application of Freudian dream analysis, that is, of the insight into the role and workings of the 'mechanisms' of repression and displacement, and, in particular (because it is virtually specific to schizophrenia), of condensation, conceptual distortion.[152]

The methodological priority of language in research into the mental world of the deranged — so self-evident perhaps to the modern psychiatric investigator — was, however, by no means taken for granted in Snell's day. Here, too, though too little noted and soon forgotten, Snell was the precursor of later developments.

Neither of these innovations, however, should be considered in isolation; they must be viewed in the context of the more fundamental methodological 'reform' which Snell's work brought to the psychiatry of the period.

Snell represents the beginning of the clinical research trend in psychiatry, focussing only on the objective, psychotic phenomenon and at the same time keeping entirely away from mechanistic theorems and from philosophical, anthropological explanations.[153]

Paradoxical as it may sound, what is interesting to us in the 'innovations' introduced by Snell is that they were not really new. The phenomenological approach to clinical psychiatry which Snell adopted in all his work was in fact, on close examination, not an absolutely isolated phenomenon. In this respect, Bodamer is correct when he points to a large number of psychiatric writers who, although they have gone down in the history of psychiatry as representatives of anthropological psychiatry (like, for example, the somaticist Nasse) or of natural science psychiatry (such as Griesinger), occasionally showed themselves, through their writings, to be practitioners of clinical psychopathological phenomenology. If, despite this, Snell's work appears as a 'solitary figure' in the landscape of nineteenth-century German psychiatry, this can only be in the sense that, as an *independent* figure of psychiatric research, it could not find a 'home' in the anthropological or natural science conceptions of psychiatry which were then current.

Thus, although *de facto* already existent in the best clinical psychiatry of the nineteenth century, the approach to psychiatry represented by Snell was only to flourish and acquire methodological legitimacy in the clinical psychiatry of the twentieth century, through Jaspers. Jaspers, in opposition to a one-sided natural science approach, defended the right to exist of a *verstehende* psychology and psychopathology, without its thereby merely [*sic*] becoming anthropological psychiatry.[154]

Thus, it is not overstating the case to say that reference to the work of L. Snell makes us aware of a real continuity in the history of clinical psychiatry which connects nineteenth- and twentieth-century psychiatry. The continuity which has continued to be preserved, despite all the theoretical and implicit, or explicit, philosophical differences in psychiatric thinking in the course of its development, is the continuity of a certain style in the approach to psychiatric reality. This 'clinical' style can be found equally well in the work of early institutional psychiatrists like Jacobi and Snell as in the work of Griesinger, Kraepelin or Binswanger. It is, indeed, the belief in clinical observation and description as the ultimate basis for testing psychiatric (psychopathological) insight which unites all these psychiatrists. This belief was strikingly expressed by H. Ey when he wrote, although without mentioning Snell by name in this connection:

For the psychiatrist who knows psychiatry, for anyone who has read the great clinical descriptions by Esquirol and by Griesinger – by Laseque and Schulte – by Falret and Westphal – by Magnan and Kahlbaum – by Kraepelin and Seglas – by Bleuler and Pierre Janet – it is the value of these clinical studies which is the foundation of psychiatric science. Hypotheses, theories, aetiopathogenic conceptions come and go, but clinical experience (*la clinique*) remains the 'pedestal' on which psychiatry rests.[155]

CHAPTER 2

THE MECHANISTIC VIEWPOINT IN
NINETEENTH-CENTURY PHILOSOPHY AND SCIENCE
(PSYCHOLOGY AND PHYSIOLOGY)

2.1. MECHANISM: TERM AND CONCEPT

The transition from anthropologically-oriented institutional psychiatry to so-called university psychiatry, which regarded itself as a natural science and which we associate with the name of Wilhelm Griesinger, is a change indicative of a comprehensive reorientation in scientific thinking in accordance with the so-called mechanistic viewpoint.

'Mechanism' is not an unequivocal term. As Dijksterhuis, looking at the history of natural science and philosophy in the seventeenth and eighteenth centuries, demonstrated, a distinction must be made at the very least between Cartesian mechanism and Newtonian mechanism. According to Cartesian mechanism, physics can only be studied by using mechanistic proofs (*des raisons de méchanique*), that is to say, with no other principles of explanation than the concepts dealt with in mechanics — geometrical definitions, such as shape, size and quantity, which mechanics, as a branch of mathematics, uses, and motion, which is its specific object. The difference between this and Newtonian mechanism lies in the fact that precisely that characteristic which, since Newton, we have been accustomed to consider as the quintessence of the mechanistic explanation of nature, i.e. the reference to the concept of force, is not found in the Cartesian definition of the concept of mechanism.[156]

What in the eighteenth century (I am thinking here particularly of Kant) and in the nineteenth century was defended and criticised as mechanism is, invariably, a scientific viewpoint which took Newtonian mechanics as its model. This is also the mechanism which we regard as the antithesis of the so-called vitalism of the nineteenth century. In short, it is mechanism as we understand the term today and this meaning is assumed in the following observations on the mechanistic ideal in philosophy and science (psychology and physiology) in nineteenth-century Germany.

Since the arguments for the mechanistic scientific view in the period we are studying involved widely divergent philosophical presuppositions it is, at the very least, desirable from the viewpoint of the history of science and philosophy, to distinguish between different variants.

37

It therefore seems to me advisable to distinguish first of all between mechanism which is somehow combined with a finalistic or holistic viewpoint in science or philosophy, and a meaning of mechanism which is not. The mechanism of Herbart, Müller, Griesinger, and Lotze — mechanism *in the broad sense* — is of the first kind, that of Helmholtz and his followers — mechanism *in the strict sense* — is of the second kind.

Fate has decreed that mechanism in the strict sense often came to be described as a position which is not only monistic but also materialistic.[157] This latter assumption is based on a misunderstanding: the mechanism (in the strict sense) which reached its peak in Germany in the third quarter of the nineteenth century is found in both a materialistic and a non-materialistic, namely positivistic, variant.

Mechanism as "the only possible and the only admissible type of rational understanding",[158] that is to say, what I have called mechanism in the strict sense, was the predominant scientific viewpoint in Germany for the greater part of the second half of the nineteenth century. In (theoretical) physics — usually considered as science *par excellence* — mechanics was regarded as the first science of reality. It was only towards the end of the century that there began to be doubts about the legitimacy of the pretensions to absoluteness of the mechanistic intelligibility model. These doubts were prompted by problems which were encountered in the fields of thermodynamics, optics and theory of electricity. As early as 1872 Mach was making an express distinction in his epistemology between the requirement of the *causal* understanding of nature and the postulate of the *mechanical* knowledge of nature.[159] Cassirer points out in this connection that these two problems are not differentiated either by Helmholtz or by Wundt, both of whom were intent on "deducing [the axioms of mechanics] as simple inferences from the general thesis of causality".[160] Mach, on the other hand, demonstrated that it is impossible to prove that strict and 'real' causality must be mechanical in nature.

This was a significant blow to the mechanism which drew its strength from precisely this non-distinction between the problem of causal understanding and the problem of the mechanical explanation of nature. The relativisation of the mechanistic viewpoint introduced by Mach was the start of a development in an instrumentalist and conventionalist conception of science with which the names of Hertz, Poincaré and Duhem, as well as that of Mach himself, are associated. The fierce conflict between the proponents of mechanics and those who championed energetics, in which this development had its roots historically, took place, however, in the last twenty years of the nineteenth century [161] and is therefore not of immediate importance to our study.

2.2. THE PHILOSOPHICAL BACKGROUND

The period from about 1834 to about 1868 — i.e. from the time when Griesinger began his studies until the time of his death — displays at first sight a confusing multiplicity of philosophical views.

In 1831 Germany's most influential philosopher of the period, G. W. F. Hegel, died. Until about the middle of the century his philosophy was predominant in German universities. But in the twenty years between Hegel's death and the sudden collapse and downfall of German idealism, conflict arose between the followers, the 'epigones' of Hegel's philosophy, which led to their splitting into a 'right' and a 'left' wing of so-called *Althegelianer* and *Junghegelianer* (Old Hegelians and Young Hegelians), respectively. A multitude of conflicting interpretations of Hegel were mooted, all stemming from a one-sided emphasis on certain aspects or elements of his philosophy. Under the powerful spell woven by this philosophy in the closesness of its systematic construction, Hegel's followers tried to keep alive the spiritual legacy of the great philosopher, but what was thus clung to as Hegelian philosophy in fact amounted to little more than systematics adapted to the philosophical preferences of its interpreters. With the exception of Marx's work and, to a lesser extent, that of Feuerbach, posterity's judgement of the literature produced by these pale imitators is hard and unflattering. It is tempting to say that what lived on in the works of these epigones in the twenty years following Hegel's death was not so much the spirit as the body of his philosophy. When the spirit had flown, all possible means were tried to breathe new life into the body — in this case with no convincing results.

The counter-forces which had already been present during Hegel's lifetime, but which had then been unable to achieve validity, now got their chance. What Schopenhauer's invectives against the philosophy of his celebrated colleague at the University of Berlin had been powerless to achieve — the dethronement of Hegelian philosophy — was now automatically brought about by the changing times. In 1841, at the request of King Frederick William IV, Schelling went to Berlin charged with the task of destroying the "dragon's teeth of Hegelian pantheism". However, despite the initial enthusiasm with which many people [162] greeted his prophecies of the dawn of a new era, interest soon waned and Schelling officially discontinued his lectures in 1846. His mission had not had the hoped-for success and Schelling was destined to outlive his fame. By the time he died, in 1854, his work had been all but forgotten. The last great representative of German idealism had been, so to speak, overtaken by the spirit of the new age.

What characterised this new age was not, however, as a persistent, almost ineradicable misunderstanding in history of philosophy textbooks will have it,[163] materialism, but the antagonism between metaphysical idealism and positivism. It was undoubtedly these two trends which were to dominate the philosophical scene after Hegel's death – positivism, which was defended by prominent scientists (Helmholtz, Du Bois-Reymond, etc., and in a different sense by Dilthey and others), and the different variants of metaphysical idealism of thinkers like Schopenhauer, Lotze, Fechner and (later) Hartmann; in comparison, so-called materialism played a minor role within philosophy itself.[164]

For the moment, however, I am less concerned with the contrast between these two schools than with seeing what links this German positivism and metaphysical idealism together and thus distinguishes them most clearly from the philosophy of the leading speculative systems of the 'older' generation. Both positivism and metaphysical idealism in the second half of the nineteenth century are related to the stormy developments which were taking place in science during this period. In the case of positivism this seems self-evident, but it is also true of the idealistic systems of Lotze, Fechner and Hartmann, which were so typical of the period, because what these metaphysicists were aiming at was precisely to justify along idealistic lines the mechanistic view of nature which had achieved supremacy with the ('mechanical') natural sciences.[165] The development of these branches of science gave food for thought, and viewed in this light positivism and metaphysical idealism can be interpreted as answers to the problem which science posed for a generation of thinkers for whom the pretensions to knowledge and the rationalist ideals of the great idealistic systems had lost their credibility and interest, and for whom the very fact of the existence of 'mechanical' natural sciences demanded a philosophical reorientation.

It is therefore the *problem of the relationship between philosophy and science* which was of crucial importance in the second half of the nineteenth century, and since it was Kant's critical philosophy which had in fact opened up the dimensions of the problem we are concerned with here, it is not surprising that during precisely this period his work had an inspirational effect, particularly among the more critically inclined scientists. In our outline of the situation we must, therefore, first of all look at some aspects of Kant's work.

2.3. KANT AND THE PROBLEM OF THE RELATIONSHIP BETWEEN PHILOSOPHY AND SCIENCE

One needs no great knowledge of the history of Western European philosophy and science to observe that, until the last quarter of the eighteenth century, so little distinction was made between the two that it is difficult to tell where the one begins and the other ends. Locke, Hume, Descartes, Malebranche, and Leibniz are to be found both in a history of philosophy and in a history of science (for example, psychology, physics, mathematics, etc.).

When, at this point, with the dawning of the era of so-called 'modern' philosophy, two different conceptions of science are seen to grow up, typifying so-called rationalism and empiricism respectively, it must be noted that in both cases the relevant conception of science was not defined on the basis of the difference between science and philosophy, but presupposed the unity of the two. This means that within each of the two philosophical camps there was a reigning conception of science which defined both science *and* philosophy: in the case of rationalism this model function was represented by mathematical (i.e. axiomatic-deductive) theory, in the case of empiricism by (empirical) physics.

Kant's 'critical' philosophy was to change all this: the symbiotic unity between philosophy and science of the pre-critical period was dissolved, and philosophy and science parted company, that is to say, their non-identity was validated.

This event is often described (not without justice) as the divorce or the division of property between philosophy and science: the former (i.e. in pre-critical philosophy) 'conjugal' symbiosis of philosophy and science is broken, or, putting a different emphasis on it, once the marital ties between philosophy and science have been broken, it is necessary to determine what is due to each of the parties in the case from the property which they had previously held jointly. Kant's 'critical' philosophy was to perform this task of 'legal' separation. The outcome of this philosophical labour is contained in his *Critique of Pure Reason*. The results of this work are well-known: (1) the domain of scientific knowledge is delimited ("the limits of our knowledge are the limits of our possible experience"), but not without (2) a transcendental philosophical foundation of this (thus limited) possibility of scientific knowledge.

Once Kant has given the problem of the relationship of philosophy and science this forceful expression, it was no longer possible for a mature philosophy to pretend that the problem did not exist: all philosophy *after* Kant is, on this point, either a (modifying) imitation or a criticism (whether or not actually expressed) of Kant – a return to the 'naivety' of pre-critical philosophy is no longer possible. This is also, of course, true of the philosophy of the period with which we are now concerned.

Thus the thinking of the metaphysical idealists (Schopenhauer, Herbart, Lotze, Fechner, etc.) and of the positivists (Helmholtz, Du Bois-Reymond, and others), who are important to our elucidation of the situation in Germany around 1850, stands in a sense in the shadow cast by Kant's critical philosophy. This is by no means to say that these thinkers conceive the relationship between philosophy and science in exactly the same terms as Kant did. On the contrary, the *metaphysical* justification of the mechanistic viewpoint attempted by one side and the other side's pursuit of an *empirical scientific* foundation of the theory of knowledge, if it did not bring them into direct opposition with Kant, must at least alienate both the metaphysical idealists and the positivists from the intentions of Kant's transcendental philosophical legitimation of (natural) scientific knowledge. It can, however, be said that throughout all these conflicting views, Kant's philosophy remained an inescapable point of reference for both parties, whether scholars criticised and dissociated themselves from Kant's interpretation of the relationship problem, or whether they wanted to make a positive reference to his philosophy and, in the call 'back to Kant', found fitting expression of their own conviction that only a reconsideration of the problem of knowledge inspired by Kant's critical philosophy could offer any prospects.[166]

Apart from the fact that this philosophy of Kant had laid the foundations for the questions being asked by the metaphysical idealists and positivists, there were two principal factors which meant that his work became important at precisely this time – around the middle of the nineteenth century. One of these factors lay outside the field of philosophy: I am referring here to the stormy development of 'mechanical' natural science, which we have already touched on, that occurred from about 1840 onwards. The most notable feature of this was primarily the speed with which the application of the viewpoint of mechanistic causation spread throughout physiology, biology, chemistry and, above all, psychology. The other factor was to be found in Kant's philosophy itself and concerns the conception of science which Kant assumed to be self-evident and which, related as it was to Newton's mechanics, was in itself enough to make his philosophy attractive and interesting to a generation of scientists who were committed to achieving the victory of the mechanistic scientific ideal.

The radical change in the relationship between philosophy and science which took place after 1850 can best be described, in the light of later developments,[167] as a shift in the treatment of the problem of knowledge, that is, as a shift away from philosophy towards science. The pre-eminence of philosophy, which had been taken for granted for centuries, was gradually lost, and the theory of knowledge became the business of the various branches of science, which now considered themselves autonomous. Philosophy, as Cassirer observes,[168] was no longer willing to accept the leading role it had played in the past and instead of

representing, on its own responsibility, a certain ideal of truth, would rather allow itself to be led by the specialist sciences and to be pushed by each of them in a particular direction.

The unparalleled splintering of the theory of knowledge which resulted from this was to become particularly obvious in the clash over psychologism and can be documented by reference to the succession of conflicting trends in the theory of knowledge which emerged in the latter years of the nineteenth century, each of which had its origins in one or other of the specialist sciences: logical formalism, psychologism, mathematicism, physicalism, biologism.[169]

The fact that in this situation, around 1850, the successful 'mechanical' natural sciences acquired a special, trend-setting significance is immediately apparent from the almost universal spread of what I shall call, for the sake of brevity, the *mechanistic 'prejudice'* of philosophy and science in Germany at that time.

The presupposition that one can only speak of science in the strict sense of the word where it conforms to the mechanistic scientific ideal can be found, during this period, in the works of philosophers and (natural) scientists of almost every school: among the metaphysicists like Herbart, Schopenhauer and Lotze, among the positivists like Helmholtz, Du Bois-Reymond, etc., among the ('naive') materialists like Büchner, Vogt and Moleschott, as well as among the various scientists: in Griesinger (medicine and psychiatry), Liebig (chemistry), Ludwig and Brücke (physiology), Virchow (cell pathology), etc..

From the historical point of view it seems useful to distinguish two phases in the process of the rise and spread of this mechanistic view of science. These phases differ from each other not only chronologically,[170] but also because in the first phase the initiative for the mechanistic conception came primarily from philosophy, while in the second phase it was principally the leading scientists of the day who defended mechanism. This division more or less corresponds with the shift in emphasis from metaphysical idealism to positivism, and thus from what I have previously distinguished as mechanism in the wider sense to mechanism in the stricter sense, as well as with a perceptible change of stress within the sciences which are of primary relevance to our study from (non-physiological) psychology to physiology (physiological psychology).

If one proceeds on the assumption that the term 'mechanism' can only meaningfully be used where the extrapolation from the example of mechanics is expressly defended and carried out, one would have to regard the psychology of Herbart as marking the start of the history of mechanistic thought in nineteenth-century Germany. As far as mechanism is concerned, however, Herbartian psychology (like all mechanistic psychology in Germany during the nineteenth century) was influenced by Kant's philosophy,[171] a fact that is all the more clearly underlined by Herbart's view that his psychology had the added significance of demonstrating the realisability of a conception of psychology which his predecessor in the Chair of Philosophy at Königsberg, although regarding it as possible *in principle*, had considered unfeasible *in practice* (and which he had indeed left unrealised in his own work).

Without Kant, the psychological mechanism of the 'philosophical' phase and that of the positivistic phase are equally inconceivable. I must therefore first say something about the significance of Kant in the mechanistic self-conception of nineteenth-century psychology, as a preliminary to all subsequent observations about individual representatives of the metaphysical-idealistic, positivist, or even (naive) materialist schools.

2.4. THE SIGNIFICANCE OF KANT'S PHILOSOPHY FOR THE MECHANISTIC SELF-CONCEPTION OF NINETEENTH–CENTURY PSYCHOLOGY

Looked at from Kant's point of view, one can conceive the problem of the possibility and meaning of a scientific (i.e. natural science) psychology as the problem of determining a meaning of 'freedom' which leaves intact the meaning of the concept of so-called transcendental freedom (fundamental to Kant's theoretical philosophy) on the one hand and that of so-called practical freedom (the basis of his practical philosophy) on the other, and which at the same time renders insightful the applicability in principle of the mechanistic concept of science to psychology.

The solution to this problem which Kant offers is to be found in his concept of *psychological freedom*. Psychological freedom, to start with, is not transcendental freedom, i.e. what, to Kant, is freedom in the real sense; freedom in the transcendental sense is defined by Kant as "the capability of a condition to start *itself*, i.e. which is not dependent on another cause according to the law of nature which determines it in time",[172] that is, as a freedom which is not of the order of the knowable, i.e. of the order of phenomena, and which therefore, unlike all phenomena, is also not subject to natural causality. It is thus, so to speak, a causality *sui generis* ('the causality of freedom'). As such, transcendental freedom is a property of our will − not in so far as it appears in visible actions, but as the so-called pure or noumenal will: will as 'intelligible' cause or *causa noumenon*.

I can have no (empirical) knowledge of the noumenal or the sphere of the thing-in-itself, because by definition it does not fall into the realm of experience (by which is meant natural science experience). What can be experienced, given the presuppositions of Kant's philosophy, are by definition phenomena (as distinct from the things-in-themselves or noumena). The phenomena in the 'phenomenal' world − i.e. the world in so far as we know it − are subject to the law of cause and effect. Everything that is causally determined in this way Kant calls 'nature'. This also includes man in so far as he is a *natural* being. It is with a view to what we would call the closed system of causal determinateness that Kant refers to the "*mechanism of nature*" (i.e. "every necessity of the association of occurrences in time, according to the natural law of causality") and he leaves no room for doubt that man in his psychological aspect is included in this: "whether one calls the subject in whom this passing in time occurs [Kant may have had Hobbes in mind] *automaton materiale*, or, with Leibniz, (automaton) *spirituale*, because the mechanism is operated by matter or by ideas respectively".[173]

In other words, if it is true that psychology as the science (by which is meant a natural science) of the "nature of the soul and the incentive of the will"[174] has to do with (psychological) phenomena, then these are phenomena in the context of cause and effect (as are *all* phenomena, according to Kant) and psychology must be conceived as a causally explaining science, a natural science of the phenomena of the mind, or what was only later (in conformity with the methodological dualism of understanding and explanation promoted by Dilthey) to be described as *natural science psychology*.

From the above it will have become clear how Kant was able to arrive at his definition of psychological freedom, which at first sight is ambiguous, not to say paradoxical. Psychological freedom concerns the independence from external factors of our acts of will, or in other words: our acts of will *being determined* by ideas and emotions (as internal, 'psychic' factors), thus something like causal determination in the realm of the psychic (psychic phenomena). Psychological freedom, as Kant conceives it, is therefore no exception to the rule of the universal determinism of natural causality: as far as my (acts of) will are determined by internal (i.e. 'psychological') factors (ideas, emotions, etc.) I am free in the sense of 'psychological freedom'.

The fact that Kant was led to use the term (psychological) freedom, when he in fact meant the causal determinateness of our acts of will, could betray one into concluding that Kant denied free will. Nothing, however, could be further from the truth.

The paradox of freedom conceived as determinateness is only insoluble where the mechanistic view is absolutised. It vanishes, however, as soon as one realises that it is relative; that is, relative to this (mechanistic) view. (This relativity is already inherent in the fact that so-called psychological freedom relates to a state of affairs which belongs exclusively to the realm of phenomena.)

The concept of psychological freedom, as we have already observed, thus does not represent for Kant 'real' freedom, that is, that freedom which he has in mind when he postulates the irrevocable categorial difference between nature and freedom. What Kant calls psychological freedom is in fact freedom *sub specie naturae*, and that, according to his own presuppositions, can never be anything but freedom *in an improper sense.*[175]

2.5. THE IMPLICATIONS OF THE NATURAL SCIENCE SELF-CONCEPT OF PSYCHOLOGY

The radical separation of the phenomenal and the noumenal world thus made it possible to maintain a concept of psychological freedom alongside and distinct from a concept of transcendental freedom and in this way to conceive of a meaning and possibility of natural science psychology in the mechanistic sense which was compatible with the defence of the concept of freedom in the proper (i.e. transcendental) sense.

The implications of this conception of psychology are important. They become clear in Kant's criticism and reinterpretation[176] of the concept of the substance of the soul from the so-called rational psychology of the metaphysical tradition. One could summarise Kant's criticism very briefly by saying that rational psychology does little more than make explicit the content of the concept of the substantial soul, and it does that by means of a number of suspect arguments ('paralogisms') which, taken together, purport to demonstrate that the soul in essence is a singular substance, numerically identical throughout time and related to the possible objects in space. Close analysis, however, reveals that nothing is proved and that what is actually involved here is an *objectivistic* exploration of the concept of soul (or subject), through which it is not possible to think of precisely that meaning of subjectivity which I must assume if I am to account for the possibility of (objective) knowledge at all (the subjectivity of the so-called transcendental subject, as opposed to that of the empirical subject).

From this understanding, it can be said that criticism of rational psychology or its objectivism opened the way to a non-substantialistic, or if you will 'desubstantialised', psychology, or, as it is sometimes described, 'psychology without soul'. This last expression is somewhat misleading on one point, because it can give the impression that it refers to a form of psychology in which the *concept* of the soul no longer has any place. What does, however, disappear as a result of Kant's criticism is only the concept of the *substance* of the soul, because the object of psychology is no longer defined as the substantial soul, but as the 'phenomenal' *life* of the soul (which for Kant is equivalent to the life of the consciousness), conceived as embracing all the phenomena of consciousness which are given to me in the 'inner sense'.[177]

On the other hand, the traditional concept of the substance of the soul is at the same time reinterpreted within the presuppositions of critical philosophy and in this reinterpretation it is to a certain extent saved. The soul (like the world and God) becomes one of the so-called transcendental Ideas of pure (theoretical) Reason, that is, one of the 'regulative' concepts, which serve as the highest viewpoints for the systemisation of the empirical material determined by the categories and which as (purely) thought-of (but not known) totalities represent the limit which this endlessly advancing experience approaches, without ever being able to reach. If, regarded in this way, the transcendental idea of the world is the conceived limit of the total empirical knowledge of external reality as this is revealed to us in the 'outward sense' and towards which we are constantly moving in the study of external reality (as far as we can know it), then in the same way it is true to say of the transcendental idea of the soul that it represents the hypothetical totality of perfect empirical knowledge of internal (i.e. psychic) reality, as this is made accessible to us by our 'inner sense'.

Thus, we see on the one hand that Kant gives up the possibility of philosophical (i.e. metaphysical) psychology, and this includes the possibility of psychological 'total knowledge', but on the other hand we find that psychology as an empirical science remains possible for him as the way in which we approach, in the endless process of experience relating to internal (psychic) reality, the limit conceived of in the idea of the soul.

<h2>2.6. KANT AND THE PROBLEM OF THE POSSIBILITY OR IMPOSSIBILITY OF SCIENTIFIC PSYCHOLOGY</h2>

As I have observed in relation to Kant's conception of psychological freedom, it is quite clear that Kant considered psychic life (the life of the consciousness) to be as rigidly determined as external natural occurrences. In this respect, psychology, to him, is thus on a par with natural science. Indeed, it is not only our acts of will or emotions which are phenomenal, but also our thinking; that is to say, looked at from the empirical psychological viewpoint, they are all subject to fixed rules. In this respect, the introduction to his *Logic* [178] is quite clear. Just as there is no disorder (*Regellosigkeit*) in nature, so there is none in the *Ausübung unserer Kräfte* (the exercise of our powers) or in the mental operations of our understanding. (Kant takes as his example the use of language, which takes place according to conscious or unconscious – but usually unconscious – rules of grammar.) The distinction which Kant makes in this connection between the necessary and the contingent rules of our thought accentuates the fact that alongside logic (which is solely concerned with the necessary rules which "are understood *a priori*, i.e. independent of any experience, because they comprise only the conditions for the use of the understanding in general, either in a pure or empirical way") there is room for an empirical ("psychology of thought") view of our thinking, which has as its object of investigation the contingent rules of our thought.

The conclusion must in my view therefore be that while Kant considered the rational psychology of dogmatic metaphysics to be impossible, he has retained the possibility *in principle* of empirical psychology, embracing our thinking, feeling and will. As he based this possibility on the fact that *all* phenomena (external *and* internal) are subject to rules, one may infer that Kant felt that if psychology as an empirical science was possible at all, then it was in the manner of natural science.

Kant's position on this point is of major importance to the history of later psychology, because his postulate of the determinedness of psychic phenomena laid the foundations for the development of scientific psychology.

The problem which confronts us on a closer examination, however, is that Kant, by adopting this position, laid the basis for a type of psychology — natural science psychology — which nevertheless was, in his view, impossible to realise in practice. This is a real crux in Kant's philosophy. When he argues that psychology as a (natural) science is not in fact possible because psychic phenomena are given one-dimensionally (in the course of time) and can therefore not be determined mathematically, he is arguing *de facto* from the specificity of psychic phenomena, which would mean that the impossibility of psychology's becoming a true natural science is a *fundamental* one because it is rooted in psychology's object of study. In this case we run up against a contradiction: psychology as a true (natural) science is and is not possible in principle. The contradiction remained implicit and unresolved in Kant's work. It is particularly when one considers later developments in nineteenth- and twentieth-century psychology that one is led to pay attention to this difficulty.

It was obvious that German scientific psychology in the second half of the nineteenth century would latch on to Kant as the man who had introduced the concept of psychology as a *science* (as opposed to philosophy). The importance of this achievement overshadowed the fact that Kant had also said that psychology was not possible as a science. The great master was mistaken; it was obvious that mathematics could be applied to psychic phenomena, psychology as a true science *was* possible. In this way, a development in psychology which Kant had not foreseen was able, running (as it were) counter to Kant, to confirm the validity of his own mechanistic principles. It is probable that from the standpoint of this later form of psychology, scholars in fact believed that Kant's verdict of the impossibility of scientific psychology concerned an actual and not a fundamental impossibility. This interpretation is, however, difficult to sustain on closer examination as it rests on an untenable assumption. As I have just demonstrated, Kant himself argued this impossibility on the grounds of the specificity of the methods and object of study of psychology; that is, he understood it to be a fundamental impossibility which could not be nullified by any actual development in psychology.

It seems to me, however, worthwhile to consider whether it is not the case that Kant tried to support a *genuine* insight into the specificity of the methods and object of study of a form of non-natural scientific, humanistic psychology he himself practised (I am referring to his *Anthropologie in pragmatischer Hinsicht*) — that is, the insight that the mechanistic way of thinking is inadequate in this form of psychology — with an argument that

he only found attractive because it would not force him to relativise or abandon the mechanistic scientific ideal. That the mechanistic conceptualisation
of the phenomena described in his "pragmatic anthropology" is inadequate,
is not explicitly acknowledged or expressed, but it is nevertheless implicitly
presupposed in the conception and execution of the work. If it is allowable to
argue on the basis of what Kant in fact does in his *Anthropology*, then we
are forced to say that in his argumentation, as far as the fundamental impossibility of mechanistic psychology is concerned, he has given a pseudo-reason,
and what was 'seen' as genuine he himself *misunderstood*.

But this 'self-misunderstanding' is also important, if only because it
managed to survive without any essential modifications until the twentieth
century, among both natural science psychologists and the early humanistic
psychologists. The heart of the matter − the so-called naturalisation of the
consciousness − was not subjected to any truly fundamental criticism until
the advent of Husserl's work in phenomenology.[179]

2.7. KANT'S INFLUENCE ON THE RISE AND DEVELOPMENT OF
NINETEENTH-CENTURY SCIENTIFIC PSYCHOLOGY

Leaving all else aside, it goes without saying that Kant's philosophy played a
decisive role in the rise of (experiential) scientific psychology. With his
postulate of the determinedness of all phenomena, and therefore also of
psychic phenomena, he allowed later scientific psychology to proceed with a
clear conscience, while the example he set with his 'pragmatic anthropology'
could be held by those of a different persuasion to strengthen their conviction
that *as well as* natural science psychology another − humanistic, *verstehende*,
etc. − form of psychology could be defended.

For a proper understanding of Kant's significance for the history of
nineteenth-century psychology it is important to bear in mind that the
conception of scientific psychology which he initiated and made possible is
the conception of psychology as the psychology of *consciousness*. This was
a logical consequence of the fact that in Kant, with the redefining of the
object of psychology as the life of the mind (instead of the substance of the
mind, as in rational psychology), the life of the mind was identified with the
life of the consciousness (embracing all phenomena of the consciousness).
So-called 'classical psychology of consciousness' did, it is true, only make its
appearance after Kant − its period was the second half of the nineteenth
century − and many thinkers (philosophers, psychologists, physiologists) of

the so-called pre-Wundtian era contributed to its existence,[180] but it is largely thanks to Kant that empirical scientific psychology of this kind could be, and was, conceived.

Certainly, the notion of a life of the consciousness which could be studied empirically — the foundation of empirical scientific psychology of consciousness — was not discovered by Kant. The British empiricism of philosophers like Locke and Hume was in this sense essentially psychology of consciousness. This psychology (philosophy) did not, however, yet perceive itself as empirical scientific psychology *distinct* from philosophy: this division of property could only take place from the standpoint of Kant's criticism and it is therefore correct, in principle, to consider Kant, rather than the British empiricists of the seventeenth and eighteenth centuries, as marking the start of the history of the conception of strictly *scientific* psychology. Not until Kant did it become possible for psychology to perceive itself as a true *science* and to become conscious of itself as science. Kant's contribution to this dawning of self-awareness was a decisive one.[181]

The 'classical' German psychology of consciousness which then came into being was still, as far as its conception of consciousness was concerned, clearly dependent on Kant's philosophy of consciousness. If Kant's philosophy had conceived transcendental consciousness as embracing all the *a priori* rules of synthesis by means of which experience is made up, it was obvious that the object of empirical scientific psychology — the phenomenal life of the consciousness, i.e. the consciousness as it appears to us in experience — could be conceived as embracing all the *contingent* rules of synthesis. In other words, if the object of examination of the critique of reason is transcendental consciousness as the principle of subjective[182] synthesis in its *a priori* universality and necessity, then the empirical consciousness which is the object of empirical psychology can be defined as the organ of subjective synthesis in its *phenomenal* aspect.[183]

Psychology according to this formula was thus essentially a science concerned with seeking out the *laws* of the life of the consciousness (and was therefore a nomological science). Such a conception of psychology was only fully realised from about 1870 onwards, after the work of John Stuart Mill had come into its own in German psychology[184] and Kantian mechanism and Mill's empiricism had been assimilated in the form of Wundt's 'apperception psychology', which was to usher in a new era in the history of scientific psychology.

Generally speaking, it can be maintained that German psychology of consciousness, which flourished in the period from about 1850 to 1900, conceived itself as a natural science psychology in conformity with the ideal of mechanism. (Even in the case of Wundt, whose psychology cannot simply be equated with mechanistic elemental or association psychology, the mechanistic viewpoint retained at least a relative validity.) The fact that Kant's philosophy must in the first instance be held responsible for this mechanistic self-conception serves further to underline the paradoxical nature of Kant's attitude to psychology. Kant's denial of the possibility of a (natural) scientific psychology and the existence of his own non-natural scientific, 'humanistic' psychology in his *Anthropologie in pragmatischer Hinsicht* are diametrically opposed to the fact that his thinking made possible and introduced a development in (scientific) psychological thought which removed the last impediments still standing in the way of the total victory of the mechanistic conception of science.

If Kant had, so to speak, given the idea of mechanistic imperialism his blessing, the actual incorporation of psychology into science was only gradually realised through the work of a series of other thinkers, of whom the most important are Herbart, Fechner, Lotze, Beneke and John Stuart Mill.

In what I have called the first phase of the genesis of the mechanistic concept of science in nineteenth-century Germany, the accent lies, as I have observed, with philosophy and psychology, in the sense that philosophy took the initiative in applying mechanism in the field of psychology. This is primarily true of Herbart. In the case of Lotze this situation has already changed to the extent that, retaining the philosophical orientation, the scientific emphasis has shifted more towards physiology (and physiological psychology, 'physiology of the soul') and its application in medicine. They will both be looked at in detail in our examination of Griesinger's mechanism, while Lotze and Schopenhauer will be discussed in the final section of this chapter.

In the work of the pupils of the great physiologist Johannes Müller – Helmholtz, Du Bois-Reymond, Brücke – and in that of Carl Ludwig, philosophy loses its precedence over science. Lotze's metaphysical idealism concedes its leading position to the positivism of natural scientists like Helmholtz, who was four years younger, and the second phase in the history of the rise and spread of the mechanistic conception of science begins.

2.8. THE ROLE PLAYED BY PHYSIOLOGY IN CONSOLIDATING THE MECHANISTIC SELF-CONCEPTION IN NINETEENTH-CENTURY GERMAN SCIENCE

In the period between 1830 and 1850 physiology gained in importance and prestige as a basic science of medicine, and every university in Germany set up physiology laboratories and established chairs of physiology.[185] This development would have been inconceivable without the work and the personality of Johannes Müller (1801–58).

The centre from which this German physiology spread was the physiological institute in Berlin, of which Müller was appointed director in 1833. This was almost the only place in Germany, and certainly the foremost one, where young, promising scientists were educated in this profession.[186] The history of nineteenth-century physiology in Germany is therefore that of a development largely brought about by Müller and his pupils. However, it must be noted here that when Müller's empirical physiology (that is, methodologically speaking, a predominantly qualitative form of physiology with a comparative anatomical and morphological orientation) had passed its peak and been superseded, towards the end of the eighteen-fifties, by an experimental form of physiology, based on physics and chemistry, which had gained particular support among Müller's pupils, it was primarily Carl Ludwig (1816–95) – admittedly not one of Müller's pupils – who determined the direction and the way in which physiology developed during the following hundred years.

Johannes Müller and Carl Ludwig, who was fifteen years younger, are therefore the two most important German physiologists of the nineteenth century; each of them great in a very different way, each representative of a stage of development in the history of nineteenth-century physiology, the latter starting in his own, new way, while the former approached the limits of his way of tackling physiology.[187]

The biological school of thought in physiology, for which Müller was the most eminent representative, had to make way of the conceptions of a younger generation of physiologists. It was the generation of his pupils, H. Helmholtz, E. Du Bois-Reymond, E. Brücke, and above all, as I have pointed out, of Carl Ludwig and his pupils, which was to apply the results of physics and chemistry, which had meanwhile made enormous advances, to physiology, and thus inaugurate the period of physics and chemistry in the history of physiology.

The contrast between Müller's ideas and the physical and chemical interpretation of the vital processes which we find in the work of his pupils and in the work of Ludwig and his followers has a philosophical background. The clarification of this is directly relevant to the question of mechanism and leads us in the first instance to an examination of the situation of European physiology at the time when Müller emerged as a physiologist.

During the first decades of the nineteenth century, European physiology was still dominated by vitalism.[188] In other words, in answering physiological questions scholars fell back on something like a principle of life or specific life force, which was held responsible for the actual course and the laws of physiological phenomena. This is also true of German physiology which, moveover, swept along on the tide of 'romantic' (idealistic) natural philosophy (particularly that of Schelling), reached its peak as 'romantic physiology' between about 1810 and 1815 and, with its predilection for sweeping generalisations and its aversion to detailed analytical investigation, stood, for the duration of its domination, in the way of the development of empirical physiology in Germany. However, empirical, vivisectionist physiology, which had no chance of survival in Germany until about 1830, was developed in France by Magendie and others. After this time, with the growing influence of Müller as director of the institute in Berlin, German empirical physiology was to become pre-eminent in Europe.

It is interesting to see how the genesis of German empirical physiology, condensed, as it were, in the growth of one person, can be traced in the mental development which Johannes Müller underwent up to 1827.[189] His 'apprenticeship', begun after he left grammar school in 1818 and formally concluded with his 'Inaugural Dissertation' (*De Phoronomia Animalium*) in 1822, is still entirely dominated by German natural philosophy. A period of study in Berlin (1823–24), which brought him into contact with Rudolphi, then professor of anatomy and physiology, and the reading of the works of Berzelius definitively 'cured' him of his natural philosophical way of thinking and set him on the track of an empirical method of investigation, based on careful observation. This, after he had qualified as a university lecturer in Bonn in 1824, took shape in publications which stand as the documentary expression of his so-called subjective-physiological-philosophical period.[190] His example here is Goethe's theory of colour, and it is Goethe's principles of investigation which in his praise of observation as "simple, unwearying, industrious, upright, unprejudiced" and in his rejection of experimentation as "artificial, impatient, eager, digressive, emotional, unreliable"[191] he is echoing here. In 1827, in the train of nervous exhaustion brought about by his attempts, sustained year in and year out, to observe his own observational activities (eye movements), there was a change in Müller's mental attitude which heralded the so-called objective-physiological-anatomical period of his development. "A deep aversion to dealing with transcendental matters, to introspection, to his own imagination, had taken hold of him . . . this is where the Johannes Müller we knew started".[192] The methodological principles which now guide him are eloquent expression of that experiential ethos so characteristic of the generation of scientific researchers inspired by Müller's work and personality. What he demands of the investigator is "that he should be untiring in observing and experiencing"; the issue at stake is "experience, which can be repeated in all cases, always giving the same results, as one is used to demanding of every good physical experiment". He also demands "real observation", i.e. "that one should distinguish the *essential* from the *accidental* in every experience", for if all our experiences were to consist of such "real" observations "all further theorising would be superfluous and theory would simply be a stating of facts, each of which being the consequence of another". And then finally that requirement which was already incomprehensible to the younger generation of researchers who had adopted the quantitative methods of the physics and chemistry school, the requirement that "one does not merely throw the experiences together when they have reached an adequate depth and the greatest accuracy, but that one tries to reach detail starting from totality, in the way nature itself proceeds in the development and maintenace of *organic beings*, provided that one has arrived by means of analysis at recognition of the detail and comprehension of the totality".[193]

The experimentation rejected by Müller in his subjective-physiological—philosophical period has here acquired its rightful place alongside observation, which — in the spirit of Goethe — is regarded as being directed towards the essence which binds together the phenomena being studied in the unity of the essential form. Together they determine the qualitative, morphological (comparative) method which Müller was always to prefer in his later anatomical and physiological investigations. This remained the case, even after he had mastered the 'modern' method of chemical and physical investigation and had adopted the standpoint that morphology "was not the ultimate goal of investigation, but only a necessary preliminary, the basis of all knowledge about life, with which the work of the physiologist, using experiment and observation, is only just beginning".[194]

It has been remarked before that a great reformer like Müller was a typical transitional figure — more so, perhaps, than the younger men like Griesinger, Virchow, Fechner or Rokitansky — so that through him, more clearly than through anyone else, the question of the relationship between old and new, i.e. of biology-oriented physiology versus physiology based on physics and chemistry, is forced upon us. Pupils of Müller, like Du Bois-Reymond, T. Schwann and others, who were in the forefront of the physics and chemistry movement, saw something ambiguous in Müller's physiological reforms because they were convinced that a commitment to the physics and chemistry approach was *fundamentally* irreconcilable with the basic *vitalistic* conviction to which Müller remained loyal to the last and which prevented him from according more than relative significance to the methods of investigation based on physics and chemistry.

I find it quite reasonable, however, to conclude that, as far as Müller himself was concerned, there was no internal contradiction here, because his vitalistic convictions did not *per se* exclude the application to physiology of the investigative methods of physics and chemistry, but only included a claim concerning the way in which the reactions of organisms exposed to certain physical or chemical influences must be *interpreted*.[195] As Du Bois-Reymond himself remarked,[196] Müller believed that the reaction of organisms to such physical and/or chemical influences was only distinguished from physical and chemical reactions in the field of anorganic nature "in that the stimulus of the organism reveals nothing but the characteristics of the organism itself, of its 'energy' ", an idea which was to find further elaboration in the context of sensory physiology in Müller's celebrated "law of the specific energies of the senses".

It was the hope of the younger generation of physiologists, who were making themselves heard around the middle of the century, that a mechanistic interpretation of the phenomena of life was viable, but the fact that much of their work remained programmatic – indeed there were far more hoped-for results of research than actual successes – must have been all the more reason for Müller not simply to abandon the possibility he cherished of a specifically biological interpretation of the phenomena of life which was consonant with his vitalistic 'prejudice'. It is no coincidence that this situation reminds us of Griesinger, who, as we shall see, all mechanism notwithstanding, stood by his statement that with regard to the phenomena of reflex regulation studied by physiology "a teleological interpretation also appears permissible".[197] One may assume, however, that Müller would reverse the emphasis and that he believed that within the framework of the teleological interpretation of the phenomena of life there was also room for a "mechanical" interpretation.[198] This goes, in any event, some way towards explaining how it could come about that Müller managed to gather round him an array of truly brilliant pupils and colleagues who not only did not share his vitalistic position but who, on the contrary, did their utmost to prove the possibility – in fact, the superiority – of the mechanistic conception.

Du Bois-Reymond's observations about the singularity of Müller's vitalism contain a valuable hint to the solution to this problem.

In contrast to the vague ways in which other vitalistis expressed themselves, he had pondered on the theory of vital force so thoroughly and expressed it so incisively and so clearly that he had actually prepared the way for those who wanted to examine this dogma critically. Out of the mist of vitalistic reveries his mistake emerges, laying itself wide open to attack. If, as follows from Müller's reflections, we have to understand the vital force as being without any specific location, divisible into an endless number of parts all equal in value to the whole, disappearing at death without any reaction, acting consciously and in the possession of physical and chemical knowledge according to a plan, *then this is tantamount to saying that there is no vital force*; thus proving apagogically its not-being. [My italics.] [199]

It can be said that what was understood by those of Müller's pupils who were oriented towards physics and chemistry as the 'vitalism' of their teacher was, in their view, significant as the most interesting formulation of the vitalistic working hypothesis. It was interesting primarily because this formulation, as a formulation, raised no essential impediments to physiological research along the lines of the physics and chemistry school of thought, so that it was possible for these pupils of Müller's to conceive their own investigative results as a (*pure* mechanistic) 'improvement' of, or advance on, Müller's scientific work. This is borne out by the fact that — in the ideas of some of Müller's pupils, in relation to Müller at least — it is not so much a case of radical discontinuity as of a feeling of having gone *a step further* along the path taken by Müller's empirical physiology, anatomy, etc. Or, in Du Bois-Reymond's words: "The modern physiological school . . . has drawn the conclusion for which Müller thus provided the premises".[200] Ultimately, Müller was considered as out of date, but — paradoxically — to a not inconsiderable extent because of his own work, which had laid the foundations for the subsequent 'progress' and *ipso facto* for his becoming outdated.

It is impossible to go in detail here into the question of the relationship between Müller and his pupils — an exciting and interesting chapter in the history of nineteenth-century natural science. The major trend in development is that gradually, with the physical, chemical and microscopic anatomy research methods gaining ground, the domain of the phenomena of life, where the vitalistic hypothesis still seemed defensible or remotely plausible, was further and further eroded. Here we must mention in the first place the anatomist Theodor Schwann (1810–82). Schwann, in accordance with the theory of organisms which he developed,[201] rejected virtually all teleological explanations which invoked a life-force governed by immanent purposes and recognised only in man (because of his freedom) principle which differed *in substance* from matter.

In a letter he wrote to Du Bois-Reymond after Müller's death, in answer to Du Bois-Reymond's question as to whether he should be regarded, in his own opinion, as a pupil or as an independent contemporary of Müller, Schwann stated that he restricted himself to the phenomena of growth in this theory "because the *anti-vitalistic principle* was already sufficiently contained in it" (my italics).[202] Schwann's cell theory, which held that the universal basic element of all vegetable and animal anatomy was the cell and that even complex animal tissue could develop only from cells, implied, however, the possibility in principle of applying the physical method of explanation to the phenomena of animal life and was therefore, in the final analysis, irreconcilable with Müller's ideas about the 'intrinsic life force' of the tissues, and in particular that of the nerves. As can be seen from the letter quoted above, Schwann was fully aware of these implications, but he shrank from publishing that part of his manuscript in which these consequences of his theory were disclosed because, he said, he was afraid "to compromise the theory itself as well as the entire trend by too detailed an explanation. The aversion to hypotheses was extremely great at that time as a reaction against the previous natural philosophical school".[203]

Without wishing to detract from the significance of the work of others among Müller's pupils (in particular, Du Bois-Reymond, Henle, Virchow, and Brücke) I believe it can safely be said that, apart from Schwann's researches in the field of cell theory, the most interesting scientific developments set in train by Müller's work are those linked with the name of Helmholtz. I am thinking here particularly of his monumental studies of the physiology of the senses.[204]

The physiology of the senses had been Müller's favoured field of research in his so-called subjective-physiological—philosophical period. The genesis and first formulation of Müller's celebrated "theory of the specific energies of the senses" dates from this time.[205] From its formulation in 1826, the theory of the specific energies of the senses was of programmatic significance in research into the physiology of the senses for several decades thereafter.[206] According to this theory, the nerves of our senses (sensory organs) have a primary 'innate energy', as a result of which they always react with the same sort of sensation to the most diverse stimuli. Thus, for example, a sensation of light can equally be the result of pressure on the eye or of an actual light stimulus. The energy specific to the eye (inherent in its nerves) is such that the eye responds to various sorts of stimuli in the same way, i.e. with a sensation of light. And the same is true, *mutatis mutandis*, of the other senses: "Pressure, friction, galvanism and internal organic stimulus, all these things cause in the optic nerve that which is appropriate to it – the sensation of light; in the acoustic nerve that which is appropriate to it – the sensation of sound; feeling in the sensory nerves".[207]

In general terms, this theory therefore states that the nature of sensory perception is not dependent on the nature of the stimulus which brings it about, but on the nature of the sensory organ which is stimulated; from which it follows that the perceptions of our senses are no more than purely subjective symbols of unknown occurrences – a conclusion which would prove to be of particular importance in Helmholtz's theory of knowledge.[208]

Müller's theory of the specific energies of the senses did not simply appear out of the blue. It had immediate predecessors in late eighteenth-century physiology,[209] but was inspired in Müller's formulation by Kant's a priori forms of Anschauung (intuition).[210]

In Helmholtz's view, Müller's law represented a step of major importance in the history of the physiology of the senses – in fact, the basis of the theory of sensory perception. According to Helmholtz, what scholars had until then surmised from the data of everyday experience and had tried to express in a way which combined truth and falsehood, or had precisely formulated only for individual sub-disciplines – like Young and the theory of colour or Bell and the motor nerves – "left Müller's hands in the form of classical perfection, a scientific achievement whose value I am inclined to consider equivalent to the discovery of the law of gravity".[211]

What concerns me for the moment is the transition from Müller's theory of the specific energies of the senses to the use which Helmholtz and a later generation of psychologists made of it. This transition is, in the final analysis, one from a form of physiology of the senses which is still natural-philosophy-oriented to a strictly mechanistic interpretation and explanation of the phenomena of the physiology of the senses. For when Helmholtz, in what he himself saw as "little more than a further development and implementation of (Müller's) theory of the specific energies of the senses",[212] developed his own theory of the specificity of the nerve fibres to explain the variations in the quality of sensation within a single sensory modality, the concept of specific 'energy' used by Müller – a concept which has the Aristotelian ἐνέργεια in the background[213] and in which, as Boring suspects,[214] the influence of earlier theories about vis viva and vis nervosa can be assumed – was de facto discredited in perceptual psychology.

It was not least Helmholtz himself who, with the law of the conservation of energy (force)[215] – formulated at about the same time by J. R. Mayer (1842), J. L. Joule (1840–45) and Lord Kelvin (1851–54) – had laid the foundations for the radical physicalisation in the fields of physiology and psychology, which — besides reinterpreting the concept of energy — helped the mechanistic scientific ideal to attain a position of undisputed priority in scientific thinking in the third quarter of the nineteenth century.

Helmholtz's "further development and implementation" of Müller's physiology of the senses was thus a definitive departure from everything in Müller's thinking which betrayed his ties to a natural philosophical background and was, generally speaking, the start of a new phase in the development of empirical physiology. It was the time when the "modern physiological school" would draw the conclusion "for which Müller . . . provided the premises".

2.9. MECHANISM IN PHYSIOLOGY. THE POSITIVIST VARIANT.

With Müller's pupils (especially H. Helmholtz and E. Du Bois-Reymond) there developed a form of physiological mechanism which, in its philosophical orientation, differed equally from Müller's old-style mechanism based on natural philosophical (idealistic) grounds and from the contemporary (naive) materialistic mechanism supported by Vogt, Büchner and Moleschott. It can best be described as the *positivist variant* of physiological mechanism.

If we regard 'positivism' as the term for an attitude of mind characterised by the fact that it aims at validating the standpoint of experience (in all senses of the word 'experience') as opposed to metaphysical speculation, this formula (whatever its shortcomings may otherwise be) has in any event the advantage of allowing room for such diverse (or, more precisely, methodologically diverse) orientations as those of the strictly scientific, humanistic and phenomenological 'positivism' which we connect with such names as Helmholtz, Dilthey and Husserl, respectively. This description of positivism as an attitude of mind[216] seems to us, moreover — and certainly if we are discussing the positivism of Helmholtz and his followers — more satisfactory than the formula put forward by Aliotta, who, writing about the first phase of positivism — positivism before 1870 — described it as "a dogmatic belief in physical science which is set up as a model for every form of knowledge".[217] This definition, applicable as it is to the positivism of Comte and Spencer, misses, in its emphasis on the doctrinaire aspect of this school of thought, what strikes us as the most essential point in the positivism of the most eminent representatives of German science (including medicine) pioneered in Germany in the eighteen-forties. If the scientific ideal of knowledge set a trend, it was certainly not primarily as a substitute for the discredited speculative philosophical systems of the idealists (which could more truly be said of the adherents of the materialism preached by Büchner and his followers), but rather as a shining example of a way of thinking which aimed at giving substance to the strict 'positivist' experiential ethos in the practice of its own scientific research. Viewed in this light, the positivism of German science from the eighteen-forties onwards bears a closer affinity to the second phase of positivism[218] (after 1870), with its self-critical intentions, than to the dogmatic positivism of the period before 1870. Partly for this reason, I prefer to reserve the expression 'critical positivism' for precisely that German positivism – influenced by Kantian critical philosophy – of Helmholtz and his supporters, which as a completely singular historical phenomenon fits neither chronologically nor systematically into the current alternative.[219]

2.10. CRITICAL POSITIVISM AND KANTIAN CRITICAL PHILOSOPHY

It will come as no surprise that a thinker like Helmholtz, who, as he himself said,[220] let slip no opportunity to impress upon his pupils that "a metaphysical conclusion is either a fallacy or a concealed empirical conclusion", held no brief for metaphysics. Helmholtz himself was nevertheless at pains to emphasize that his critical rejection of metaphysics did not imply a condemnation of philosophy: "Even without metaphysics, there still remains a large and important field for philosophy, i.e. that of the knowledge of higher and lower mental processes and their laws". This philosophy, "the real science of philosophy",[221] is none other than that of his own (psychological) epistemology.[222]

It was obvious that Kant's philosophy (i.e. his critique of reason) would play a part in the crystallisation of this figure of thought, which was characteristic of Helmholtz, if only because of the general intellectual situation around 1850. The ceaseless strife and dissension within the ranks of the Hegelians had seriously weakened the prestige of philosophy, and many, disappointed by idealism, sought support in 'naive' materialism, which found acceptance in a wide circle with the works of Moleschott, C. Vogt and L. Büchner, which appeared between 1852 and 1855. The great progress in the field of natural science had led to the occurrence of an uncritical veneration for science, a sometimes superstitious belief in science, which – although popular – was hard to reconcile with the spirit of scientific criticism and self-criticism.

It is, therefore, by no means surprising that in these circumstances the idea of a return to Kant's philosophy as a paradigm for a critical self-limitation of scientific reason was able to strike a responsive chord among more critically-minded scientists, and was in tune with their attempts to give a positivist (i.e. not metaphysical or materialistic) interpretation of science with a view to fostering scientific research itself.[223] The idea behind this was that a science which was in essence disguised metaphysics and which struggled, in its conceptualisation, under the weight of a metaphysical mortgage could not possibly arrive at 'real', scientifically justified results. It was, scientifically speaking, of the greatest importance to keep the limits of scientific knowledge clearly in mind; in other words, according to good (i.e. 'critical') positivist conceptions, science is not, and cannot be, metaphysics.

It is in this positivist interpretation of scientific endeavour that Kantian critical philosophy was able to achieve a special significance for leading researchers both in science (Helmholtz, Du Bois-Reymond) and in the humanities (Dilthey). What these thinkers found attractive in Kant's philosophy was not the idea of a possible revival of transcendental apriorism in the strictly logical interpretation which was to be characteristic of so-called Neo-Kantianism, but its anti-metaphysical tendency. It is precisely this contrast with the Neo-Kantian philosophy which came to the fore rather later that serves to throw light on the special character of the critical-positivist figure of thought developed by Helmholtz and on the limits of its dependence on Kant's critique of reason. Helmholtz did not make the sharp *division* of the (transcendental) logical and the psychological points of view which was taken for granted in Neo-Kantian theories of knowledge. The fact that in his work, despite his Kantian use of words (I am referring here to his repeated use of the term 'transcendental'), even the *distinction* between the two points of view was not and could not be expressed, is related to his empiristic orientation. There is no room in his thinking for Kant's synthetic *a priori* judgements or principles – a genetic psychological explanation of the philosophical axioms is possible and the principle of causality must be reinterpreted as a hypothesis – and thus, among other things, the possibility of using the apriority of the axioms of Euclidean space theory as an argument in support of the transcendentality of our spatial perception is lost. To put it another way, it is only if one conceives the apriority of these axioms as transcendental in the meaning which Kant attached to this term that it coincides with the apriority of space. What Helmholtz, somewhat misleadingly, calls 'transcendental' is, however, often no more than '*a priori*', and refers in fact to what we would call the structure of the (psychological) subject of knowledge.

All this serves to restrain us from taking Helmholtz's cry of 'back to Kant' too literally. To Helmholtz, Kant was primarily of importance as an ally in his 'positivist' position against speculative idealism and later also against dogmatic materialism, which rashly concluded from the collapse of that idealism the complete impossibility of philosophy. The return to Kant has no other use here than to serve as a reminder of a philosophical theory which through its criticism (which, according to positivist ideas, was worth taking to heart) of metaphysics' pretensions to knowledge and its related limitation of the realm of our knowledge to the area of possible experience (i.e. scientific experience and therefore ultimately mediated by external observation), had laid the foundations of 'real' *scientific* philosophy; or, in other words, for 'the real science of philosophy', i.e. Helmholtz-style epistemology.

Kant's epistemology could only be a starting-point for Helmholtz. As a theory of knowledge it was, so to speak, not 'scientific' enough, insufficiently 'permeated with science', and therefore not in a position to satisfactorily justify the developments which had come about since Kant's time in the areas of physics, mathematics, physiology of the senses, and psychology. The 'solution' to the problem – the linking of this Kantian starting-point with the results of post-Kantian science – as it was to be realised in Helmholtz's epistemological writings (that is, in his theories about optical and acoustic perception and about geometrical knowledge) therefore only in fact came about because Helmholtz interpreted Kantian epistemology in an empiristic sense. In this (highly un-Kantian) empiristic interpretation, the Kantian theory of space, for example, was able to play a part in the formulation of a framework which would make possible a synthesis of the physical, psychological and geometrical aspects of the problem of (visual) perception. Indeed, it was only this psychologising interpretation of Kant which made it possible for Helmholtz to link his 'Kantian' starting-point to the results of post-Kantian developments in science – one thinks here in the first place of the insights gained by his teacher, Johannes Müller, in the field of the physiology of the senses.

The aforegoing will have made it clear that the appeal to and the orientation with Kant's philosophy (specifically, his epistemology), as we encounter it in positivists like Helmholtz, had a fundamentally different meaning from that which the later school of so-called Neo-Kantians wanted to bring out. The positivists were concerned with defending the standpoint of science (Griesinger would say the "standpoint of empirical knowledge") against the claims of 'naive' materialism and metaphysical idealism. In this case, Kant became important as the man who gave us the consciousness of the *limits* of scientific knowledge and was able to remind us of the dangers of the (metaphysical) violation of these boundaries. But among the positivists this appeal to Kant was coupled with a tendency towards a *'positive' scientific* foundation of the theory of knowledge, and this implies a clear departure from the transcendental–philosophical intentions of Kantian critical philosophy. It was, however, precisely the renaissance of this transcendental-philosophical intention, the rehabilitation of the standpoint of critical reflection, which concerned the Neo-Kantians, who emblazoned the rallying cry "back to Kant" on their banner and aimed at a specifically *philosophical* foundation of the theory of knowledge in opposition to both naive materialism and positivism.

The orientation with Kant's theory of knowledge thus proves to have been the starting-point for two distinct epistemological attitudes in the ten to fifteen years after the middle of the century. We see here the origin of an antithesis which only came to light in the course of time. This antithesis was based on two conflicting motives – the concern for the 'scientificality' and the concern for the (specific) 'philosophicalness' of the theory of knowledge – which, it is true, remained latent as long as the common opposition to (metaphysical) idealism and (naive) materialism held sway, but which could not remain concealed in the long run. The positivist school, which began with Helmholtz's empirical epistemology, would in time, by way of the so-called empirio-criticism of Avenarius and Mach, result in the so-called neopositivism of the Vienna Circle.[224] The second school, that of 'critical' reflection, was, on the other hand, to culminate in the influential Neo-Kantian movement. Together, these two schools of thought would represent the two dominant positions in epistemology in the first decades of the twentieth century.

2.11. THE MECHANISM OF HELMHOLTZ, DU BOIS-REYMOND, BRÜCKE, AND LUDWIG

2.11.1. *H. Helmholtz*

The mechanism we shall discuss now – mechanism in the strict sense – found its main expression in the works of Helmholtz. Helmholtz's often-quoted words from his introductory observations on the law of the conservation of force, 1847, are a striking illustration of this mechanistic standpoint.

It can be stated as follows ... the task of physical, natural science consists of reducing the phenomena of nature to immutable, attracting and repelling forces, their intensity depending on the distance. The solvability of this problem is also the condition for the complete comprehensibility of nature ... Theoretical natural science will have to harmonise its viewpoints with the established demands concerning the nature of simple forces and their inferences if this science does not want to stick fast halfway along its road to understanding. Its business will be completed as soon as the reduction of the phenomena to simple forces is completed and at the same time it can be proved that this reduction is the only possible one of which the phenomena admit. Then this would be established as the essential conceptual category for the understanding of nature; objective truth could then also be assigned to it.[225]

In this regard I should like to make the following observations. Firstly, what Helmholtz formulates here is a programme, that is to say, the reduction of the multiplicity and qualitative diversity of natural phenomena is a postulated, not an empirically proved reduction. It is precisely on this point that this *methodological* mechanism of Helmholtz is distinguished from the mechanism of popular, 'naive' materialism. As Cassirer rightly remarks,[226] this form of materialism only seemed to stem from natural science but was in reality a survival and a late descendant of dogmatic metaphysics.

The fact that we are dealing with Helmholtz's mechanism in the context of mechanism in *physiology* will come as something of a surprise, particularly to those who are used to regarding the 'discoverer' of the law of the conservation of energy (force) as one of the greatest physicists of his time and not primarily as a physiologist. One must, however, realise that in the period we are now discussing prominent scientists like Helmholtz, J. R. Mayer, J. von Liebig, etc. started as physiologists or physicians before they made a name for themselves as physicists or chemists.

In this respect, physiology was a matrix for developments in other scientific disciplines besides physiology. We can see this link with a physiological or medical background clearly reflected in the scientific careers of the researchers of the time. It is true that Helmholtz declared in his speech *Das Denken in der Medicin*, delivered to a group of physicians in 1877, that his main interest was not pure physiology, but physics,[227] but this does not alter the fact that before the period from 1871 until his death in 1894, during which he was primarily concerned with problems of physics (electrodynamics, aerodynamics, thermodynamics, electromagnetism, etc.), he had worked as a physiologist with great success for about twenty-two years (from 1849 to 1871). And Robert Mayer,[228] who formulated the principle of the conservation of energy in 1842, even earlier than Helmholtz, and who is also credited in the annals of physics as the founder of the mechanical theory of heat, considered himself, according to a remark he made to Preyer in 1864, as both a physician and a physicist.[229]

This connection between physics and physiology was not based solely on the fortunate circumstance that a number of important scientists managed to combine the two disciplines in a sort of personal union, but was primarily connected with the fact that some fundamental insights in the field of physics were developed from what were originally physiological questions. This is, for instance, true of the formulation of the law of the conservation of energy (force), both in Helmholtz's case and in Mayer's.[230] Thus, for Helmholtz, it was the relationship between muscle activity and heat, i.e. the problem of energetics in muscle activity, which was the starting-point for his work on the problem of energetics in general, as his early study *Ueber die Erhaltung der Kraft* (1847)[231] bears witness. In this way, the formulation of the principle of the conservation of energy could be said to symbolise the start of the definitive 'physicalisation'[232] of physiology, and since, according to Helmholtz's conception, this principle could be reconciled without more ado with the mechanistic view of nature, he saw it, as did many others of like mind, as supporting the mechanistic idea.[233]

2.11.2. E. Du Bois-Reymond

In Emil Du Bois-Reymond (1816–98) the mechanistic *Weltanschauung* had a fervent champion who, as well as being a celebrated researcher in the field of physics-oriented physiology,[234] was also, like his friend and fellow-student Helmholtz, who was five years younger, interested in the epistemological and scientific theory aspects of his own discipline. The figure of thought which emerges from his more 'philosophical' publications can be summarised under two main headings: (1) criticism of vitalism, and (2) criticism (but now in the Kantian sense of the word) of the limits of mechanistic natural science.

The first point of view was expressed in his study *Ueber die Lebenskraft* (1848).[235] Not without a certain satisfaction, Du Bois-Reymond remarked almost forty years later, looking back on "those still somewhat crude products of my *Sturm und Drang* period", that his criticism of vitalism had achieved a certain significance in the history of German science inasmuch "as it remained the last demonstration against vitalism which nowadays, in our country, has really disappeared from the scene, as I had wished and predicted".[236] And it must indeed be admitted that while the similar criticism of predecessors like Berzelius (1839), Schwann (1839), Schleiden (1842, 1843) and Lotze

(1842)[237] had had no effect worth mentioning, after Du Bois-Reymond's attack on the vitalistic theory of life and life forces, vitalism as a separate school of scientific (physiological) thought was finished for a very long time. I believe that the unexpectedly great effect his writing had – which surprised even Du Bois-Reymond himself – cannot primarily be explained by the author's keen, polemic style nor by saying that his treatise was the final blow which made the vitalists, whose spirit had already been sapped by earlier criticism, capitulate. This effect was, in my view, due much more to the fact that the criticism formulated in Du Bois-Reymond's treatise was based on a state of affairs in physics-oriented physiological research which at the time his predecessors were writing could not be expressed in the criticism of vitalism because it did not then exist or was too new and unfamiliar to play a significant part. I am referring here to the situation which arose after the work of Helmholtz (1847) and Robert Mayer (1842, 1845) on the principle of the conservation of energy (force) had laid the foundations for a radical 'physicalisation' of physiology.

What distinguished Du Bois-Reymond's fight against vitalistic ideas from that of his predecessors and was largely responsible for the impact of his criticism of vitalism was the fact that he applied – for the first time – the conclusions of Helmholtz's theory of the conservation of energy (according to which "the sum of the available vital forces and energies is constant") to the criticism of vitalism.[238] In fact, Du Bois-Reymond's 'achievement' on this point is simply that he expressed in an eloquent and telling fashion the *implications* of the innovations started by Mayer and Helmholtz, at a moment when the accumulation of the criticism of vitalism and the growing series of research results achieved by the physical school in physiology (to which Du Bois-Reymond's own *magnum opus* on animal electricity certainly made an important contribution) must have been raising serious doubts in scientific circles about the viability of the vitalistic paradigm.

Certainly no less important than his criticism of vitalism is Du Bois-Reymond's 'critical' determination of the limits of mechanistic natural science, contained in his lecture *Ueber die Grenzen des Naturerkennens* (1872), which was widely discussed at the time. He elaborated somewhat further on the theme of this lecture in *Die sieben Welträthsel* (1880).[239] When Du Bois-Reymond wrote these pieces it was no longer necessary to fight for the cause of mechanism in physiology, as it had been in 1847 when he wrote *Ueber die Lebenskraft*. It was no longer the applicability of 'modern' physics and chemistry in dealing with physiological problems which had to be defended; the problem was now how the scientific physiology which had thus been

reformed could be kept from scientifically unfounded excesses in the area
of metaphysics such as the materialists were guilty of. The 'naive' meta-
rialists of the eighteen-fifties, although still productive, had, it is true, been
past the peak of their influence for some considerable time, but Haeckel was
achieving prominence and spurred more critical minds to adopt a position.
In 1847, in the flush of his mechanistic zeal, Du Bois-Reymond was still
proclaiming that "if the difficulty of dissecting were not beyond our powers,
analytical mechanics would in essence *suffice to include* even the problem
of personal freedom", but from the agnostic standpoint which he defended
twenty-five years later, this assertion made in his youth appeared to him to
be incorrect.[241]

The issue here was to determine the limits of our knowledge of nature or,
more precisely, of scientific knowledge or the knowledge of the material
world (*Körperwelt*) with the help of, and in the sense of, theoretical natural
science. The point that Du Bois-Reymond wanted to defend was, in short,
that these limits are not those of our *actual ignorance*, but are those of a
fundamental inability to know. That is to say, they are limits which cannot
be passed even in the hypothetical case of a perfect knowledge of nature
such as we can impute to a 'Laplace mind'[242] in an experiment in thinking.
The human mind is distinguished from the Laplace mind only in degree. Put
another way, the Laplace mind represents the highest conceivable level of
knowledge of nature, and therefore anything which cannot be penetrated by
the Laplace mind must *a fortiori* remain an enigma to our much more limited
intellects. Thus in natural science (analytical mechanics) we think in terms
of matter and force, but the *essence* of matter and force cannot in turn be
fathomed within the framework of mechanistic explanation and thinking:
at this point we come up against insoluble paradoxes.

Another enigma which cannot be solved through scientific thinking is
that of the consciousness and its genesis. Even an 'astronomical' knowledge
of what goes on in our brains cannot explain the genesis of consciousness
even on the most elementary level, i.e. that of sensation. The relationship
between "specific movements of certain atoms in my brain on the one
hand, and on the other the, for me, original, not further definable, undeniable
facts: 'I feel pain, feel pleasure; I taste something sweet, smell the scent
of roses, hear the strains of the organ, see something red' as well as the
direct consequence of this: 'Thus I am'" remains a mystery.[243] In short:
"Our understanding of nature became stuck for ever, trapped between these
two boundaries, i.e. on the one hand the inability to comprehend matter
and force, on the other the inability to understand how spiritual processes

can spring from physical circumstances".[244] The essence of force and matter, the consciousness — to which Du Bois-Reymond later added in his *Die sieben Welträthsel* (The Seven Mysteries of the World) the origin of movement, the first genesis of life, the apparently intentionally functional structure of nature, rational thought and the origin of language closely allied to it, and finally the problem of the freedom of the will[245] — are areas of reality which remain forever closed to scientific knowledge. These are "mysteries of the world", therefore, which demand not simply a confession of our actual ignorance (*ignoramus*) but also an acknowledgement of our fundamental inability to penetrate them: *ignorabimus*.

From this it seems that with his 'critical' self-limitation of mechanistic science Du Bois-Reymond has defeated his own object, because — as Cassirer has pointed out[246] — in his criticism of (mechanistic) materialism he is forced to confirm implicitly what he appears explicitly to deny and oppose, namely the presupposition that mechanism "is the only possible and the only permissible type of understanding at all, beside which there can be no salvation for natural scientific knowledge" (Cassirer). This means that the dogmatism of the mechanistic materialists and the scepticism of Du Bois-Reymond are, as it were, two of a kind. I believe that it is on precisely this point that the critical *positivist* interpretation of mechanism as a programme, as we encounter it in Helmholtz, differs from Du Bois-Reymond's "positivisme *manqué*":[247]

What distinguishes Du Bois-Reymond from Helmholtz — the implicit absolutisation of mechanism as a *Weltanschauung* — does, however, bring him closer to his friends Ernst Brücke and Carl Ludwig, physiologists who had little inclination towards philosophy. No discussion of the role of the mechanistic conception of science in physiology can be complete if it fails to mention these two researchers who, although they were not philosophers, contributed through their experimental work in physiology to the consolidation of the mechanistic 'prejudice', i.e. the implicit 'philosophy' which Du Bois-Reymond explicitly declared to be untenable.

2.11.3. *E. W. Brücke*

Ernst Wilhelm (Ritter von) Brücke (1819–92) can be described as the most versatile physiologist of his time.[248] He himself did not appear to consider this versatility as a virtue, to go by a letter he wrote to Du Bois-Reymond in 1853, in which he compared his own characteristic manner and quality of work with those of his friends: "You work *multum non multa*, Helmholtz *multum et multa* and I *multa non multum*".[249]

The fact that this judgement of his own work is too modest is of less importance to us than the fact that physiology in the second half of the nineteenth century had in him a passionate advocate of the physics and chemistry approach. Something Du Bois-Reymond once wrote has almost become a maxim:

Brücke and I have sworn that we shall make the truth prevail: i.e. that the only forces active in the organism are common physical–chemical ones. If these have so far not been sufficient to provide an explanation, either one has to look for their nature and the way in which they work in actual cases, by means of physical-mathematical methods, or one has to assume [the existence of] new forces which are of the same order as physical-chemical ones, inherent in matter, and which can always be reduced to repelling or attracting factors.[250]

Brücke's scientific work, as E. Lesky remarks,[251] was indeed a sustained search for and uncovering of those physical and chemical forces 'inherent in matter' as they occur in the various processes of secretion and resorption of the digestive and urinary tracts, in the molecular movements of plant and animal cells, and in the muscle spasms which appear in reaction to stimuli.

As we have already observed, the significance of Brücke's work lies in what he accomplished in the field of experimental physiology, and this makes him (like his friend Ludwig, of whom the same can be said) if not an apostle of the mechanistic view of science like Helmholtz and Du Bois-Reymond, then certainly its loyal disciple, who through the practice of his scientifically varied work effectively confirmed the assumptions of this mechanism. In this connection, it is certainly relevant to recall that Freud, who worked in Brücke's physiological laboratory in Vienna between 1876 and 1882, later spoke of Brücke as his "most highly revered teacher"[252] and "the greatest authority I have ever met".[253] From Freud's research it can be seen clearly enough[254] that the mechanistic scientific ideal embodied in Brücke's physics-oriented physiology — or more specifically and rather more unkindly expressed, his "mythology of energetics"[255] — set the pattern for the formation of psychoanalytical theory.

2.11.4. *C. Ludwig*

In the history of physiology in the second half of the nineteenth century, pride of place must be given (as K. E. Rothschuh has demonstrated) to Carl

Friedrich Wilhelm Ludwig (1816–95). If Johannes Müller was the most important figure in the first ('biological') phase of empirical physiology, Carl Ludwig was certainly the most prominent and influential representative of the physics and chemistry school of thought in physiology.

His two-volume work *Lehrbuch der Physiologie* (1852, 1856) marks the caesura between the physiology of a bygone age and a new period in the history of this science. The programme formulated in this work – "Scientific physiology has the task of determining the functions of the animal body and of *deriving these, of necessity, from elementary conditions of the latter*" [256] – implied the idea of a form of physiology which was essentially physics and chemistry applied to living organisms.

This programme took shape, in the years that followed, in a staggering number of experimental physiological research results, which gained a world-wide reputation for Ludwig's institute in Leipzig between 1865 and 1895 and made it, as the Mecca of experimental physiology, an internationally sought-after place for the education of aspiring physiologists. Ludwig had a large number of students (more than two hundred) who, spread far and wide, disseminated his ideas in the theory and practice of scientific research. All this certainly played a major part in the consolidation of the conviction that mechanism "is the only possible and the only permissible type of understanding at all, beside which there can be no salvation for natural scientific knowledge".

2.12. MATERIALISTIC MECHANISM (VOGT, MOLESCHOTT, AND BÜCHNER)

We have already demonstrated that mechanism does not necessarily go hand in hand with materialism (in the metaphysical sense). The mere recalling of Descartes, Kant, Herbart and Lotze, among the philosophers, is enough to bear out this point. We have, however, also seen that among the scientists, such as Helmholtz and Du Bois-Reymond, the defence of the mechanistic idea was in no way regarded as a commitment to materialistic metaphysics. Helmholtz and Du Bois-Reymond were, according to their own *self-conception*, positivists not materialists, and if one does not want to conceive 'materialism' in such a way that the specific distinction between positivism and materialism disappears, it is as well to keep this distinction clearly in mind. [257]

While it is true that it appeared to us that Du Bois-Reymond's agnosticism involved the implicit confirmation of a dogmatic metaphysical standpoint which conflicted with the positivist standpoint he defended explicitly, this nevertheless does not make him, in my view, a materialist (assuming that we only call someone a materialist if he is prepared explicitly to endorse the thesis of [metaphysical] materialism.)[258] Equally, as we shall see, it is impossible simply to rank Griesinger among the materialists because, when he describes his standpoint as 'materialistic', he wants this clearly distinguished from the standpoint of the "dull and shallow materialists".[259]

The fact that the rise, spread and consolidation of the mechanistic concept of science played into the hands of the physics-oriented materialism of Vogt, Moleschott, Büchner and others is as obvious and as true as the fact that not every form of mechanism is (or was) materialistic mechanism. Historically, the situation was that the consistent support of the mechanistic point of view – by both the idealists (Lotze) and the positivists (Helmholtz, Du Bois-Reymond, etc.) – *laid the foundations* for that 'naive' materialism which came into being in an almost explosive fashion at the beginning of the eighteen-fifties.[260] The criticism of the 'supersensory' concept of a life force, particularly that of Lotze (1842) and, finally, Du Bois-Reymond (1848), could be interpreted by these materialists – entirely contrary to the intentions of the authors (or in any case to those of a metaphysical idealist like Lotze) – as putting paid once and for all to the dualistic belief in the distinction between a sensory and a supersensory world.[261]

This point is not unimportant as it helps draw attention to the fact that the fundamental motive in this materialistic movement was not that which the term 'materialism' suggests at first glance. The materialists we are now discussing were anti-idealistic or anti-'spiritualistic' – not because they were against the spiritual (in the sense of spiritual values), but because they were *anti-dualistic*. The irresistible attraction which the mechanistic conception exerted on them lay mainly in the fact that it was pre-eminently suitable as a vehicle, a medium for a *monistic* view of life and of the world.[262] If an anti-anthropological element was concealed here, then it was only in regard to Christianity-inspired anthropology, for as a result of this monism all those dualisms which constitute the Christian view of man and the world, particularly those of body and soul, man and nature (world), world and God, were – of course – denied. The cause of scientific anthropology (anthropological biology), on the other hand, was not only unharmed by this monism but was even effectively supported by it. Thus it seems to me that it is not overstating the case to say that what was partly at stake in the keen polemics of Carl Vogt and Rudolf Wagner was, for Vogt and his followers, the meaning and possibility of 'true' *scientific* anthropology. A look at this controversy will serve to clarify this point. The main factor which touched off the so-called materialism conflict[263] was a lecture given by the Göttingen physiologist and comparative anatomist Rudolph Wagner (1805–64) at the conference of natural scientists held in Göttingen in 1854. This lecture, *Ueber Menschenschöpfung und Seelensubstanz*, and the pamphlet which followed hard on its heels, *Ueber Wissen und Glauben mit besonderer Beziehung auf die Zukunft der Seelen* (1854), revealed to a wide circle the war which for several years had been waged more or less under cover by the Christian conservative Wagner against the innovator Vogt. Vogt, in the face of this extreme provocation, responded with his *Köhlerglaube und Wissenschaft* (1854), which attacked Wagner in such a personally damaging and essentially annihilating fashion that, looking back on it and bearing in mind the popularity of Vogt's polemic pamphlet, one is amazed that Wagner had the temerity to continue to publish his controversial views.[264]

Wagner's concern – apart from the personal feud – was to demonstrate that the conclusions of scientific research (and in particular anthropological and physiological research) were entirely consonant with the insights derived from the Bible concerning the origin of the human race and its diversity, and the (immaterial) substance of the soul. "There can be no doubt", he says in his Göttingen lecture, "that all historical Christianity, deeply connected as it is to the creation of man, stands or falls with the acceptance or denial of the descent of all people from one couple; the simplest, most modest faith in the Bible, as well as the entire edifice of the dogmas of our church collapse, and our scientific theology, as far as it identifies itself with the church, is losing its foothold".[265] The result of Wagner's research concerning human descent – the thesis that all human beings are descended from just two people can neither be proved nor refuted – leaves the religious truth of the Bible untouched; that is, if the Bible teaches us that all human beings are descended from a single couple we can accept this religious truth with a clear scientific conscience. In the context of his defence (which was, in fact, a religious one) of the thesis of the existence of an individual, immortal soul, Wagner speaks of faith as "a new organ of the mind, a new way towards understanding, in addition to thinking, natural reason" and expresses the opinion that the human soul is "a product of combining the divine spirit with matter into an individually independent being", for which reason, as he said himself, he "has to accept, on dogmatic and metaphysical grounds, a providential relationshp between body and soul, and the resurrection of the (transfigured) body".[266]

It is indeed hard to understand what this confession of faith on Wagner's part had to do with the aims and accepted practice of the conference of natural scientists. Ueberweg speaks, with justice, of "a pointed misuse of a scientific rostrum to put across religious views",[267] and it can be assumed that the resentment this caused among the majority of those present only increased the sympathy with which Vogt's criticism was received. It is certainly not true to say that they disputed Wagner's right to his religious conviction; what really rankled was that he had tried to settle a scientific question by using religious arguments. The issue here was an inadmissible combination of religious and scientific points of view, which — if it were allowed to stand — would mean the end of *all* scientific discussion. Because of this, Vogt's *Köhlerglaube und Wissenschaft*, besides underlining the right to existence of strictly *scientific* anthropology and physiology, acquired the significance of an appeal to the scientific conscience in general.

One cannot, however, overlook the fact that where, as in the case of Vogt and his supporters, the defence of the cause of strictly scientific anthropology was inspired by an ethos of scientific rationality which prided itself on being free from Christian religious ideas, a philosophical conception of man which was not recognised as such came into its own. This was the conception of *homo natura*.

A point which has not received due merit in the current discussions of this trend in nineteenth-century thinking is that interest in this sort of a concept of man was a major factor in the thinking of the much-reviled materialists of the eighteen-fifties.[268] It is not only in Vogt's work that we can detect this motivation. In the work of J. Moleschott (1822–93) we also find indications of this which could not be clearer. Moleschott wrote in his memoirs that anthropology (in the sense of the science of man as a "culture-bound product of nature")[269] lay at the heart of his scientific endeavour:

anthropology in that sense, for which my father sowed the seeds and Ludwig Feuerbach set the objective, ranked and still ranks with me as my life's work. For its sake I occupied myself with medicine and medical care, for its sake I studied the theory of life, for its sake I was dedicated to philosophy, which I saw only in it. Therefore not 'philosophy, law and medicine, and unfortunately theology too', but anthropology, only anthropology in all its aspects, without theology and teleology, without religious mania and efficiency theories, but with religion, with that religion that sees man as a dependent being, conditioned by nature, who comprehends as his duty the task of raising his natural conditioning increasingly towards a cultural conditioning, instilling to him, with the admiration of nature, the urge and the art to rule over it.[270]

What was true of Vogt and Moleschott was certainly no less true, albeit in a different way, of the third figure in the materialistic triumvirate, Ludwig Büchner (1824–99). With his book *Kraft und Stoff* (1855) Büchner created what can justly be regarded, in many respects, the Bible of naive materialism. This book, revised and expanded several times, went into no less than twenty-one editions before 1904 and was translated into fifteen languages. Partly because of the simple language in which it was written, which everybody could understand, it was more instrumental than any other work in the materialist literature in bringing the new gospel to a wide circle. If the motivating force behind Vogt and Moleschott was primarily a *scientific* one, in which the humanitarian ethos and the existential ideal of the 'free-thinker' are realised, where Büchner – more of a populariser than a researcher – is concerned the emphasis lies more on proclaiming the new materialistic or monistic *Weltanschauung* which, based on the solid foundations of modern scientific knowledge, would replace the old theological and philosophical *Weltanschauung*.

The old, idealistic philosophers had believed that great importance had to be attached to the differences between man and the animals, and many scholars had been convinced that these differences proved the existence of an eternally unbridgeable gulf between men and beasts. According to Büchner, on closer examination these differences proved, without exception, to be relative and were to be understood as having arisen from a gradual process of developing, perfecting and self-educating. "Therefore man does not stand *outside* or *above* nature, but wholly and entirely within it, and therefore the gross and grave mistake – that all nature was only created for him and for his use and benefit – has to be considered, once and for all, as obsolete, in the same way as the earlier erroneous notions about the significance of our small earth as the centre of the universe were discarded for ever by science".[271] The idea of a long and laborious education of theory and life which brought man, after he had overcome innumerable setbacks, to "that pure lucidity of free, unprejudiced thinking . . . in which, nowadays, all scientific minds are moving or ought to move" reminded Büchner of the poet Titus Lucretius Carus, whom he was fond of quoting in this context. Lucretius believed that through his celebrated didactic poem, preaching the ancient atomist doctrine of Leucippus and Democritus, he could free man, trapped by delusions and religious images, from all fear.[272]. This comparison with the Roman poet – suggested by Büchner himself – is indeed a very fitting one, and it also throws light on a characteristic trait of Büchner's endeavours. What the ancient atomic theory was to Lucretius, contemporary science was to Büchner: not so much science or a scientific theory as a (metaphysical) theory of reality whose revolutionary significance for a view of the world and an outlook on life had to be brought to the attention of their contemporaries through the spoken and the written word.

This aspect of a philosophy of life and the world, which is most evident in Büchner's work, is, however, also to be found in Vogt and Moleschott, where it bears the characteristic stamp of the materialist movement in general. I mention this point because it is of importance for a more or less exact determination of the significance and the nature of the influence which materialism can be considered to have exerted on the scientific and philosophical thinking of the generation which emerged around 1850: the generation of Griesinger and Lotze, Helmholtz and Du Bois-Reymond, Virchow and Meynert, etc.

Generally speaking, it has to be said that, despite all the enthusiasm which the materialistic *Weltanschauung* managed to arouse in a wide circle during the second half of the nineteenth century, the works of these materialists have left only faint traces in the history of philosophy and science. Of all philosophers of any standing, it was only Feuerbach, hailed by the non-dialectic materialists as their spiritual father,[273] who openly sympathised with the philosophical implication of materialist writings. It was Feuerbach, too, who in a book review (which, as Moleschott later remarked, created more of a stir than the book itself) coined the phrase: "man is what he eats",[274] a phrase which, with Moleschott's "No thought without phosphorus", and Vogt's "Thought bears the same relationship to the brain as bile does to the liver or urine to the kidneys", is still in vogue in the historiography of nineteenth-century philosophy as a pithy summary of the materialistic conviction.

2.13. SCHOPENHAUER'S AND LOTZE'S CRITICISM OF MATERIALISM AND ITS RELEVANCE TO THE IDENTIFICATION OF THE SELF-CONCEPTION OF THE SO-CALLED 'MATERIALISTS' OF THE EIGHTEEN-FORTIES

The fact that mechanism or reductionism in nineteenth-century Germany does not necessarily imply materialism needs no further demonstration. It now remains for us to put forward arguments in support of the thesis that, from the eighteen-forties onwards, what was understood *in the first place* as materialism among scientists who called themselves 'materialists', was a methodological position which corresponds with what we should nowadays be more inclined to describe as naturalism (and which, as we shall see, was also criticised by Schopenhauer under this name as a sort of less strict form of materialism).[275] The contemporary criticism of materialism[276] has a particular — heuristic — significance in this context, and two critics from the idealist camp — Schopenhauer and Lotze — deserve the historian's special attention.

Anticipating our argument for a moment, it can be said, partly on the basis of the following discussion of Schopenhauer and Lotze's criticism of materialism,[277] that so-called naive materialism (from the eighteen-fifties onwards) was, at least in Germany, a secondary and later form of materialism which was in fact not representative of the scientific self-conception as we encounter it among the leading scientists from the eighteen-forties onwards. As I shall demonstrate, the philosophy of F. A. Lange, viewed in this light, can be seen as an attempt to counter the naive metaphysical materialism of the eighteen-fifties by providing a philosophical justification for the original, methodological meaning of materialism. I shall discuss Lange's philosophical position in detail in Chapter 4. It is now necessary first to look at Schopenhauer's criticism of materialism and subsequently, somewhat more briefly, at Lotze's.

2.14. SCHOPENHAUER'S CRITICISM OF MATERIALISM (IN THE PROPER SENSE) AND NATURALISM

In the first instance, it is Schopenhauer's distinction between and criticism of materialism and naturalism which is of relevance to our discussion. Schopenhauer's views on this subject are found mainly in his principal work, *Die Welt als Wille und Vorstellung* (1819), particularly in the second part, which appeared in 1844 and contained supplements to Books I–IV, which made up the first part.[278] As well as this chronological note, some history of science background is important for the correct identification of Schopenhauer's criticism of materialism (which already had, as we see, essentially been formulated in 1819). First of all, it must be remembered that early nineteenth-century materialism in Europe was primarily physiological materialism and that it was, moreover, in the first place a French affair. The development of German materialism in physiology did not take place until the end of the eighteen-thirties. What distinguishes this German materialism from its French (physiological) predecessor is, first and foremost, that where French physiology, with the odd exception,[279] defended a vitalistic form of materialism, German 'materialism' from about 1838 onwards was characterised by a pronounced anti-vitalistic mechanistic orientation. In this context, M. Schleiden and T. Schwann, the founders of the natural scientific cell theory, are especially important.[280] Both Schwann and Schleiden were, however, in the final analysis, more mechanists than materialists (in the metaphysical sense): Schwann was a very religious man,[281] and Schleiden mounted a fierce attack on materialism in modern German science.[282] This, taken together with the fact that in both cases the weight of their criticism was aimed at opposing vitalism, serves further to underline the fact that the newer, specifically German brand of 'materialism', which became prominent in scientific physiology from about 1838 onwards, cannot be interpreted as materialism *in the proper sense*, precisely because the defense of the mechanistic idea was coupled with rejection of materialistic metaphysics. This later German 'materialism' is therefore not the same as the materialism that Schopenhauer criticised.

What Schopenhauer criticised as materialism in 1819 and 1844 is metaphysical materialism, which was inherent in French physiology as represented by Cabanis and Bichat;[283] in fact, the same materialistic physiology which, because of its vitalistic character, was to suffer so intensely at the hands of mechanistic German science in the middle of the nineteenth century. To a certain extent, the German 'materialists' of the generation of Schleiden, Schwann, Griesinger, etc. agreed with Schopenhauer in their criticism. The difference is that, while Schopenhauer had the greatest admiration for French physiology, as physiology, and only condemned the materialistic metaphysics inherent in it, the young, German 'materialists' regarded this physiology as scientifically out of date because of its vitalistic character, and at the same time (whether from religious, philosophical–idealistic or positivist motives) withheld their approval of the strict materialist position or even expressly dissociated themselves from it.

After these preliminary remarks, we turn to the question of what Schopenhauer understood by materialism and naturalism, and the manner in which he opposed them. For Schopenhauer, to whom the Kantian distinction between appearance and the thing-in-itself is the beginning and unquestionable starting-point of all serious philosophy, materialism must be defined first and foremost as the philosophical position which identifies the thing-in-itself with matter: "anyone who accepts matter as an independently existing entity must also be a materialist, i.e. must make it the principle for explaining everything. (Anyone who denies that matter is thing-in-itself is *ipso facto* an idealist.) However, anyone who conceives matter as the explanatory principle of everything that is, will also be unable and unwilling to understand the subject of knowledge as anything other than a product or modification of matter. This apparent absurdity to which the materialistic thesis leads should nevertheless not cause us to reject materialism unconditionally: materialism, too, has a certain right to exist, because the materialist thesis 'that which knows is a product of matter' is as true as the idealist thesis 'that matter is merely an idea of that which knows'. Both the materialist and the idealist theses are only relatively valid; materialism and idealism are equally one-sided.

Materialism's one-sidedness is expressed in its forgetting the subject: "Materialism [is] the philosophy of the subject that fails to take itself into account".[285] This obliviousness of the subject is the reverse side of the materialistic postulate:

... the fundamental absurdity of materialism [lies in the fact that], starting from what is *objective*, it accepts something *objective* as the final basis for explanation. This can either be *matter* in the abstract, *existing* only in *thought*, or as *substance* already formed and empirically given, for instance chemical elements and their closest compounds.[286]

To say that materialism starts from the objective (in the above sense) is to say that it conceives this as existing by itself and as absolute, and — as we have already observed — once this postulate has been accepted, it is only consistent to try to explain everything that is, including organic nature and even the subject of knowledge, in terms of that objective (matter in all senses of the word).

Materialism's great error is that it is blind to the fact that what it itself holds to be *directly* given (i.e. that 'objective') is only given to us indirectly, through the intermediary of the subject of knowledge and the forms of knowledge inherent in it; or, conversely, that it tries to explain the only thing that is given directly to us, namely the data of awareness of self or "the inward self, the *subjective*",[287] in terms of that which is given indirectly (i.e. matter).[288] The root of all evil (by which he means 'untruth') which is lodged in materialistic philosophy — as Schopenhauer repeatedly impresses upon us — is nothing other than the "assumption that matter is a simple and absolute datum, i.e. something existing, independent of the knowledge of the subject; in other words that it is really a thing-in-itself".[289]

According to Schopenhauer, matter is 'really' not thing-in-itself. The only real thing-in-itself is the 'Will' perceived by us in our awareness of self. This Will,[290] which is the essence of things, reveals itself first of all on the level of experience in the empirical knowledge of the form, quality and functioning of material things. We conceive these forms and qualities, which are phenomenally given, as expressions of matter acting in a specific way, which although it is not itself perceived (or perceptible) is thought by us to be the (unknown) substratum of the actions experienced and of the specific properties of material reality on which these actions are based. Matter conceived as substratum (thing-in-itself) is, however, according to Schopenhauer, not what it is taken to be (by the materialists): it is not the thing-in-itself, absolute, that is, independent of anything else in existence or essence, but is based on the 'subjective forms of our intellect';[291] it is, in the terms of Schopenhauer's metaphysics, 'Idea', that is to say, subject-dependent, relative.

What Schopenhauer calls 'Idea' is always an objectification of the Will and he is therefore able to describe matter as "the objectified *causality* itself (namely the Will)". With this objectification the Will, in itself invisible, appears, it becomes visible.[292] Therefore he can also say: "matter [is] merely the *visibleness* of the Will or the bond between the world as Will and the world as Idea".[293] This indicates at the same time that matter belongs to two worlds and in what sense this is so: considered as a product of the functions of the intellect, matter belongs to the world of the Idea; as a manifestation of the Will, however, it belongs to the world of the Will. Therefore (says Schopenhauer) every object as thing-in-itself is Will, and as appearance is matter.[294] Summarised in a short formula, what Schopenhauer's criticism of materialism boils down to is that materialism makes the mistake of taking as an absolute object something which in reality is subject-dependent objectification (namely of the Will).

Schopenhauer, however, wanted to make a distinction between the materialism we have been discussing — *real* materialism or materialism *in the proper sense* — and a looser conception of materialism. This distinction is based on a different interpretation of what matter is. In the first case, matter is seen as nothing but "the *vehicle* of the qualities and natural forces which act as its attributes";[295] matter, then, is thought of as the substratum, in itself invisible and possessing no properties, of visible qualities and actions. In the second case — materialism in the improper sense [296] — matter is conceived as:

real and *empirically* given matter (i.e. substance or rather substances) ... provided as it is with all physical, chemical, electrical characteristics, also with those that, out of themselves, raise life spontaneously. The real *mater rerum* then, from whose dark womb all phenomena and forms emerge, only to fall back into it again some day.[297]

In other words, matter here is not, as in the case of real materialism, the sought-for, invisible basis of material phenomena; it is a total concept of the *given* material reality. In this case, however, we are not dealing with real *materialism*, but simply with *naturalism*, that is, "absolute physics . . . . physics sitting on the throne of metaphysics";[298] a form of physics which believes it can do without metaphysics, because it proceeds (perhaps tacitly, perhaps unwittingly) on the assumption that it can take the place and fulfil the role of metaphysics.[299] Needless to say: "Physics that states that its explanations of things – in particular from causes and in general from forces – really sufficed and thus exhausted the essence of the world, would actually be *naturalism*".[300]

The distinction between this naturalism, or absolute physics or materialism in the improper sense, and real materialism lies in the fact that naturalism is more 'superficial' than (real) materialism, because in its search for explanation it does not penetrate to the metaphysical dimension but restricts itself to the surface-level of the physical. Naturalism is ultimately "a point of view . . . that impresses itself on man naturally again and again, and which can only be eradicated by deeper speculation".[301] The fact that such a "fundamentally false view" as naturalism forces itself of its own accord on man and must be removed in an artificial manner, can be explained by the view "that the intellect originally is not meant to teach us about the essence of things, but only shows us their relations with respect to our Will, that is to say, it (the intellect) is the mere medium of motives".[302] In other words, the 'superficiality' of 'natural' naturalism is based on the fact that the intellectual penetration of the reality which confronts us goes no further, in the first instance, than is strictly necessary from the viewpoint of controlling nature. Where, in contrast, there is 'deeper speculation' and the mind is not prepared to stop at controlling nature, but goes on to a real, i.e. metaphysical, explanation of nature, naturalism will deepen into materialism (in the real sense).

From the philosophical standpoint, a transition from naturalism to materialism – as, for example, was realised at the beginning of Western-European philosophy in the transition from the naturalism of Ionic natural philosophy to the materialism of Leucippus and Democritus – means a step forward. It is the transition from 'physics without metaphysics' to metaphysics. This point is not unimportant: a one-sided form of metaphysics like materialism is evidently preferable, in Schopenhauer's view, to the standpoint of physics without metaphysics, or naturalism. Philosophically speaking, this naturalism without metaphysics is an untenable position, and Schopenhauer therefore does his utmost emphatically to underline the fundamental "inadequacy of pure naturalism". The standpoint of pure naturalism or of physics (as opposed to metaphysics) is that of merely immanent knowledge, that is, the standpoint from which I can only talk about phenomena (as opposed to the thing-in-itself).[303] From the viewpoint or the purpose of metaphysics, this means an essential lack, since metaphysical explanation is concerned precisely with discussing what is concealed 'behind' the world of phenomena. It is clear that if one restricts oneself to the phenomenal world one will, in tracing causal regression (from cause to earlier cause to still earlier cause, etc., until one comes up against a law of nature which in turn finally leads to a force of nature) ultimately have to stop for a "force of nature, which remains as the simply inexplicable",[304] while moreover (as "our contemporary fashionable materialism" demonstrates) one also makes the mistake of "[taking] the objective, without more ado, as simply given, in order to deduce everything from it without regarding the subjective at all, through which, even in which, the former has its sole existence".[305] In short:

One will never be able to manage with naturalism or with the pure physical approach; it resembles a calculus example which never tallies. Causal series without an end or a beginning, primal forces that cannot be studied, space without limits, time without a beginning, endless divisibility of matter – and all this also depending on a cognitive brain in which it solely exists just as a dream and without which it disappears – form the labyrinth in which it ceaselessly leads us around."[306]

There are, therefore, basically two counts on which 'pure naturalism', in Schopenhauer's view, is essentially deficient: (1) in its 'positivistic' self-limitation to the realm of the empirical it cuts itself off from the dimension of metaphysical problems, and (2) in as far as this naturalism contains implicit metaphysics, it is the one-sided, 'bad' metaphysics of materialism.

What makes Schopenhauer's definition and criticism of naturalism of such interest to our study is that we can recognise in it, without much difficulty, that German 'materialism' which took hold among German physiologists from about 1838 onwards, and which distinguished itself from the stricter materialism of French physiology precisely through its refusal to embrace materialistic metaphysics.

The supposition that Schopenhauer, in formulating what he saw as halfway materialism and called naturalism, was characterising and criticising the self-conception in force among German physiologists at the time becomes highly likely, not only on the history of science grounds we have previously mentioned, but also because of the result of our review of Lotze's criticism of materialism in 1852, which leads us to the same conclusion.

2.15. LOTZE'S CRITICISM OF MATERIALISTIC METHODOLOGY

Lotze's criticism of materialism stems from his great work on psychology.[307] The position defended by Lotze is that of an interactionism which implies a *fundamental* separation of mind and body, and since it was precisely at this time, particularly in psychology, that the influence of materialistic monism was being expressed in criticism of more traditional dualistic conceptions of the mind–body relationship, it was necessary to refute "die Einwürfe des Materialismus"[307] with counter-arguments.

What strikes one in the first instance about Lotze's rebuttal of these arguments is (1) the fact that it dates from just *before* the materialism conflict which was brought into the open by the Wagner–Vogt polemics, that is to say, before this movement achieved its greatest notoriety, and (2) that it criticised materialism as a *methodological* position.

The misleading methodological desire for unity of principle has never expressed itself in any theory more passionately than it has in those materialistic theories which have appeared every now and then throughout the ages, but it is particularly encouraged at present by the rapid progress of natural science and is coming to the fore to an increasing extent and with growing confidence.[308]

It is reasonable to assume that Lotze's criticism of materialism, characterised as it was by its emphasis on the methodological aspect, can be seen as – indirect – evidence of the self-conception held by the German 'materialists' in the eighteen-forties – not only for the chronological reasons we have advanced, but also, and no less, because around 1850 this methodological materialism must have appeared to be the *only* expression of what was then known as 'materialism' which was *scientifically respectable and could be taken seriously philosophically.*[309] This does not alter the fact that Lotze's ontological criticism of the presuppositions of materialistic methodology is of course equally applicable to the methodic materialism of his period, which was his primary target (and which corresponds approximately with what Schopenhauer understood by naturalism), and to the naive materialism of Vogt, Moleschott and Büchner, which came to the forefront several years later. In both cases it is the same mistake which is the issue, with all its methodologically disastrous consequences: "These (materialistic) theories are concerned not only to avoid the existence of a psychic principle itself, but in particular to absorb psychology entirely into natural science; to assimilate its basic ideas with those principles which have long been contained there in the practice and continuous expansion of science".[310] Among the most fanciful excesses of this misplaced enthusiasm for natural science, according to Lotze, is a methodological precept which we often encounter as a positive formulation: "It must always be our aim, it is said, 'to snatch away as much ground as possible from the territory of the soul. One should everywhere try to push back this immaterial principle as far as possible and try to reduce the phenomena to the only proper basis of physical forces' ".[311] Lotze emphatically rejects the imperialistic claims of this methodological materialism: "Psychic phenomena are *not* identical or analogous to physical ones, and can never be considered as conjunctions of these".[312]

If we want to explain the transition from the physical to the psychic, we must assume a substratum which is not similar to matter, on which external stimuli act. The notion that matter could function as substratum of both physical and psychic properties is ruled out, since the unity of the consciousness does not admit of this explanation: "The soul *cannot* be viewed as the resultant of anything whatsoever, but only as a unity, because its separate actions are not distributed over different subjects and its overall condition cannot be considered as *the sum of the motions in a complex system*".[313]

2.16. SCHOPENHAUER AND LOTZE

Comparing Schopenhauer's criticism of materialism (and naturalism) with Lotze's, we find that in Schopenhauer's elucidation the accent is placed on the non-metaphysical, that is, on the inadequate character of naturalism viewed from the *metaphysical* viewpoint, while in Lotze's case the emphasis is on the *methodological consequences* of this, which were, in his view, untenable.[314]

Lotze (1852) goes further than Schopenhauer (1844) in so far as he casts doubt on the methodological rationale of the positivistic self-limitation which had been denounced by Schopenhauer as "naturalism" or the "purely physical way of reflection".

Summarising, we can conclude that both criticisms reflect the actual existence of a trend in natural science thinking *before* the major spread of naive-materialistic ideas, which – in as far as it thought about it at all – regarded itself primarily as a methodological position.

Schopenhauer's metaphysical criticism and Lotze's methodological criticism stood in the way of an explicit philosophical *justification* of methodological materialism (or naturalism). It is therefore only natural that when methodological materialism was finally justified in the work of the Neo-Kantian F. A. Lange, this implied at the same time a (critical) abstention from metaphysical claims (in this case, therefore, against Schopenhauer as a metaphysicist) as well as a defence of that figure of methodological materialism which Lotze had opposed with ontological arguments.

This concludes our exploration of the background. Our task now is to place the thinking of W. Griesinger, the founder of scientific psychiatry, in the context of the philosophical and scientific ideas of his time, and to interpret it in this light.

W. GRIESINGER AND THE MECHANICIST CONCEPTION OF PSYCHIATRY (FROM ABOUT 1845 TO 1868)

In the discussion of L. Snell's "clinical-psychopathological phenomenology" at the end of the first chapter we touched briefly on a subject which we must now look at in depth. This is the fact that the period in which 'scientific' psychiatry came into being – between the appearance of the first issue of the *Allgemeine Zeitschrift für Psychiatrie* in 1844 and Griesinger's publication of the *Archiv für Psychiatrie und Nervenkrankheiten* in 1867 – was dominated by the opposition between anthropologically oriented psychiatry and natural scientific psychiatry.

We have seen that the psychicists and the somaticists must be considered as representatives of anthropologically oriented psychiatry and we have observed that men like Damerow and Zeller, who could not be classified as psychicists or somaticists – or at least not without qualification – also belonged to this school.

By about 1850 the era of psychological–anthropological total visions in psychiatry had already effectively come to an end. Carus's great works, *Psyche* (1846), *Symbolik der menschlichen Gestalt* (1853), and *Organen der Erkenntnis der Natur und des Geistes* (1856) went unremarked in the *Allgemeine Zeitschrift für Psychiatrie* and evoked no response elsewhere.[315] Now and then the odd anthropological note was struck,[316] but Bodamer is certainly right when he says that after Wernicke had made his appearance (in the eighteen-seventies) the anthropological unity in psychiatry was, for the time being, out of the question.[317]

The period which saw the peak of anthropological psychiatry in the works of men like Damerow, Flemming, Roller, Jacobi, Zeller, etc. was, however, also the period in which the natural scientific school of psychiatry grew and spread. This school was ultimately to gain the upper hand. Its spokesmen, psychiatrists of a younger generation like Wernicke, Meynert, etc. were already clearly alienated from the anthropological tradition in psychiatric thinking. Indeed, the genesis of the natural scientific viewpoint, which took place from the eighteen-forties onwards, and the simultaneous downfall of anthropologically oriented psychiatry are clearly related events.

The period we are now discussing was – in terms of the history of science – dominated by the conflict between natural philosophical thinking and scientific thinking and was, therefore, markedly transitional in character. This character forms a common bond which links together, despite all their mutual differences, its typical representatives — Johannes Müller, W. Griesinger, J. L. Schönlein, K. von Rokitansky, R. H. Lotze, G. T. Fechner, etc. It was these transitional figures who made their mark on this period, and since we are concerned with identifying the general trends of this period in the psychiatry of the time, it is obvious that Wilhelm Griesinger, as the man whose name is indissolubly linked with the genesis of the natural scientific viewpoint in psychiatry, must occupy the centre of the stage.

The period mentioned in the heading of this chapter (about 1845–68) ends with Griesinger's death in 1868. To say that with Griesinger's death a stage in the history of psychiatry came to an end is to say nothing controversial. The date — about 1845 — which we have taken as the beginning of this period is, however, more open to debate. What decided us upon this date was the consideration that, although the natural scientific approach in medicine (psychiatry) had been eloquently defended by Griesinger in a number of articles in the *Archiv für physiologische Heilkunde* from 1842 onwards, it was not until the publication of his *Pathologie und Therapie der psychischen Krankheiten* in 1845 that it appeared on the scientific–psychiatric scene in *systematic* form. It was, as it were, the first major response to the influential institutional psychiatry, which had created it own organ a year earlier with the establishment of the *Allgemeine Zeitschrift für Psychiatrie*.[318] It is quite remarkable that the discussion between the up and coming scientific psychiatry and classical institutional psychiatry, which was begun in that first major work by Griesinger and was to be continued in the years which followed, did not lead to the establishment of an *alternative* scientific psychiatric organ until 1867 — that is to say, a whole generation later. Griesinger died in the following year, but the dialogue with anthropological institutional psychiatry was by then, in fact, at an end; "university psychiatry"[319] had carried the day and was to achieve unprecedented growth in the following decades.

In contrast with current studies of Griesinger's work, which look at it one-sidedly from the perspective of the university psychiatry that flourished after his death and which thus lose sight of the factors which — in terms of philosophical presuppositions — connect Griesinger's psychiatry with his 'predecessors' and separate it from his followers, I shall argue that Griesinger's work can be regarded as an exemplary manifestation of the link between old and new.

3.1. GRIESINGER'S 'APPRENTICESHIP' (UP TO 1844)

When Griesinger, at the age of sixteen, left school and went to study in Tübingen he encountered in the psychiatry lectures there the Schellingian physician and philosopher Eschenmayer. However, he had no time for romantic medicine, and once wrote: " ... I prefer to read Müller's Physiology rather than to have outdated views dictated to me".[320] This reference undoubtedly relates to the first volume of Johannes Müller's pioneering work, the *Handbuch der Physiologie*, which had appeared in 1833, the year in which Müller became a professor at Berlin.

In a way, Müller's work symbolised a farewell to 'romantic' physiology and at the same time the beginning of a development which was to lead to the dominance of the so-called empirical trend in physiology. This empirical trend, in contrast to the 'romantic' aversion to the particular, to observation and analysis, emphasised the indispensability of experiment and of experience and, after 1815, it prevailed among the then rising generation of researchers. It originated in France, and it was France, in the person of François Magendie, which was to lead Europe in the field of (empirical) physiology until the fourth decade of the nineteenth century – until, in fact, Germany, with the appearance of Johannes Müller (1801–58), took over this leading role.[321]

It therefore comes as no surprise to learn that Griesinger, whose sympathies so clearly lay with the new trend of empirical physiology, visited François Magendie while studying in Paris in 1838. Indeed, one can scarcely imagine a greater contrast than that between the 'romantic' Eschenmayer and the down-to-earth scepticism of someone like Magendie. It is, however, typical of Griesinger that, despite his antipathy towards speculative physiology and medicine, he did not subscribe uncritically to the extreme of experimental, vivisectionist physiology represented by Magendie. If I am right in my assumption that the anonymous person to whom the criticism at the beginning of Griesinger's early article *Theorien und Thatsachen* (Theories and Facts) (1842)[322] was addressed was Magendie (or someone of like mind), then we must go so far as to say that Griesinger subjected this position to criticism which was no less fundamental than his criticism of 'romantic' physiology and medicine. In order to be able to appreciate this hypothesis one must realise that Magendie is noted in the history of physiology as a sceptic who was highly critical of the vitalists' tendency to try to explain all unknown processes of life in terms of an equivalent number of life forces.[323]

Magendie's criticism was aimed not so much at the vitalistic idea *per se* as at the groundlessness of vitalistic explanations. His own pretensions concerning the realisation of a programme of physics of the phenomena of life (as expressed in his *Phénomènes physiques de la vie* published in 1842) therefore went no further than to state that he had investigated the phenomena of life sufficiently to be able to demonstrate that physical laws were operative at least *in some of their aspects.*[324] Although he expressed a pragmatically-argued belief in determinism as a guiding principle, the most striking thing about Magendie was his extreme caution in the face of theory in any shape or form. It is known that he even resisted drawing general conclusions from his experiments, and once described himself to his illustrious pupil C. Bernard as a "rag-gatherer" who, rake in hand, collects everything he finds.[325]

It is difficult not to identify this absolutisation of the anti-theoretical standpoint of 'facts', which was so characteristic of Magendie, with what Griesinger denounced in the above-mentioned article as a form of positivism which misunderstood itself.

Facts! Nothing but facts! exclaims a positivism which does not have the faintest idea that science has to employ negation at every point to make the next step possible. It does not want to understand that analysis always has to precede each reconstruction of concepts. The little world of medicine lives after all on ideas, like the great world of science and life. Just because medicine has said goodbye to hollow speculation it does not mean to say that we can happily assume that thinking has been superseded. Facts, hard-won through observation, experimentation and the microscope, need intellectual analysis and synthesis.[326]

It was precisely such great researchers as Müller and Henle who demonstrated most strikingly "that it is, above all, those who themselves have wrestled, in laborious research, with the details of the matter, who are capable of a theoretical definition of concepts and competent to provide one".[327] It is to be hoped that science will be enriched by new facts discovered through precise observation, but, he adds, "we [are], however, also convinced that these [observations] in all cases immediately have to be supported by a theoretical investigation, and that the attempt to familiarise oneself with the explanation and interconnection of phenomena is the task of anybody who devotes his efforts to science".[328]

The principle of science, of the scientific attitude, is that of criticism, "Negation".[329] The scientific standpoint must be seen as having to be defended on two fronts against the threat of misunderstanding. It must be defended on the one hand against those who "vociferously try to make the world believe that they are the positive ones"[330] (Griesinger is alluding here to representatives of 'romantic' physiology and medicine[331]) and on the other hand against the atheoretical (not to say anti-theoretical) attitude of those who lack the spiritual warmth and resolution which is necessary if one is to be able to commit oneself to any sort of theoretical standpoint. In the first instance Griesinger is criticising what he sees as an inadmissable dissociation (non-identity) of theory and practice, that is to say, of physiological and pathological theory and medical experience, and in the second instance he is condemning an illegitimate severing of theory and facts such as that which was *de facto* advocated by Magendie and his followers.

The interesting thing about this criticism is that Griesinger evidently felt that he was dealing with two different figures of *a single inadequate self-conception of science*, which he lumped together under the title 'eclecticism'.[332] This title is perhaps an unfortunate choice, as it appears to describe only one of the figures referred to. However, the context makes it clear that his criticism of eclecticism is directed against a scientific attitude, lacking in credibility, which lies at the root of both manifestations: science which is properly understood is science which proceeds from endeavours to achieve penetration, interaction, of facts and theory, of theory and practice. The negation which is contained in eclecticism (in the sense referred to above) is thus negation in a false sense; it is not the negation which is the principle of science: "Negation, as it occurs in eclecticism, is not an advancing force, pushing and stimulating from within; it is simple denial, whereas mind and will would have belonged to affirmation."[333] To Griesinger, science in the true sense of the word exists in a sense through the tension of the non-identity of theory and practice, of theory and facts, and its purpose lies in achieving the mutual penetration of these opposites.

We can thus already see, contained in this early plea on behalf of science, the philosophical motives which prompted Griesinger as a scientist to choose a path which diverged from what he felt was the over-speculative romantic physiology and medicine of his compatriots and from the too 'positivistic' physiological materialism in France.[334]

The critical intention of the major article *Herr Ringseis und die naturhistorische Schule* (1842)[335] can also be understood when viewed against this background. It is said that Griesinger caused a sensation with this article in the medical circles of his day.[336] The article was provoked by an attack made on the so-called natural history school led by Griesinger's former teacher Schönlein, by the Munich physician Joh. Nepomuk von Ringseis in his *System der Medicin* (1841). In this work Ringseis applied the doctrine of the fall of man and of redemption to medicine in a manner reminiscent of the pathogenetic notions found in the psychicistic psychiatry of Heinroth and his adherents. Like them, he did not hesitate to draw conclusions from this relating to medical therapy: before a cure is attempted, doctor and patient must "cleanse themselves of sin" through the means offered by the church. Griesinger's reaction to this therapeutic recommendation can stand as a summing-up of his attitude towards the 'religious' medicine represented by Ringseis: "In vain, Herr R.! The bell tolls for us in vain. We respect the church as long as it makes no incursions into foreign territory, but we expel dogma from medicine and we do not want science going to the Capuchin friars for confession!" [337] It is essentially the same criticism as that levelled by Jacobi against the disastrous mixture of religious and medical points of view which, in his opinion, characterised the work of the psychicist Heinroth. It was, however, by no means Griesinger's intention, as one discovers when one reads on, to take sides in the dispute between Ringseis and Schönlein. The controversy served him only as a welcome opportunity, through criticism of the protagonists of outdated medical ideas, to advance the reformation of physiology and pathology − something which he saw as the major task of the researchers of his generation. Thus we can see how the criticism of Ringseis was balanced by criticism of Schönlein and his school of thought.

In Schönlein, leader of the so-called natural history school, in whose work the "ontological conception of disease" which had traditionally dominated German medical thinking reached its zenith, [338] Griesinger was criticising a figure of medical, pathological research with which he had become familiar while studying in Zürich (in about 1837) and which he had − initially − considered highly estimable. This criticism is not simply negative, but undertakes, on close inspection, a *reformulation* of Schönlein's idea of an ontological theory of illness. An ontological theory of disease, as interpreted by Schönlein, generally provides an answer to the question of the nature of various diseases. This means that the attention of the nosographer is directed towards those factors in the symptoms of a disease which are invariable and constant. However, as Griesinger observed, [339] because diseases are not constant or unchanging, but are in fact variable processes, a systematic description of disease in this sense is only possible if such a description is limited to the peaks of the disease process, that is, to those conditions in which the disease concerned reaches its highest stage of development and finds its fullest expression. In this way one arrives at a system of clinical pictures which simply and solely reflect diseased conditions which have fully developed in this manner. The application of the method used is governed from the outset by a model of botanical classification: the disease is regarded and described as a natural history datum, a product of nature, and categories which are valid for organisms are applied to this.

Griesinger's greatest objection to Schönlein's version of an ontological theory of disease is that it gives pictures only of mature diseased conditions, while a great gulf continues to exist between the causes of the disease and the complex of symptoms described, and the questions of the nature, the genesis, the physiological properties and the development of the disease remain unanswered.[340] He sees this as the consequence of methodological one-sidedness and writes in this context of an "excess of the *analytical–descriptive* method . . . in which the inside of the processes has more or less been lost sight of".[341]

His proposal is, therefore, to correct this one-sidedness: confronting the ontological view of disease is the "purely phenomenological concept of illness . . . as was postulated in the principle of Broussais;[342] its path is as yet a little-used one and still leads to new truths. However, the two opposites will find a real intermediary in an *all-round physiological conception* which is now possible considering the standpoint lately adopted by factual physiology"[343] (my italics).

It is important to note that the "all-round physiological conception" advocated by Griesinger was not so much the promise of yet another physiological *system* as the formulation of a research programme. The superiority of this "all-round physiological conception" over other physiological conceptions lay, in his view, in the fact that it avoided the *methodological* one-sidedness which was inherent in these other conceptions.[344] Thus we see that the ultimate goal of the criticism of Schönlein's essentialism was to clear the way for the formulation of a *methodological idea* which was to give direction to the new research programme of the "all-round physiological conception". The new physiological pathology, like Schönlein's ontological theory of illness, was to endeavour to define the 'nature' of disease. There is, however, one difference: this nature which the new physiological pathology is on the way to defining is not, as it was in the earlier ontological pathology "a short definition, an abstract, fitting all individuals who have a disease; [in modern pathology] the nature, the concept of disease emerges only from the entire history of the disease, from the knowledge of each and every detail of the pathological processes, i.e. the gradual progression of quantitative and qualitative dysfunctions".[345] This means in fact that the nosological essences, 'natures', which are being sought amount in this case to much the same thing as the ideal, conceived ends of physiological–pathological research. Griesinger is therefore also able to write:

In this sense we know nothing as yet about the nature of any disease. Nonetheless, indeed all the more, its investigation has to be the constant aim and goal of all our observations of nature.[346]

The (new) physiological way of looking at things, he adds, can therefore offer no system — that system does not yet exist.

Thus we see that, while Griesinger scarcely takes Ringseis' theory seriously, he looks more kindly on Schönlein's natural history school in as far as it gives physiological pathology the goal of the definition of the essence, the nature of diseases and the processes of disease. The great failing of this school was that it had a false conception of what this objective implied: "What medicine must now try to do is always to keep in step with real and factual physiology in pathology. In this way a *natural history* of the process of a disease will be obtained, but in an entirely different sense from that given by the natural history school".[347]

The 'new' empirical physiology of Magendie, Müller and others, from which Griesinger took his inspiration, put medicine on a new footing, and it is obvious that anyone who hoped in this period to advance medicine in one of its many forms had to start by making clear the relevance of this new physiology to (general) pathology and to medicine in general. Their common aim of validating this new physiological point of view in medicine led three men who had been students together — Wunderlich (internist), Roser (surgeon) and Griesinger — to found the *Archiv für physiologische Heilkunde* in 1842.

In this context it is as well to point out that — surprising as it may be to anyone used to thinking of Griesinger exclusively as a psychiatrist — virtually all the articles by the celebrated founder of clinical psychiatry up to the year before his death, that is, before the establishment of the *Archiv für Psychiatrie und Nervenheilkunde* in 1867, appeared in a clearly general medical — or at any rate not specifically psychiatric — journal like the *Archiv für physiologische Heilkunde*. Griesinger, indeed, took a keen interest not only in psychiatry but also in the problems of general pathology, and in particular in questions relating to the problem areas of internal medicine.[348] In the eighteen-forties and -fifties, the fundamental possibility and purpose of the reformation of medical thinking *in general*, aimed at by Griesinger and his followers, was certainly at stake and had to be defended. Griesinger, because of his universal medical orientation and interest, was eminently suited to assist in the realisation of the general objectives of this journal, and it is therefore by no means odd that he should have remained faithful to it[349] for so long, nor that he should have published in it not only several important psychiatric studies but also a not inconsiderable number of pages on general pathological and other matters. I should also point out that, as far as his psychiatric work was concerned — of which the most important appeared in the first and second editions of his *Pathologie und Therapie der psychischen Krankheiten* (1845, 1861[2]) – Griesinger in fact devoted the major part of his psychiatric and scientific life to accomplishing a fairly *general* task. This was to demonstrate the viability of a new figure of psychiatry, i.e. that of so-called 'scientific' psychiatry, which had become possible on the basis of the new physiology. It is in line with this that the connection with the problems of general pathology is preserved in Griesinger's psychiatric work, and his psychopathology – with the exception of a few later studies in the field of specific pathology – is essentially and substantially general psychopathology.[350]

The year 1842 was important in the history of physiological and medical thinking because it saw the more or less simultaneous appearance of a number of publications which heralded the start and growth of a new era in science, which was to become a fact in subsequent generations.[351] The period with which we are now concerned is, however, as we have already observed, a period of transition and of preparation, and in the proclamation of the new order, which makes the year 1842 so memorable, the antithesis to the 'old', i.e. romantic-idealistic and vitalistic physiology, is still constantly present. The new order must therefore be understood in terms of this antithesis. This is the common factor in the developments which then set in.

3.2. LOTZE AND GRIESINGER

The year 1842 saw not only the foundation and first issue. of the *Archiv für physiologische Heilkunde* already referred to, with Griesinger's aggressive and sensational contributions, but also the publication of the first part of the celebrated *Handwörterbuch der Physiologie mit Rücksicht auf physiologische Pathologie*[352] by Wagner of Göttingen. The most important article in this, *Leben, Lebenskraft*, was written by a man who was later to be a colleague of Wagner's, the philosopher Rudolph Hermann Lotze. In that same year Lotze also published his book *Allgemeine Pathologie und Therapie als mechanische Wissenschaften*.

There can, I feel, be little doubt that the work of the philosopher and physician Lotze must stand as the leading expression of the 'transitional thinking' of the period. It is therefore certainly an inexcusable omission that, to date, the question of the relationship between Lotze's work and that of Griesinger has been systematically disregarded in the history of science literature. By looking in more depth at Lotze's publications in 1842,

18

mentioned above, and at an article by Griesinger, published in 1843, in which he gives his opinions of "the most recent development in general pathology",[353] I shall try to make it clear that Griesinger, in respect of his view of science, in fact followed the self-conception of the medical sciences expressed by Lotze.

The fact that Lotze was still generally reckoned to be a physiologist until 1855[354] may come as a surprise to present-day philosophers, for whom the name Lotze is associated exclusively with the figure of an all-but-forgotten philosopher who is remembered only as the forerunner of twentieth-century philosophy of values, through the work of pupils (or critics) like Brentano, Stumpf and Husserl, and whose place in the history of science is primarily due to his contribution to the psychology of perception.[355] Lotze must, however, as a glance at his work and the progress of his university career confirms, be described as a philosopher-physician, judging from his principal interest.[356]

While Lotze may well continue to be remembered by posterity primarily or solely as a philosopher, it is undoubtedly true that to the generation of scientific researchers to which Griesinger belonged[357] Lotze was known above all as the author of a number of fundamental works specific to medical thinking (general physiology, pathology and therapy, and medical psychology). Chief among these is the work published in 1842 which we mentioned earlier, *Allgemeine Pathologie und Therapie als mechanische Wissenschaften*, which created a great stir when it appeared and ushered in a new era in the science of general pathology.[358]

The new (general) pathology necessarily assumes a new (general) physiology, and Lotze's reflections on "Leben, Lebenskraft" (Life, Vitality) — central concepts of (general) physiology — can therefore be usefully applied in the clarification of his general-pathological position.[359]

After what has been observed above about Griesinger's first articles in the *Archiv für physiologische Heilkunde*, it must strike one — although it will come as no surprise — that Lotze perceives the *problem situation* in essentially the same way as Griesinger and defines his attitude towards it in a similar fashion: the physiology of an earlier period (by which is meant 'romantic' physiology) has lost its credibility because its many attempts to explain the phenomena of life have all been unsuccessful. With speculative trends retiring from the scene, however, a deep mistrust of theory in any shape or form has taken hold, and this has led to a far too uncritical acceptance of 'stray ideas'. Against this rejection of any sort of 'theory', Lotze defends the necessity of what we would call philosophical fundamental research into physiology or, formulated in his own terms, the necessity "of giving a well-founded account of the conditions and the general concept of life, from the nature of knowledge and of things, as given to us by knowledge".[360]

The controversial aspects of physiology are not the facts which everyone knows, but the explanations put forward for them, which in their turn are based on "previous convictions and misconceptions of what the actual task of natural science in general is, of the means through which it fulfils this task, and on the incorrect connection, expression and interpretation of the facts to which these preconceived opinions imperceptibly tempt one". In other words, what Lotze is aiming at here is a criticism of the general explanatory basis of the (natural) scientific undertaking. Without a preliminary theoretical clarification of this kind it is impossible to develop a sound concept of physiology as a (mechanical) natural science.

To summarise, Lotze's criticism of the earlier physiology (that is, not 'truly' scientific physiology in the sense of the 'mechanical' natural sciences) amounts to the following. This earlier physiology was guilty of mixing points of view, questions or methods of looking at things which should be kept strictly segregated. "There are three relationships in particular, which are metaphysical conditions of all connections between things, and which provide the inspiration for *three* different, precisely delimitable, methods of research which are, however, so often confused, much to the detriment of science; i.e. the relationships of *reason* and *consequence*, of *cause* and *effect*, and of *end* and *means*".[361] Accordingly, a distinction must be made between ways of looking at things or formulations of questions in which a thing or occurrence is only of interest (1) as a particular case of a general law, (2) as the actual cause of an actual consequence, or (3) as the means to an end. In the first case we are dealing with a problem of subsumption in the second case with a problem of the application of the general to the (real) particular, and in the third with the demand for a goal or purpose of what is or what happens.

The critical point of Lotze's argument can be formulated as follows: (1) logical deducibility (of the particular from the general) does not necessarily imply the applicability to reality of the cause and effect relationship found,[362] and (2) scientific explanation of what is or what happens is not an answer to the teleological question of the reason why something is or happens (and vice versa).[363]

This last point — essential to a 'mechanical' conception of physiology such as he has in mind — is summarised by Lotze in the formulation of the following methodological requirements to which, in his view, any theory about the phenomena of life must be subjected:

never give an end as the cause of the realisation and of the quality of a phenomenon; never use the indication of ends and of causes as two coordinated principles of explanation, to be used arbitrarily according to circumstances; never believe that the statement of the end can excuse one from also indicating the causal instrumentation by which the end has been reached.[364]

Lotze's criticism of the self-conception of 'earlier' science (in this case, physiology) can be summarised as criticism of a double non-distinction, i.e. that between (1) ideality and reality ("ideal relationships are not (necessarily) real relationships"), and (2) mechanical–causal (= [natural] scientific) and teleological (= philosophical, metaphysical) questions. The consequence of this for Lotze's self-conception of science is that science is, by definition, the science of reality; the questions it formulates are mechanical–causal and all science is *ipso facto* 'mechanical' (natural) science. I shall not go into the detailed argumentation with which Lotze backed up these views, but will look at the general considerations with which he introduced his *Allgemeine Pathologie und Therapie als mechanische Naturwissenschaften.*

It goes without saying that, where the relationship is as close as that between physiology and pathology, the general views expressed in the article *Leben, Lebenskraft* will recur here – albeit more specifically directed at pathology. Although the fundamental standpoint remains the same, the specific application to general pathology which we find here nevertheless holds a certain interest, because it in fact expresses the perspective in which the special pathology of mental illness must be placed. This was the field of research which, while it was not the only one in which Griesinger worked, certainly occupied a privileged place in his attention.

Lotze rejects any form of pathology which begins by giving a definition of the concept and essence of disease, thus in fact determining before any research is undertaken what is to be regarded as disease "as if all aspects of the matter to be subjected to investigation could be given by the content of such [a definition]".[365] This is exactly the same anti-essentialistic criticism which Griesinger levelled at the ontological medicine of the so-called natural history school[366] led by Schönlein. Unlike Griesinger, Lotze goes on to show that (and how) the confusion of points of view, which we have already discussed, begins with essential definitions which lay down the law in this way. In other words, essentialistic pathology is criticised for its unacceptable *methodological* consequences.

In natural science, in this case pathology, one must proceed on the assumption that there are *general laws* which make possible the interaction of the various processes and their substrata. "If there are disturbances in the

regular course of the phenomena of life, and if there is a general science to
deal with these disturbances and their consequences [general pathology, of
course, as Lotze conceives it], then that science must be able to indicate step
for step how the disturbing influence works, point for point, from one
sutstratum to the other, according to general rules, and derives from them the
degree of its success".[367]

He is here concerned with finding laws of disease on which the very
possibility of disease and how it comes into being are based. The formulation
just used makes it clear that in the case of these 'laws' Lotze is thinking of
something different from what we would nowadays be inclined to think of
at first. The general 'laws' of (general) pathology he is looking for are not
empirically acquired generalisations arising from an intensive interaction of
pathological experience and theoretical reflection, but are the result of an
analysis of fundamental pathological concepts whose sole purpose is "to bring
the fundamental concepts of pathological disturbance into line with physical
possibility" and which to this end serves the "mechanical theory as a guiding
rule".[368] In other words, Lotze's (general) laws of disease are, more or less,
(particular) natural philosophical principles.[369]

Scientific interest (in pathology) is not, however, limited to the general
laws of disease: no less important is knowledge of "empirically given *forms*,
among which the phenomena of life act as particular instances of those
general *laws* and of the special effects which the actions of the latter must
have, by virtue of the given constitution of living bodies".[370] Without this
empirical information the knowledge of scientific laws remains abstract: it is
only through application to (pathological) reality that this error is rectified.
The possibility of application only exists, however, if I have knowledge of the
empirically given forms in which the generalities of these laws are examplified.
If the former – 'nomological' – approach thus provides "in all cases the basic
principles of pathology by reducing it to its simple, general conditions", it is
the empirical knowledge of forms which enables the bridge to be built to
(pathological) reality.

As well as (1) the 'nomological' and (2) the empirical approach, the
teleological approach (3) also has a right to exist.[371] With teleological lines
of research one is, in fact, in a sense leaving the field of science – in this case
of physiology and pathology as 'mechanical natural sciences'. The question of
the essence of life and disease, determinative for physiology and pathology,
respectively, acquires another meaning here than it assumes in the nomological or empirical perspective. In the teleological approach to physiology
and/or pathology one is seeking the essence of life and disease in the sense of

"the meaning and significance, the values, the purpose of this whole business which we call life and disease".[372] The phenomenon of nature interests us in this case only "in view of the meaning . . . or the idea, for the representation of which it is at all present in the phenomenal world".[373] For physiology this is the idea of life, for pathology that of disease.

One should not think (according to Lotze) that this teleogical viewpoint, which — strictly speaking — represents comparatively the highest standpoint of speculation,[374] is irreconcilable with the supposition of mechanical theories or that it renders them superfluous: ". . . as soon as it needs a play of masses against one another for its appearance, [every idea] necessarily [presupposes] a well-ordered mechanical interconnection . . . which provides it with the material forces for its realisation".[375] The idea which is meant in the teleological approach and the phenomena in which this idea is presented on the level of experience lie, so to speak, on different planes: the order of the ideas and the order of the phenomena, the sphere of the ideal and that of the real are not the same.

Thus Lotze, in clarifying this point, can make use of the following, rather nice comparison: "In the same way as everything has to submit to the law in a state, without the law itself having to give a mechanical impulse to the musculature of its citizens, the idea of life will in fact also be the ruling law of its phenomena, but it will never itself actively interfere in the interconnection of these phenomena".[376] This same thought occurs elsewhere in Lotze's work, expressed in a slightly different form: one must guard against "taking the idea of life, which is always only a *legislative* power, to be an *executive* one".[377] So much for Lotze and his philosophical views on the fundamentals of pathology.

As we have observed, Griesinger's work falls entirely within the bounds of the self-conception of natural science in general and pathology in particular developed by Lotze. The fact that, on this fundamental level, there was complete agreement of conception between himself and Lotze was acknowledged by Griesinger himself in such an unambiguous and clear fashion that it is indeed remarkable, not to say astonishing, that in the history of psychiatry, and more especially in the interpretation and analysis of Griesinger's position in pathological thinking, nobody has paid any attention to this relationship.

In the article mentioned earlier, *Bemerkungen zur neuesten Entwicklung der allgemeinen Pathologie* (1843)[378] several pages are devoted to Lotze's *Allgemeine Pathologie und Therapie*, which the generally highly critical Griesinger treats in an unusually complimentary way. It is above all Lotze's discussion of fundamentals in the general section which gains his unqualified admiration, as well as the way in which Lotze takes a stand against the undisciplined and pretentious character of old-style pathology. In Griesinger's view, Lotze has generally hit the nail on the head and (he adds) "I am all the more delighted to agree with all this, because I recognise my own viewpoints in it, but so splendidly and sensibly put across that I could scarcely have done it better myself".[379]

Indeed, Lotze's book, which was published in 1842, must have appeared to Griesinger as valuable support for his efforts to place medicine on a "new"[380] footing. It is true that much of what Lotze broaches in his book had already been familiar to a younger generation for some time, but Lotze must take the credit for having expressed so clearly what everyone else perhaps "saw unclearly and imperfectly".[381] This, I suspect, will also have been true in some measure of Griesinger himself; his criticism of ontological medicine and his defence of the "mechanical" viewpoint are in tune with Lotze's methodological separation of the scientific and the teleological view. It is thus with a certain amount of satisfaction that Griesinger quotes Lotze in this context, where the latter insists that the "legislative power" of the idea of life cannot be concieved as "executive".

Notwithstanding all of this, it is as well to realise that the link which joined Griesinger and Lotze chiefly consisted − on the negative side − of their common opposition to the anthropologically oriented psychiatry of the psychicists and the somaticists,[382] and − on the positive side − of the commitment both felt to the idea of medicine as a "mechanical natural science". In the early years of the 'reformation' of medical thinking in Germany, in particular, this agreement will have been the most important and it is therefore quite justifiable to deal with the comparison of Lotze's standpoint and that of Griesinger in the context of the discussion of Griesinger's early work. This cannot, and should not, blind us to the fact that, as was to become clearer later on, the idea of medicine as a 'mechanical natural science' could be reconciled with various metaphysical presuppositions; more particularly, with various interpretations of the mind-body relationship. It was on this point that Griesinger will not have been able to find the final justification of his own assumptions in Lotze's philosophy, nor will Lotze have been able to find whole-hearted support for his philosophical insights in Griesinger's work.[383] While Lotze believed explicitly in a dualistic interactionism,[384] Griesinger's sympathies, as I shall demonstrate in more detail later, clearly lay in the direction of an identity-theory conception of the mind−body relationship.

For the moment it is not necessary to pursue this theme further. What

concerned me in the comparison of the views of science of Lotze and Griesinger was to find a basis for a reconstruction of Griesinger's *self-conception*
of psychiatry. This in turn will help us to develop some ideas about the
systematic unity of his thought, which more often than not, is lost sight of
in the varied, often conflicting interpretations which subsequent generations
have placed on his work.[385]

For the same reason, there is no need to look in depth here at the study
Ueber psychische Reflexactionen; Mit einem Blick auf das Wesen der psy-
chischen Krankheiten (1843)[386] or the follow-up article *Neue Beiträge zur*
Physiologie und Pathologie des Gehirns (1844).[387] The interesting thing
about these two articles is that they contain the fundamental conception
of Griesinger's psychiatric theory, that is to say, they are the first form in
which Griesinger's idea of truly natural scientific psychiatry was realised.
Their importance in the context of the question we are dealing with —
the self-conception of Griesinger's psychiatry — can best be discussed in
connection with his *Pathologie und Therapie der psychischen Krankheiten*
(1845).

3.3. GRIESINGER'S PSYCHIATRY IN THE PERIOD 1845–68

The year 1845 saw the publication of the book to which Griesinger owes
his reputation in the field of psychiatry, *Die Pathologie und Therapie der*
psychischen Krakheiten für Aerzte und Studirende. The title of this work
immediately reminds one of Lotze's textbook *Allgemeine Pathologie und*
Therapie, which we have already discussed. Bearing in mind that the special
pathology and therapy developed by Griesinger fit into the horizon of Lotze's
conception of pathology and therapy as 'mechanical' natural sciences, it is
certainly not unreasonable to assume that Griesinger was inspired by the title
of Lotze's work.

In this work, for which the way was paved by the psychiatric studies of
the preceding years, classical institutional psychiatry receives its first 'major'
rebuttal from a representative of the 'new' physiological medicine. This
marked the emergence of an alternative conception of psychiatry in its first,
systematic form, and curiously enough it came from the pen of a young
doctor who was probably better known at the time as a practitioner of
internal medicine rather than as a psychiatrist.[388] He was, moreover, a man
who in all likelihood preferred to regard himself as a general physician rather
than as a specialist[389] and who, in fact, could boast of no more than two

years' psychiatric experience gained – and this is not unimportant – during a sort of probationary period (1840–42) in one of the most renowned centres of German psychiatry at the time: A. Zeller's institution in Winnenthal.

The figure of psychiatry developed by Griesinger has been preserved by posterity in the form of a number of dictums which are often quoted but usually partly or wholly misunderstood, such as 'mental diseases are diseases of the brain' and 'insanity is only a complex of symptoms of different, anomalous cerebral conditions'. Because such pronouncements were taken out of context they easily led to misunderstandings.

My aim, in the section which follows, is (1) to say something about the basic model of Griesinger's psychiatric theory (an answer to the question of why Griesinger referred to pathology, etc., as 'mechanical' natural science), in order (2) to develop, based on the character of Griesinger's "mechanical" conception of psychiatry which we have thus derived, an interpretation of his *self-conception* of psychiatry which permits a clarification of his relationship with the assumptions of institutional psychiatry, on the one hand, and the presuppositions of (in this case, mechanistic) materialism on the other. In this way I hope to make clear what links Griesinger with his (anthropological psychiatric) predecessors and contemporaries, and what separates him from some of his followers, who determined the further course of so-called university psychiatry.

3.3.1. GRIESINGER'S PSYCHIATRIC THEORY AND THE MECHANISTIC CONCEPT OF SCIENCE

One reason why Griesinger will have found Lotze's views on the foundations of medicine attractive was that Lotze's position provided insight into the *compatibility* of the mechanistic ideal of science and a teleological interpretation of reality. The question of how 'mechanical natural science' could go hand in hand in one way or another with the teleological view of the phenomena of life, which was traditionally relied on in philosophy and the biological sciences and supported by everyday experience, was the great problem which the philosopher-physicians (biologists, etc.) of the generation of Lotze, Griesinger, Johannes Müller, *et al.* tried to clarify.

Griesinger was thus allied to Lotze on this problem of the compatibility of 'mechanical' and teleological views. This does not alter the fact that in the final analysis this compatibility was seen by the two men in a different way. Lotze's standpoint was fairly explicit from the outset (1842), while

Griesinger's conception remained largely implicit and can only be brought out by means of a reconstruction of the systematic unity of his work.

In Lotze's case the problem is solved in the sense that the 'mechanical' and the teleological points of view are distinguished as the perspectives of (natural) science and of philosophy (metaphysics) respectively. This is not the case, as a closer examination shows, in Griesinger's work. Although Griesinger — more a scientist than a philosopher — was primarily interested in demonstrating that the concept of the disciplines of medicine as 'mechanical' natural sciences, defended by Lotze, also held good in the specific case of psychiatry and, certainly at first, the question of the possibility and right of existence of a mechanical conceptualisation of psychiatric phenomena occupied the forefront of his attention, we must nevertheless say that the viewpoint of a (possible) teleological interpretation of the 'facts' was never completely out of his thoughts. This means that Griesinger, at least implicitly, granted the teleological viewpoint a certain right to exist within the scientific perspective, so that it seems to fulfil (within science) a complementary function, as it were, to the 'mechanical' point of view.

This point must be emphasised now because in the current interpretations of Griesinger's mechanistic theory it is either overlooked or its fundamental significance has not been sufficiently considered, so that these interpretations fail to arrive at an adequate definition of his position in the history of psychiatry. The significance which must be attributed to this preliminary observation will, I hope, gradually become clear in the following discussion.

The fact that Griesinger's psychiatry fits without difficulty into the limits of the self-conception of general pathology and therapy as 'mechanical' sciences, put forward by Lotze, is not surprising when one sees that this psychiatry, as the concrete expression of the mechanistic ideal of science [390] which was gradually spreading in the eighteen-forties, is something of a synthesis of (earlier) mechanistic views in physiology and psychology, like those formulated by Johannes Müller and J. F. Herbart respectively.

The basic model of Griesinger's psychiatric theory, which contains this synthesis, is in fact already present in the article *Ueber psychische Reflex-actionen* (1843) and is expounded most fully in the second edition of *Pathologie und Therapie der psychischen Krankheiten* (1861). Since we are not concerned with a detailed analysis of the content of Griesinger's psychiatry, but with the (psychiatric) *self-conception* which he endorsed, I shall restrict myself to those elements in the two works which are of importance in determining that self-conception. These are (i) the extension of the mechanistic point of view into the domain of psychopathology, as defended

by Griesinger in the 1843 article, and (ii) the (relative) validity of the teleological point of view within the mechanistic perspective, for which there are
several points of contact in both the earlier and the later work.

(i) The major article *Ueber psychische Reflexactionen: Mit einem Blick auf
das Wesen der psychischen Krankheiten*, which appeared in the second year
of publication of the *Archiv für physiologische Heilkunde*, immediately
reflects in its opening words the difficult situation in which medical thinking
found itself at the beginning of the eighteen-forties: there is the express wish
to break the ties with speculative philosophy and to give medicine a new
foundation in a 'really' scientific physiology.[391]
    The distinction created here between philosophy and science must not,
however, cause us to overlook the extent to which the intelligibility of
Griesinger's scientific programme and of the conceptualisation adopted by
him depends on certain philosophical presuppositions. A case in point here is
the 'naturalistic' ontology that is presupposed when Griesinger differentiates
between the organic and the psychic only specifically, not generically. In a
broader sense of the word, in which 'organic' equates with 'nature', the
psychic is no more than a certain mode of being of the organic. Viewed
from this angle, the study of psychic phenomena is *a fortiori* a matter of
natural science and psychology, natural scientific, or more precisely, physiological psychology.[392]
    It is precisely this naturalistic-ontological presupposition that creates a
clear conscience for the parallelism of organic and psychic phenomena, of
organic nature and psychic nature, and makes it possible to apply the same
(general) natural laws and concepts on both the organic and the psychic
level.
    One must also understand Griesinger's germinal psychiatric thought –
the extrapolation of the (neuro-) physiological concept of reflex action in
the nervous system to the field of the psychic – against this background.
The justification for this extrapolation is rooted in the structural agreement
between the organic and the psychic. Griesinger can therefore state in his
article *Ueber psychische Reflexactionen* that his aim is "to emphasise the
parallels between the workings of the spinal cord (with the *medulla oblon-
gata*) and those of the brain, in so far as it is an organ of psychic phenomena
in the strict sense of the word, and to prove those parallels to exist by reference to normal and abnormal phenomena".[393]
    The concept of reflex action in the sense used by Griesinger comes from
the (neuro) physiology of M. Hall and J. Müller and as such is the central

concept in a natural scientific physiology which regards itself as 'mechanical'. The concrete expression of the mechanistic concept of science, as Griesinger realises it in his psychiatric theory, in fact thus comes about because the psychiatric theory is conceived as a form of reflex theory.[394]

From the viewpoint of a mechanistic (neuro)physiology, like that which Griesinger invokes, the area of the phenomena of life to be studied is perceived and conceived in terms of centripetal and centrifugal actions and the transition from the former to the latter. Such centripetal and centrifugal actions take place in the spinal cord (and therefore, according to Griesinger, also in the brain in so far as it is an "organ of psychic phenomena"). The concept of reflex action functions in this context as a concept aimed at understanding the transition from centripetal to centrifugal actions in the spinal cord, etc..

Griesinger's starting-point is the consideration that, on closer examination of the comparison of the actions of the spinal cord and the brain, it appears that the two are not subject to different laws and that, where the transition from centripetal to centrifugal excitations is concerned, there exists a "curious harmony" between the more or less conscious actions of the brain itself — so-called ideas and strivings — and between the phenomena of sensation and movement of the spinal cord.[395]

His concern is now twofold: (1) to demonstrate that there exists a parallelism which extends to the details "between the vital manifestations of the medulla, perception and movement, and between those of the brain, ideas and strivings", and (2) to contribute to a correct conception "of the relationship of the consciousness to the other [read: not conscious] reflexes" by studying the reflexes within the consciousness.[396]

Explanatory note. The spinal cord and the brain, that is the so-called "central organs", are the organs which regulate the reflexes through which the organism (animal, man) realises the goal of maintaining its own existence. The process of reflex regulation is set in motion by external stimuli (so-called 'centripetal actions, stimuli'). These stimuli are converted in the 'central nervous organs', by means of a process of dispersal, into a sort of excitation-mean which, spread throughout the organism, maintains and regulates the tone:[397] that is to say, the spinal cord is responsible for regulating the tone of muscles, cell tissue and blood vessels, and the brain for the so-called psychic tone.[398]

The ultimate destination (!) of the centripetal impressions is, however, not the aforementioned dispersal, but is, on the level of the spinal cord, "to become the source of stimulation for motor activity and in this way itself actually to change into movements, to find its destination in these".[399] The same is true, *mutatis mutandis*, on the level of the brain.[400]

This presupposes that the centripetal impressions which have the significance of 'impressions of sensation' on the level of the spinal cord, have undergone a further change on the level of the brain and are there converted into ideas (to Griesinger the brain is the 'organ of ideas').

(ii) The reflex regulation which brings about the transition from centripetal sensory impressions to (purposeful) movement plays the major role in animals where brain development is (very) slight: the movements of animals are "effective by nature, i.e. intended for the preservation of individual existence". In animals this "effectiveness" is not based on a free choice of the will − animals neither have nor need this − but lies in the neural organisation of the animal organism itself: "all its aims in life [are] organic aims".[401]

In man's case things are different, because he has a relatively highly developed brain. The reflex regulation of the human organism is therefore not restricted to that of the spinal cord reflex activity, but takes place on the level of the spinal cord and the brain. The ultimate goal of the reflex regulation of the brain, in which idea changes into striving and finally into action, is "to realise our spiritual ego", because it is "the act that is our destination, that liberates our inmost depths".[402]

This short summary is enough to shed light on a characteristic point in Griesinger's thinking − a point that is important to our argument − namely the fact that, for Griesinger, the 'mechanical' view left room for a teleological interpretation. It is important to realise that we are not simply and solely concerned here with a viewpoint which is wholly *implicit* in his mechanistic theory, i.e. a view which Griesinger himself was not aware of or did not acknowledge, but with a conscious way of looking at things, acknowledged in its own relative right, which − apparently − was not considered to affect the universality of the mechanistic concept of science.

On the one hand, it is indeed true that when Griesinger, in developing his fundamental model on the basis of reflex actions on the lowest level ("as it were at the aphelion of the consciousness"[403]) ascends to the consideration of the psychic activities which are coupled to 'arbitrariness' and 'consciousnesss' − ideas and impulses of will or strivings − and thus constructs a sort of hierarchy of functions of the central nervous organs (spinal cord, brain), a design of this kind can only be understood as the result of an extrapolation, which was certainly original at that time, of the physiological reflex action model, that is, as a very specific concrete expression of the *mechanistic* concept of science. Griesinger's formulations, however, make it clear that at the very least he does not *de facto* exclude the possibility of positing teleological questions. To put it even more strongly, it is not only true that he speaks of the suitability, "purpose", "aim", "destination", of reflex regulation, it is also true that he expressly underlines the right to speak in this way when he observes that "this teleological conception of the fact [i.e. of reflex action] seems permissible if one remembers how tone and muscular movements are absolutely indispensable for the attainment of the vital purposes of animal organisation".[404]

In this context it must be pointed out that when, in the article in question, Griesinger states that one of the objectives of his undertaking is to contribute to a "correct view of the reaction of the consciousness to the other reflexes"[405] and amplifies this point by pointing to the "apparently" great dependency of the reflex regulation of the tone of muscles, blood vessels, etc., on the consciousness, that is to say, the *content* of conscious ideas, he is *de facto* — whether consciously or not — overstepping the bounds of the mechanistic perspective.

The meaning which must ultimately be attributed to this observation in relation to the self-conception of Griesinger's psychiatry which we are trying to reconstruct — are we dealing with an internal inconsistency which undermines the *unity* of Griesinger's conception? — can only become sufficiently clear in the following pages if we look at his psychiatric theory in its most mature form, that is, as we encounter it in the second edition of his *Pathologie und Therapie der psychischen Krankheiten* (1861).[406]

3.3.2. THE BASIC PATTERN OF GRIESINGER'S 'PHILOSOPHY': NATURALISM ON THE BASIS OF IDENTITY THEORY

Although we can agree with Ackerknecht[407] that all dominant trends in modern psychiatry can be traced back to parts of Griesinger's work, Griesinger is almost exclusively[408] remembered in present-day psychiatric circles as the founder of so-called neuro-psychiatry, that is, as the man who laid the foundations for the fusing of psychiatry and neurology. This is also true of Jaspers who, although the one to underline the significance of Griesinger as a descriptive author,[409] nevertheless primarily remembers him as the author of the neuropsychiatric 'dogma' of the identity of diseases of the mind and diseases of the brain ("Geisteskrankheiten sind Gehirnkrankheiten") and in that context mentions him in the same breath as Meynert and Wernicke.[410]

His criticism makes it clear how he interprets the thesis that "diseases of the mind are diseases of the brain", i.e. as a thesis which holds that *the* (ultimate) cause of a disease of the mind is a disease of the brain, and which thereby implies the (ultimate) *identity* of physical occurrences in the brain and mental illness. An identity thesis of this kind describes, however, in his view, "perhaps a possible aim, compared with actual possible research and real experiences, but then it is an aim of research which is infinitely far away. It does not denote, however, an object of research".[411] In brief: I can perhaps usefully *endeavour* to explain (causally) mental disease on the basis of brain diseases, that is, regard the identity thesis as a guideline, as the *idée directrice* of research (into the pathology of the brain), but I cannot proceed on the assumption that such identity simply *exists* somewhere and can thus be an object of research in itself.[412]

There can be no doubt that Griesinger (if he could) would have rebutted such criticism with the question: "how can research directed towards the idea of the identity of diseases of the mind and diseases of the brain be at all meaningful unless, in some cases at least, experience has proved the *fact* of this identity?" Indeed, where Jaspers, given his psychophysical parallelistic attitude, could at best attribute a pragmatic significance to the neuropsychiatric identity thesis, a justification of this thesis (as a guideline for neuropsychiatric research) in the spirit of Griesinger would unhesitatingly refer to the *fact* [*sic*!] of this identity as something that had already been established by experience.[413]

It must, however, be observed in this respect that what Griesinger saw as an empirical fact was criticised by Jaspers — apparently — not only as the πρῶτον ψεῦδος of Griesinger's research programme, but as the metaphysical dogma of the *entire* neuropsychiatric school. Given Jaspers' own dogmatic assumptions this is understandable and consistent, but in terms of the history of science its effect is to cloud the issue, because the difference between Griesinger's careful treatment of the identity thesis as neuropsychiatrically regulative and the uncritical ontologising of it by some of his most influential followers remains obscure, and a closer connection is suggested between Griesinger's psychiatry and that of his immediate successors than can be proved historically.

In fact, it was only with the realisation by Meynert and Wernicke of the programme conceived by Griesinger that this tendency to ontologise the scientific, neuropsychiatric theory took hold — and here Jaspers' criticism is eminently applicable. What then came into being was namely a figure of naive-realistic thinking which, made over-confident by the resounding successes of anatomical and physiological research into the brain (think, for example, of Wernicke's aphasia theory!), was able to imagine that it possessed the key to the solution of all the riddles of (in this case disturbed) mental life. Here, indeed, what to Griesinger only had the significance of a "guiding principle" (Lotze) seems to have been elevated to a dogmatic truth.

3.4. GRIESINGER'S THESIS OF THE IDENTITY OF MENTAL DISEASES AND DISEASES OF THE BRAIN

Mental dieases or madness (*Irresein*) says Griesinger in the opening paragraph of his *Pathologie und Therapie der psychischen Krankheiten* — disturbances in the life of ideas and the will — are symptoms. The first step towards a knowledge of the symptoms is their location. The decisive question for psychiatry is thus: "To which organ does the phenomenon of madness belong? What organ must necessarily and invariably be diseased where there is madness?". Griesinger's answer to this question is: "Physiological and pathological facts show us that this organ can only be the brain; we are therefore bound to recognise primarily, in all cases of mental disease, *diseases of the brain*".[414]

When we read these words against the background of what we know of the rival school (or schools) of institutional psychiatry, it immediately becomes clear that the thesis of the identity of mental diseases and diseases of the brain is not primarily opposed — as is often tacitly assumed — to a romantic 'psychicistic' conception of mental disease, nor implies a materialistic denial of the specific nature of psychic phenomena, but is directed against the 'somaticism' of Jacobi.

As we have already seen,[415] Jacobi defended the thesis "that *all* phenomena of psychical *disease* ... can only be considered as symptomatic, as accessories to states of disease formed and developed elsewhere in the organism", that is, so-called mental disease has the status of a symptom of one somatic disease or another. Griesinger is also convinced that mental diseases must be regarded as *symptoms* of "deeper-lying" somatic diseases. Where Griesinger's opinion differs fundamentally from Jacobi's is, however, that he believes that mental diseases must be regarded as (symptoms of) brain diseases and not as (symptoms of) other somatic diseases,[416] hence: "mental diseases are *diseases of the brain*" (and not diseases of the tissues, organs, skin, bones, etc.). This standpoint does not exclude the (causal) relevance of diseases other than brain disease to the existence of mental diseases, but does limit it. Only diseases of the brain can be considered as primary, *direct* causes of mental diseases, other somatic diseases (or suscepti- bilities) can only *indirectly* play a part in the pathogenesis of mental diseases through stimulus or impairment of the brain.

The figure of psychiatric aetiology defended by Jacobi was by no means the rule among the somaticists, and the idea of mental diseases as diseases of the brain was not only not new, but — as far as I have been able to discover — was not even exceptional. Thus we find a figure akin to Jacobi's "symptom- atological" view in the work of Combe (*Observations on mental derangement*, 1831) in which mental diseases are described as symptoms of diseases of the brain, and we encounter similar ideas among the German somaticists who, although impressed by Jacobi's work and personality, did not follow him on the point of aetiology in that they placed diseases of the brain in a privileged position.[417] Viewed in this light, Griesinger's dictum appears at first sight to be not much more than a repetition of an aetiological figure which was already current in somaticist circles.

It would, however, be wrong to infer from this that Griesinger was con- cerned with nothing more than defending a return to, or repetition of, precisely this somaticistic conception against Jacobi. Griesinger was, it is true, an adherent of the 'somatic theory', but he was most decidedly not a somaticist in the sense of the word which implied ties to the Aristotelian tradition. The new look which psychiatry acquired through him was thus not determined, to all intents and purposes, by a fundamental agreement with the assumptions of the anti-Jacobi wing of the somaticist school, but by philosophical presuppositions which differed essentially from those of the two major variants of somatic psychiatry which we have previously distinguished.

This change in the philosophical foundation of psychiatry put the thesis of the identity of mental diseases and diseases of the brain — in fact (as we have observed) commonplace in the somaticistic tradition — into a different perspective, and gave it, although the formulation remained the same, a different meaning. A look at some aspects of Griesinger's psychiatric work will serve to elucidate and confirm the above.

To begin with, we must remember that Griesinger developed the pro- gramme of his own psychiatric undertaking from his opposition on the one hand to the anti-theoretical scepticism of men like Magendie and on the other to the 'ontological' medicine of Schönlein and others. Theory is possible, and even indispensable, in medicine, in this case psychiatry, but it cannot be a theory of the type found in ontological medicine, that is to say, a 'theory'

which confines itself to describing ideal types, divorcing the symptoms of
the disease from the way in which they occur; it must be a theory which also
'explains' these things. True enough, followed to the letter, the task remains
that of penetrating the 'nature' of diseases, but 'nature' here no longer stands
for the essence to be hunted down by analytical description, but in fact for
an ideal limit of all-round empirical scientific research which pursues the ideal
of an explanation in the terms of 'mechanical natural science'.

Since Griesinger was concerned with a pathology and therapy of mental
diseases as "*mechanical* natural sciences", and since psychiatry conceived
in this way was still very much in its infancy at that time and was more of
an ideal than a reality,[418], it seems obvious that in the first instance his
"mental diseases are diseases of the brain" should be read as the formulation
of a comprehensive *working hypothesis*. Not only the scientific problem
situation at the time, but also the context and incidental remarks made by
Griesinger point in this direction. Indeed, according to Griesinger, it is the
(pathological, physiological) *facts* which impose this hypothesis on us,[419]
and the thesis (or hypothesis) that mental diseases are brain diseases (i.e.
are based on diseases of the brain) should not therefore be seen, in his view,
as we have already said, as proceeding from a philosophical (aprioristic)
conviction concerning the identity of body and mind, but as the result of
an *empirical* consideration.[420]

It is difficult to overlook the fact that the "empirical standpoint" which
Griesinger expresses here against the "abstract" (i.e. lacking empirical content) apriorism of his more philosophical colleagues – Schönlein, Heinroth,
Jacobi, etc. – contains an aprioristic element. It is, however, more important to observe that Griesinger's apriorism, unlike the ontological apriorism
which he opposed, was the apriorism of a methodological idea. The thesis
of the identity of mental diseases and diseases of the brain, as he understood
it, should therefore not primarily be regarded as the confession of faith of
a metaphysical materialist, but as the formulation of the *idée directrice* of a
line of psychiatric research which had pledged itself to follow the anatomical
and physiological method (with regard to the brain).

Of scarcely less importance for an adequate definition of the fundamental
figure of Griesinger's psychiatry is the fact that Griesinger was convinced
that mental diseases were only a sub-class of the class of "cerebral affections".[421] This means that psychiatry is a part of cerebral pathology. What
makes this insertion of psychiatry into the framework of cerebral pathology
possible is that all mental diseases *ex hypothesi* are attributable to diseases
of the brain. However, given the existing embryonic level of development of

cerebral pathology, according to Griesinger it is — for the time being — impossible to say that the class of mental diseases coincides with the class of diseases of the brain. It is exactly this deficient state of affairs in cerebral pathology that will have prompted Griesinger to acknowledge the meaning and right of the symptomatological view in psychiatry as well as the (symptomatological) distinction made therein between diseases of the brain (of course, in the narrowest sense) and mental diseases.[422]

In principle (thus we can clarify Griesinger's ideas), "madness is only a complex of symptoms of anomalous conditions of the brain", "mental diseases" are (nothing but) "diseases of the brain", and any claim to independence which psychiatry may make is unjustified because psychiatry is totally swallowed up in cerebral pathology. "But, although at some more distant period this may perhaps be looked for, any attempt at such a fusion would at present be premature and quite impracticable."[423] Even though, with these words, Griesinger stamps the desired annulment of the at present (still) independent and chiefly symptomatologically descriptive psychiatry as an almost unattainable ideal of the future, he nevertheless feels that psychiatry is capable of contributing to the task of cerebral pathology "if the intimate fundamental union which exists between psychiatry and the remaining cerebral pathology be only constantly kept in view — if in the one, as in the other case, the same exact anatomical physiological method be as far as possible pursued . . . ".[424]

If anywhere, then certainly here, it becomes clear that to Griesinger the thesis of the identity of mental and cerebral diseases had the significance of a guiding rule of psychiatric scientific research. This, too, is the only interpretation in which his plea for (continued) recognition of the symptomatological difference between mental diseases on the one hand and cerebral diseases on the other, his defence of the independent treatment of psychiatric problems, and the existence of a (still) independent psychiatry, can be reconciled with his theses about the (ultimate) identity of mental diseases and diseases of the brain, of psychiatry and cerebral pathology.

It is certainly a striking characteristic of Griesinger's thinking[425] that, for all his dedication to the *idea* of a pathology and a therapy of mental diseases as mechanical natural sciences, he remained fully alive to the gap between the ideal and the reality of psychiatry, and to the limits which had been or must be reached in carrying out the mechanistic programme, without allowing this to betray him into concluding that his idea was impossible or meaningless. The same empirical attitude which prevented him from indulging in uninhibited 'speculative' theoretical construction in science (compare, in contrast, Wernicke and Meynert's 'brain mythologies') led him to reserve judgement on the traditional metaphysical problem of the mind—body relationship. It is important to refrain from philosophical interpretations, explanations of "that inexplicable unity (of soul and body)"; all such hypothesis of "attenuated fluids which mediate between them, 'fluids subtle enough to be reckoned spirit', even by the system of pre-established harmony according to which body and soul never act on, but always along with, one another — these hypotheses are, empirically considered, as difficult to sustain as to refute. How a material physical act in the nerve fibres or cells can be converted into an idea, an act of consciousness, is absolutely incomprehensible; indeed, we are utterly unable even to pose the question of the existence or nature of the *mediating processes* existing between them. *Everything* is still possible here . . . ".[426]

Thus we see here clearly expressed what 'materialism' means to Griesinger: the most attractive *hypothesis*, scientifically speaking, because it presents "fewer difficulties, obscurities and contradictions (particularly in respect of the origin of the life of the soul), than any other". Griesinger feels that it is scientifically perfectly admissible to postulate, leaving entirely out of account those possible, but entirely unknown, mediating processes, the ultimate identity of mental activities on the one hand and the body, more specifically the brain, on the other, in the sense that they are related to each other as function and organ. Our ideas and strivings must be seen as the expression of the activity or energy which is specific to the brain, that is they must be understood in the same way as one usually regards the transmission in the nerves and the reflex action in the spinal cord as the functions of the nerves and the spinal cord. Scientifically speaking, there is nothing to stop one (by way of a hypothesis) from considering "the soul primarily and pre-eminently as the sum of all cerebral states".[427] What Griesinger calls here his "materialistic hypothesis" is, as the reader will have observed, nothing but the *regulative idea* of mechanistically conceived psychiatry which we have already discussed.

A further clarification of Griesinger's position provides us with an idea of his attitude towards (metaphysical) materialism. This materialism (like 'spiritualism' and thus, generally, any philosophy at all) is not in a position to give "*definite information* regarding what takes place in the soul . . . And even if we did know all that takes place within the brain when in action — if we could penetrate into all the processes, chemical, electrical, etc., of what use would it be? Oscillation and vibration, all that is electrical and mechanical, are still not a state of the soul, not ideas. How they can be transformed to these is, indeed, a problem which shall remain unresolved to the end of time; and I believe that if today an angel from heaven came and explained all to us, our understanding would not even be able to comprehend it." [428]

Put in a somewhat different way, what it says here is that, given the ideal, entirely imaginary case that, at a certain moment, through whatever felicitous circumstances, we were to be in possession of *total* knowledge of our brain, it would be of no avail in finding an answer to the question of the genesis of psychic phenomena (mental state, ideas). The limit which is *here* placed on psychiatry as a "mechanical natural science" is no longer an actual one, dictated by the realisation of what is "presently possible" for scientific psychiatry (Binswanger), but a fundamental one. The limitation in question is founded in human nature and its cognitive possibilities. That is to say, the motive in the background apparently derives from Griesinger's philosophical view of man. A similar motive appears to be at work in his rejection of the (naive) materialistic *Weltanschauung* as incompatible with the human self-conception to which he felt committed. [429] Only the "empirical view" *is compatible* with the recognition of the value and true nature of man as a spiritual-moral being, because a characteristic of the "empirical view" is that it remains free of totalitarian pretensions to knowledge such as those held by (metaphysical) materialism. [430] A philosophical conception which leaves no room for "the most general and most valuable facts of the human consciousness" is equally unsuited to Griesinger's purpose as a philosophy which, absolutising the spiritual side of man, explains matter as a side-effect of the spirit. Humanness encompasses both aspects, it is body and mind (soul). While we may want to *believe* that there is ultimately a relationship of identity between the mental [431] and the physical, we cannot *know* for certain (in the manner of metaphysics), because this is a possibility which is not given to us humans.

What remains after the (possibility of) metaphysical total knowledge has been taken away is the meaning and possibility of the (empirical) scientific undertaking, in this case the psychiatric undertaking, conceived as an endeavour which is guided by the idea of the identity of the mental and the physical. 'Science' is thus tacitly interpreted as 'mechanical natural science' − that, too, is implicit in 'the empirical view' − and this definition implies, in its application to (natural) scientific psychology, that attention is only paid to the form of mental life. [432] Therefore Griesinger can also write that the "empirical view" (i.e. in psychology) "[leaves] the actual contents of the life of the human soul intact in all their richness". Since a psychology of this kind was conceived by him as an extension of the physiological reflex theory, it is natural that what he puts forward as 'psychological' considerations is expressed in the language of 'physiology'. As a natural science this psychology is directed towards the organic aspect of mental phenomena, or should we rather say, to stay within Griesinger's notion of things: this psychology concerns itself with the study of so-called mental phenomena, understood as a subclass of organic phenomena (in the broader sense). Although the actual relationship of mental phenomena conceived in this way to organic phenomena (in the narrower sense) is still largely unexplained, the postulated subsumability of the mental into the concept of the organic (in the broader sense) implies a structural identity of the two classes of phenomena.

The decisive reason to describe Griesinger as the advocate of an identity theory in the philosophical sense lies, however, in the fact that he proceeded on the assumption of an ultimate identity of the content of our mental life and its 'form'. It is indeed true that, in Griesinger's view, questions such as how the relationship between the form of mental life and its content must be regarded, where psychology as such expressly limits itself to the form and relinquishes the content, or how the genesis of the individual, qualitative abundance of mental life can be explained at all in scientific terms, were still insoluble at that time.[433] However, this does not alter the fact that Griesinger, with blind faith in 'mechanical natural science', did not hesitate to advise his supporters — advocates of the 'empirical view' — to "patiently await the time when the questions concerning the connection of the contents of the life of the human soul, with its form, shall have become physiological instead of metaphysical problems".[434] Since reflections about the contents of our mental life would be considered by Griesinger as prescientific and, in a broad sense of the word (such as we have already encountered in Lotze), of a 'teleogical' nature, the statement quoted above is doubly interesting: it tells us something, albeit indirectly, about the way Griesinger saw the relationship between the mechanistic and the teleogical perspective and it makes it possible to contrast his views on that matter with those of Lotze. The following observations will make clear what I mean.

If we can assume that Griesinger implied that the situation of a total knowledge to which the ultimate identity of body and soul is revealed is none other than that in which the τέλος of the total 'mechanical' scientific explanation is achieved, then we must say that the irreducible difference between the two approaches recognised by Griesinger, although definitively *insoluble* from the standpoint of man (who is limited in his capacity for knowledge and explanation), must be described from a 'higher' standpoint, from the 'angelic' perspective, as *provisional* and *not ultimate*. At this 'higher' level of total 'mechanical' explanation and explainableness the 'teleological view' conclusively vanishes from sight, no longer has a right to exist, is 'abolished'.

On precisely this point then, there would have to be assumed to exist a fundamental difference between the philosophy of Lotze and the 'philosophy' of Griesinger, since it is unthinkable that Lotze, for whom the recognition of the inviolable (!) right and priority of the teleological approach was indissolubly bound up with the idea of the primacy of philosophy over science, could have agreed with a figure of thought in which the eventual abolition of the teleological approach and thereby of philosophy (= metaphysics) in favour of science was regarded as the 'higher' standpoint.[435] The salient features of Griesinger's naturalism can best be expressed in a summary of agreements and differences with Lotze's position in philosophy of science. First, what links Griesinger to Lotze, what causes the mechanistic self-concept of his psychiatry to fit without more ado into the perspective of the self-conception of 'mechanical' natural science put forward by Lotze, is the identification of the scientific approach (in the narrow sense) as a 'mechanical' approach.

The teleological approach, which Lotze considers as philosophy (metaphysics), is, however, regarded by Griesinger as a sort of (prescientific = 'philosophical') left-over within science, which will be cleared away in time, i.e. with the perfecting of our 'mechanical' natural scientific knowledge. A (further) distinction between the two positions would then lie in the fact that Griesinger saw this *'factum'* of the non-identity of the 'mechanical' and the teleological approach as non-ultimate, provisional. From a higher, superhuman, 'angelic' standpoint, that non-identity would have to be described as 'seeming'; and the abolition of the teleological view Griesinger had in mind must be thought of as the abolition of philosophy and its replacement by science. To Lotze, however, the philosophical foundation of teleology and thus of the possibility of philosophy in the *factum* of the unity of the consciousness ruled out the possibility that the mechanistic viewpoint as the viewpoint of science could ever reveal itself to be the final, 'highest' point of view.[436]

However this may be, the fact that the mechanistic conception functioned for Griesinger as a regulative idea and not as a metaphysical dictate makes it in any case understandable that Griesinger's psychiatry (psychology), despite the sought-after limitation of psychological ('physiological') research to the form of the life of the human soul, as opposed to any (psychological) consideration of its content, *de facto* left room for what was later to be known

as a 'descriptive and analytical' or *verstehende* ('understanding') psychology.[437] In fact he practised this ('unscientific') psychology with such great
skill that K. Jaspers did not hesitate to place Griesinger, in the history of
clinical psychiatry, in the company of the great descriptive writers, to which,
in his view, psychiatric authors like Esquirol and Kraepelin also belonged.[438]
    Thus it came about that in the methodological conflict which was carried
on at the beginning of the twentieth century between advocates of natural
scientific psychology and humanistic psychology, Griesinger's psychiatry
provided an example for both sides.

3.5. GRIESINGER AND HERBART

If the interpretation developed in the preceding paragraphs is broadly correct,
it is possible to arrive at a more detailed definition of Griesinger's relationship
to Herbart than one has so far been able to find in works on the history of
psychology and psychiatry.
    The fact that Herbart, and not Lotze or any other philosopher (Schopen-
hauer, for instance), has attracted the attention of historians of psychology
and psychiatry when they wanted to explain the antecedents of Griesinger's
theory is understandable but misleading if one − recognising the (relative)
right of this interest − is tempted to conclude from it that Herbart was the
philosopher who determined Griesinger's thinking, or, to express it more
precisely, that it was primarily Herbart's *philosophy* (metaphysics) which
influenced Griesinger as a 'philosopher'. It must be stated emphatically that,
contrary to any suggestion of this kind, the text of Griesinger's work does not
support such a conclusion. Instead we must say that what Griesinger − within
certain limits − appears to have borrowed from Herbart is not his philosophy
but his psychology, but without the metaphysical groundwork which, for
Herbart, was indissolubly linked to it. This is precisely what we can expect of
a thinker for whom the mechanistic view has the function of a (working)
hypothesis or regulative idea.
    In the following discussion of the relationship between Griesinger and
Herbart I shall look in more depth into the, in my view, fundamental differences between Griesinger and Herbart in so far as they can be ascertained
in regard to the problem of the mind−body relationship and − on the level
of the presuppositions of the two views of psychology − become visible in
divergent (philosophical) conceptions of the I. In contrast to the more usual
history of science approach, I am less concerned with the question of the

influence which Herbart's psychology had on that of Griesinger than with fixing the limits within which, *given the philosophical premises of the two*, it is necessary to take into account the possibility of such an influence.

The fact of a (certain) influencing of Griesinger by Herbart is, in itself, beyond question. I have already remarked[439] that the concretisation of the mechanistic concept of science effected in Griesinger's psychiatric theory came about through a synthesis of the physiological reflex theory of Hall and Müller and J. F. Herbart's mechanistic psychology. Griesinger himself, moreover, expressly acknowledged the significance of Herbart's psychology for his 'ego psychology' in his *Pathologie und Therapie der psychischen Krankheiten* (1845).[440] And when, after Griesinger's death in 1868, people were evaluating the scientific achievement of the great man, a leading Neoherbartian like Lazarus, not without pride, pointed to the Herbartian influence in Griesinger's work.[441]

3.6. HERBART'S METAPHYSICS AND GRIESINGER'S 'EMPIRICAL STANDPOINT'

I should like to make a few preliminary observations in order to place the philosophy of Johann Friedrich Herbart (1776–1841) in context.

In terms of the history of philosophy, Herbart belongs to a generation of thinkers who, in different ways, opposed the post-Kantian idealism of their time. In Herbart's case,[442] this criticism was aimed at the problem of (the interpretation of) the Kantian 'thing-in-itself'. Jacobi's famous words ("without thing-in-itself one cannot get into Kantian philosophy, with it one cannot stay there") had provided a concise description of the problem which his work had created for his (later) contemporaries. There were two ways in which scholars took a stand concerning the problem created by Kant's philosophy. While post-Kantian idealism (led by Fichte) aimed, in the name of the spontaneity of the mind and its aprioristic forms, to eliminate from the Kantian system the last realistic remnants – Kant (still) referred to 'affections' emanating from the 'thing-in-itself' – Herbart opted for a unique sort of realism which defended Kant's thing-in-itself against the idealists, but at the same time interpreted it in a way which brought his philosophy closer to pre-Kantian philosophy than to that of Kant.

Herbart differs fundamentally from Kant in his attitude to (the possibility of) metaphysics. Kant's radical separation of the empirical world and reality-in-itself (in Kantian terms, of the phenomenal and the noumenal world) is abolished: the semblance of the empirical world points to the existence of reality-in-itself ("so much appearance, so many indications of reality"). If, however, this reference function is to be fulfilled, it is necessary to clarify the concepts of non-philosophical thinking which are geared to this empirical world, and full of internal contradictions, through a philosophical method of concept analysis. The objective of philosophy is thus none other than the bringing about of this concept clarification.[443]

The basis of this conception of the task of philosophy is the metaphysical conviction that the ultimate reality can be conceived in one system, free of internal contradictions, which expresses the structure of reality, of being. This reality consists of a plurality of entities (so-called 'reals' [*Realen*]), each of which is, in itself, absolutely singular, of positive quality, independent of all the others, indivisible, and therefore also non-spatial and immutable. The remarkable thing about Herbart's metaphysical views — and this was proving a stumbling-block as early as Lotze — is that these totally unrelated 'reals' can still relate to one another: an interaction arises in any confrontation of opposing qualities. This interaction threatens to bring about a disturbance, (*Störung*) to which the, by nature immutable, 'real' can only react by an action of self-preservation. Thus we see emerging the figure of a (real) world of immutable and mutually unrelated 'reals', that is, a world without movement and change, which is, as it were, duplicated by a world which is governed by the dynamics of disturbance and self-preservation. According to Herbart's conception there is, of course, no question of a real duplication here, but even if one distinguishes the first world as the world of (real) being from this second world of appearance, and bases the phenomenal qualities and changes in the empirical world (i.e. the world of appearance) in the *relations* between the reals and in the *changes* in these relations, it still remains to be explained, as Lotze rightly observes, "how it is that this appearance is formed in us, which is indeed a reality and which is not [merely] an appearance for a third observer".[445]

Be that as it may, this general metaphysical conception lies at the root of Herbart's psychology. The bridge between this metaphysical idea and psychology is built on the hypothesis that all so-called acts of self-preservation (*Selbsterhaltung*) of the soul are ideas.[446] That is to say, it is not known what form the acts of self-preservation of any real but that of the soul take, but in the case of the real of the soul alone Herbart is prepared to assume that we can say something. The acts of self-preservation with which the soul-real reacts to (the threat of) disturbance are ideas, which are either directly caused by the contact of the soul with other reals, or, in accordance

with a law of inertia, continue to survive in the soul eternally, once they have
come into being, but are thereby involved in all sorts of dynamic relationships
to one another which are subject to continual change.

The life of the soul, as the object of psychology, is thus established as the
dynamics of ideas[447] and the traditional mental faculties – thought, feeling,
will, etc. – are accordingly understood as so many modifications of this
life of ideas.

There is no need here to go any deeper into the details of Herbart's metaphysics and psychology. For our purpose – defining the difference in philosophical orientation between Herbart and Griesinger – the most important
thing is to make explicit a certain implication of Herbart's philosophical ideas.
Herbart distinguishes appearance and being, the empirical world (as conceived)
and reality (as conceived), from one another in accordance with a logical criterion: the first is internally contradictory, the second is without contradictions.
From this it is clear that when Herbart invokes experience in countering speculative thinkers, this appeal is a qualified one. Given his metaphysical conviction concerning the relationship between the empirical world and reality-in-
itself, experience can only interest him as the *necessarily false starting-point* of
philosophical (metaphysical) research. In other words, experience has no
independent significance for knowledge, it is not a positive source of truth.

Aside from the fact that Griesinger, in more 'philosophical' pronouncements, showed signs of sympathising with a form of concept analysis ('concept reconstruction'), which is perhaps reminiscent of Herbart literature,[448]
it must be clear, after all we have said about his views, that Griesinger's
positivism, his passionate defence of the 'empirical standpoint', do not
fit in at all with the disqualification of experience which is contained in
Herbart's philosophy.

This point is by no means contradicted by the fact that the title of Herbart's major psychological work refers to a science of psychology which is
based not only on metaphysics and mathematics, but also on experience.[449]
On the contrary, the overall impression left with the reader of Herbart's
psychology is that these generally extremely abstract reflections have very
little to do with experience and are anchored only with very slender threads
to what we would be prepared to recognise as experience or the empirical
world. This can, however, scarcely come as any surprise against the background of Herbart's metaphysics, in which the empirical world, or more
precisely the empirical world conceived through non-philosophical thinking,
is the sum and substance of appearance, which must be destroyed through
philosophical concept analysis.[450]

A further confirmation of the difference between Herbart and Griesinger which we have assumed is provided by the comparison of Herbart's endeavours to elevate psychology to science with Griesinger's parallel efforts to make the psychiatry of his time scientifically respectable: the parallel cannot blind us to the fact that the emphasis of each man is very different – indeed, in a sense one can say that Griesinger takes up where Herbart leaves off.

Thus (to start with Herbart) the pointed description of 'psychology *as science*' in the title of his major psychological work must be seen against the background of Kant's assertion that psychology could not be a (true) science. Herbart sees here a possibility for psychology which Kant had ruled out.[451] Psychology is possible as a (natural) science because, and in so far as, mathematics can be applied to it. In fact, Herbart and Kant are in agreement on the *criterion* for a science, but they disagree on the question of whether psychology can meet this criterion. The way in which mental phenomena (i.e. phenomena of the consciousness) are given in the 'inner sense' – one-dimensionally, in time – excludes, according to Kant, as we have already observed, the possibility that psychology can be a science with the same structure as (mathematical) natural science. And it is precisely this point which Herbart disputes when he defends "mental research . . . which is equivalent to the investigation of nature".

Griesinger's position is very different from that of Herbart. It is not solely the application of mathematics which has the power to make psychiatry a science (i.e. a 'mechanical' natural science), but also the connection with the methods and results of mechanically oriented (cerebral) anatomical and physiological research. Herbartian psychology was thus only of interest to Griesinger in a synthesis with the physiological reflex theory of Hall and Müller, that is to say, only after Griesinger had seen the possibility of integrating Herbartian psychology into a pathology of mental diseases based on (cerebral) physiology. With this integration it can now be seen in what sense Griesinger goes, as it were, a step further than Herbart, in what sense (as we have observed) Griesinger takes up where Herbart leaves off: the importance which Griesinger attached to empirical physiological and anatomical research, the idea of the need to give psychology a physiological 'substructure', was entirely foreign to Herbart, who always held to the primacy of psychology. More significant, however, is the fact that, with the introduction of Herbartian psychology into an empirical scientific undertaking which did not consider itself as an *ancilla metaphysicae*, but rather understood itself in terms of its antithesis to philosophy (metaphysics), precisely that difference in the relationship between Griesinger and Herbart is introduced which is of

importance to our argument. If I am right, then we must indeed say that
Griesinger *de facto* adopts Herbart's idea of mechanistic psychology (as an
element in a more comprehensive *programme* of psychiatry as a 'mechanical
natural science'), but not the metaphysics on which this mechanistic psychology is based.

This conclusion accords with our earlier findings. It is certainly true that
when Griesinger — incidentally — speaks of the limits of 'mechanical natural
science' as *actual* limits, which could *in principle* be abolished were research
to be carried on for long enough, this proves the force of the natural scientific
'prejudice' in his thinking. But that does not alter the fact that his *self-conception* as a scientist contained only a commitment to natural scientific
method (or methods), and did not also imply a commitment to Herbart-style
metaphysical realism or to the metaphysical materialism of the generation
of Vogt, Moleschott and Büchner. This means that the reasons which preclude
us from identifying Griesinger's standpoint with that of the 'flat and shallow'
materialists whom he loathed are, in fact, the same reasons which prevent
us from seeing in him, without more ado, a Herbartian 'realist'.

3.7. GRIESINGER'S 'EGO PSYCHOLOGY': ASSIMILATION OF HERBARTIAN ELEMENTS

The relationship between Griesinger's psychology and Herbart's psychology
has been dealt with frequently and extensively in the history of psychology
and history of psychiatry literature. A striking feature of this is that this
theme has been tackled (to a greater or lesser extent) more or less exclusively
in terms of historical research into the antecedents of Freudian psychoanalysis.[452] It is, indeed, not unreasonable to assume that Herbart's influence on
Freud, which scholars have tried to substantiate on the grounds of, amongst
other things, terminological similarities (e.g. terms like *Hemmung, verdrängt*,
and *Schwelle*), is not based on direct knowledge and derivation, but was
mediated by Griesinger's work.[453] It has also not escaped the historians that
such 'modern' concepts as the role of the unconscious, ego structure, frustra-
tion and wish-fulfilment in symptoms and dreams are — as Ackerknecht
says — developed in Griesinger from Herbartian 'suggestions'.[454] Certainly
the points of contact between Griesinger's psychology and Freud's are too
striking to be put down to coincidence. In my view there can thus be little
doubt that Griesinger's work was an important impulse in the development of
psychoanalytic thinking. This is, in fact, borne out by the rare but significant

references to Griesinger's work which are found in Freud.[455] None of this alters the fact that the preoccupation with Freud in the examination of the relationship between Herbart's psychology and that of Griesinger has led to a certain one-sidedness in interpretation[456] which cannot be countenanced. On the other hand, however, we are not about to carry asceticism in historical speculation as far as does the author of *Die Psychologie Johann Friedrich Herbarts und ihre Bedeutung für die Psychiatrie des 19. Jahrhunderts*, who — apparently — feels that it is enough to remark that Herbart's psychology was of great importance to Griesinger but otherwise has nothing to say about Griesinger, as if he were discussing an insignificant psychiatric author who deserves only to be mentioned in this context but does not merit any more central treatment.[457]

In the discussion of the relatonship between Griesinger's psychology and that of Herbart which follows, I shall try to give substance to the theses that (1) Griesinger not only has a metaphysical orientation which differs from that of Herbart, but that (2) despite his mechanistic language, as a psychologist too he cannot be described without qualification as a Herbartian, because (and in so far) as the standpoint of a *psychology of content* does make itself felt. In connection with this last point in particular, I shall pay special attention to the significance of the ego concept in the psychology of the two men.

3.7.1. *The Ego Concept in Herbart*

Reference is sometimes made, not without justice, to Griesinger's 'ego psychology', and one could also, without objection, speak of Herbart's psychology in the same way. This term reminds one in the first place of a movement in the history of psychoanalysis in America in the nineteen-forties and -fifties, which we connect with the names H. Hartmann, F. Kris and R. Loewenstein — the so-called New York group.[458] Taking as their starting-point what Freud wrote in one of his last articles about an original Ego–Id unity, in which the structure and course of development of the Ego were preformed before the Ego came into being,[459] these psychoanalysts aimed at giving the Ego a greater autonomy, where Freud in his earlier writings had always underlined the dependence of the Ego in respect of the Id. The 'classical' psychoanalytical conception of the Ego as the 'descendant of the Id' thus found itself opposed by a significantly different conception, because Hartmann and his colleagues maintained that the Ego was not derived from the Id, but that the Ego and the Id developed — independently of each other

– from dispositions inherent in the human organism. This means that the term 'ego psychology', which was used of this conception, has acquired a negative connotation for present-day representatives of psychoanalysis, for whom the (meta-psychological) subordination of the Ego to the Id is the cornerstone of Freud's theory: it represents a movement of defection, an aberration from the 'true' doctrine.

When we talk of Griesinger's psychology as "ego psychology", this is primarily because this psychology is essentially a psychology of the ego. The fact that this ego psychology, moreover, uses a number of terms which, to our ears, have a depth-psychology meaning – these are in fact concepts from Herbart's (ego) psychology – must therefore not be misunderstood. Analogues of psychoanalytical concepts like Id, Ego and Super-Ego, which one can find in Herbart, represent to him different idea masses in the soul, that is to say, the principle of the unity of the soul[460] is not breached. Herbart's psychology (like Griesinger's) is essentially a psychology of consciousness, and the concept of the subconscious refers to those ideas which are no longer or not yet conscious, but which can in principle become conscious (comparable with Freud's 'preconscious'). The structural unconscious as a part of what cannot *in principle* be made conscious – Freud's great 'discovery' – was unknown to Herbart and his followers. The analogy between Herbartian 'ego psychology' and that of Hartmann *et al.*, which perhaps gave rise to the use of this term,[461] is thus that the former was not yet, and the latter no longer in possession of the psychoanalytical 'truth', according to which the principle of the ego plays a subordinate role in our mental housekeeping. The point of reference in the application of this term is in any case orthodox psychoanalytical theory.

The psychoanalytical viewpoint should not, however, be made the important one in an explanation of Herbartian ego psychology. Precisely what 'ego-psychology' implies here can only be made clear by placing Herbart's psychology against the background of the (philosophical) psychological conceptions from which he wanted to dissociate himself, and considering how it is related to similar nineteenth-century psychologies.

As far as this last point is concerned, it must be remarked[462] that, generally speaking, the problem of the ego came to a head in nineteenth-century psychology on two questions; firstly, whether the ego is a psychological fact and, secondly, if so, can this psychological fact be reduced to other psychological phenomena. The psychology of Herbart and his followers can be seen as one in a series of theories which were developed in the course of the nineteenth century, in which an (exclusively) empirical psychological interpretation of the ego goes hand in hand with an attempt to reduce this ego to other (more elementary) psychological phenomena.

Thus we see first of all that Herbart adopts a position against the distinction between absolute and empirical ego which was current in the philosophy of the period. The philosophy (what is meant is, of course, not Herbartian philosophy) which believes that it is enough to define the ego, in extremely abstract terms, as the identity of subject and object "seems to distance itself from the datum, by rejecting *temporal* observation",[463] that is to say, it goes to the extreme of assuming an ego concept from which everything which is individual is eliminated, a *pure* ego (*reines Ich*). This concept of a pure ego is, however, a contradictory concept, because an ego without an element of individuality is not an ego. The ego conceived in the concept of the pure ego is completely void and unreal, or, in Herbart's words: "Of all the distinctions encountered in the *real* ego, according to whether man feels himself depressed or delighted and either advanced or exhausted in his efforts, the ego as a metaphysical principle neither knows nor contains the slightest thing"[464] (my italics).

It is thus obvious, after the untenability of the concept of a *'reines Ich'* has been demonstrated, that the real ego should be sought where this ego is given as something individual, that is, in self-perception. The empirical ego, which is what is involved in this case, is 'the ego of common sense', i.e. the ego as we conceive it in our everyday thinking (non-philosophical thinking, not yet refined by Herbart-style concept analysis); it is the individual ego which is referred to in the question 'who am I?'. That ego, however, "contains only purely *accidental* attributes [my italics], as is disclosed by the analytical judgements [of the answers to the question 'who am I?'] in the same way as the ideas of sense objects are decomposed by judgements into nothing but predicates whose subjects are blindly assumed for a long time, but ultimately turn out to be lacking".[465]

Preposterous as it may be to postulate with traditional philosophy a pure, that is, absolutely universal, non-individual ego, however, it is equally problematic to conceive the individuality of that ego as purely *coincidental*, that is to say, varying according to the circumstances in which someone finds himself at a given moment in his own life history. If I were to take this conception seriously, I should have to assume a multiplicity of egos for myself, because if it is established that the ego is individual (and what else can it be?) and varies with the circumstances of my life (and this, too, is indisputable — our (self) perception bears witness to it), then it follows that I must be a different I (ego) every moment, that is to say, in that case I am an aggregate of egos.[466]

In the final analysis, the current philosophical and everyday ego conceptions support each other in the illusion that the ego is something like an immutable substantial substratum, and they must therefore, if they are thought through logically, come up against the problems we have just described.[467]

What Herbart's solution in fact amounts to is that, to start with, he abandons the idea of an immutable, substantial ego. When, before this, the empirical ego, in its opposition to the philosophical concept of a 'reines Ich' (which Herbart criticised), is described as the 'real' ego, this is using a meaning of 'real' which does not imply the above-mentioned substantiality. Using the terms of Herbart's own metaphysics, only what belongs to the order of being, i.e. of the 'reals', is real. However, that 'empirical' ego, i.e. that ego which is the object of our self-perception, does not belong to this order, because it belongs, like all changeable phenomena, to the world of change, the order of appearance. The reality of the empirical ego (thus I interpret Herbart) is not ultimate, because in the final analysis it is tied to the perspective of the everyday experience which has not yet been illuminated by philosophy.

What Herbart's conception (stated positively) boils down to is that our ego is not something constant, something unchangeable. What we perceive at every moment of our individual life histories as our self, our I, is determined by the nature of the series of ideas which are crossing in the ego at the moment of self-perception. "Depending on the nature of the series of ideas, which meet and intersect in the ego, and on the way these are stimulated at that particular moment: that is what determines how man sees himself at that particular moment." The ego "constantly ... fluctuates; now it is something sensory, now rational, now strong, now weak; now it appears to lie on the surface, now at an unfathomable depth".[468] The identity of the ego, or the ego-ness which teaches us by experience (by which we may understand self-perception), "advances, as it were, in one and the same man from his childhood until old age, on different and heterogeneous feelings, wishes, deeds, thoughts, external relationships, which he gradually considers to belong to his 'self' as time passes",[469] and does not therefore have the (metaphysical) status of an ultimate, immutable being, but is the (changeable) *product* of those crossing series of ideas mentioned previously.

This can only mean that the psychological fact of the ego (the empirical ego) is not an original, irreducible datum, but must be understood as something based on more fundamental mental occurrences; more precisely, on the 'mechanism of ideas'. We are, in short, regaled with what ultimately proves to be an ego concept which is purely mechanistic, because it is based on association psychology: an ego concept which puts a definitive end to every form of psychology or philosophy in which the ego concept is still deemed to contain any element of creativity or spontaneity. In the case of philosophy, this point is brought out in Herbart's resolute rejection of the aprioristic forms (of time and space, the categories) and of Leibniz-style innate ideas, and in the case of psychology in his repudiation of any form of faculty psychology.[470] There is nothing under the philosophical/(empirical) psychological sun, belonging to the mental, which, according to Herbart, resists being understood as a product of the mechanism of ideas, and which does not gain in value in this mechanistic-psychological 'translation'. The gain lies in the fact that this psychology, which aspires to be strictly scientific, does at least *explain* something, whereas the earlier psychology of the faculties stops at the dubious result of an unmethodical concept formed on the shaky basis of our self-perception: the highest genera of ideas, feelings, desires.[471] Even so-called apperception (in the case of man, self-awareness) — a concept borrowed from Leibniz — proves, in Herbart's hands, to be susceptible to a mechanistic interpretation, and in this translation acquires a key position in Herbart's psychology, where it serves to bring to the surface and to resolve the contradictions in Fichte's concept of the pure ego, described earlier.[472] Apperception in the Herbartian sense refers to a relationship between idea masses, of which one, already existing (the so-called 'apperception mass') represents the perceiving party, and the other, new idea mass represents the perceived. Apperception is thus, more specifically, the assimilation of the new idea mass by, and on the basis of, the old, already existing 'apperceptive mass'.[473] To Leibniz, apperception is an occurrence through which the perceptions (ideas) which come into the consciousness themselves become conscious.[474] For Herbart, however if ideas are to be apperceived, that is, to become conscious themselves, this can only happen if, in their turn, they become the object of *other* ideas. That means that in his view they can never *of themselves* become the object of ideas. It is on precisely this crucial point that Leibniz and Herbart part company. The standpoint of mechanistic association psychology, which Herbart in all his radicalness wants to defend, does not permit of an element of spontaneous activity being preserved anywhere in the mental; even apperception must be conceived here as a process which is essentially in agreement with the regularities of the postulated mechanism of ideas. Leibniz, whom Herbart so admired, was in Herbart's view entirely correct in his idea that the soul "[generates] all its ideas *out of itself*" (my italics), but the prestabilised harmony "according to which the soul must produce its ideas not only *out of* and *through* itself, but also *of itself*, without an external cause, has its weak sides", and must, according to Herbart, be corrected on that point through the insights of Locke's (association) psychology: or rather, the issue is to reconcile Leibniz and Locke.[475] It is, however, clear that in the synthesis envisaged by Herbart the viewpoint of association psychology has the upper hand, and it can therefore come as no surprise that his concept of apperception, in which the apperceptive activity was reinterpreted as — not to say reduced to — a reactive occurrence in which the system of ideas reacts to new ideas coming 'from outside', came up against solid opposition from later generations, both in philosophical circles and among psychologists.

As early as 1846, Lotze, who had succeeded Herbart in the chair of philosophy at Göttingen, in his article *Seele und Seelenleben* (in Wagner's *Handwörterbuch der Physiologie*), had raised objections to the total mechanisation of the consciousness, and showed himself, despite his defence of the mechanistic viewpoint *in science*, to be too strongly bound in his *philosophical* presuppositions to the Leibnizian intuition of substance as "a being capable of action" not to take exception to Herbart's mechanistic psychology. It is consistent with this that Lotze at the same time rehabilitated the categorisation of faculty psychology, which had been condemned by Herbart as unscientific, and did his utmost to demonstrate the untenability of Herbart's extremism.[476]

As far as the reaction of the psychologists is concerned, it should be remembered that apperception psychology — traditionally the sworn opponent of association psychology — had a powerful champion in Wilhelm Wundt (1832–1920), who did not hesitate to brand Herbart's apperception concept as the fundamental error of his psychology,[477] and who reinstated the element of spontaneous activity, considered fundamental to apperception, in his own apperception psychology.

However, this is not the place to delve more deeply into the history of philosophy and history of psychology 'after-life' of Herbart's psychology. What we were concerned with was defining Herbart's ego concept in the context of a description and placing of his ego psychology. We have seen that the only ego which Herbart is prepared to call the 'real' ego, in a specific sense of the word, is the empirical ego. This is the ego that we become conscious of from the standpoint of 'common sense', as the object of our 'self-observation'; the ego which (mistakenly, from that standpoint) is regarded as the unchangeable substratum of our individual ideas, feelings, thoughts, etc., and which, as such, is also meant in all individualising statements which answer the question 'who am I?'.

We have, however, also established the fact that in Herbart's psychology
the ego has a derivative character in the sense that that ego — and everything
which traditionally is also considered as belonging to the realm of the ego
(consciousness), such as awareness of identity, apperception or consciousness
of self — is regarded as the changeable product of the mechanism of ideas.
This mechanistic interpretation of the ego concept, with which (natural
scientific) psychology in fact unmasks the ego concept of 'common sense'
as an illusion (and this is a point which we have not yet discussed), is linked
with Herbart's metaphysical interpretation of the soul as a 'real'. The mechanistic translation of our psychological categories and the associationist
explanation of the facts of consciousness may well represent the 'actual'
reality of our consciousness against the standpoint of 'common sense'.
Metaphysically speaking, however, this can only relate to the reality of
appearance, because the history of the inner life conceived as the mechanism
of ideas belongs to the order of change, that is, it does contain an 'indication'
of being (of the world of the reals), but is itself not of the order of that being.
This means that empirical psychology — that is, that psychology which tries
to grasp the reality of the life of the consciousness in terms of a mechanism
of ideas — still needs a metaphysical complement. To formulate it another
way: the results of empirical psychology or, as it is called by Herbart, taking
up the pre-Kantian tradition, must be complemented by rational psychology.
For this reason, too, the subtitle of his major psychological work refers to a
science of psychology which is based not only on experience and mathematics, but also on metaphysics. This is also why, in the then widely-used
Lehrbuch zur Psychologie, he deals with 'rational psychology' as well as
with 'empirical psychology'.

It is only in rational psychology that we find the (unjustly reviled, according to Herbart) concept of soul through which we acquire a "correct
knowledge of ourselves": the contradictory ego concept of common sense
must be transformed into that concept of soul.[478] But what is the soul, in
fact? Well, the soul is (as we have already observed) a real. This means it is
"a simple being; not only without parts, but also without any multiplicity
in its qualities".[479] It is not reality "somewhere" or "some time", it has
"absolutely no aptitude or capability either to receive or to produce anything", and it follows from this that it is neither a *tabula rasa*, on which
foreign (outside) impressions can make their mark, nor a substance which
manifests itself as original self-activity, in the Leibnizian sense. "It has originally neither ideas, nor feelings, nor desires; it does not know anything about
itself nor about other things; nor does it contain any forms of perception and

conception, any laws for willing and acting; nor even the remotest preparations for all these".[480] And Herbart therefore concludes: "The simple *what*
of the soul is completely unknown and will always remain so; it is as little
an object of speculative as of empirical psychology".[481] Although we can
therefore say nothing about that completely unknown *what* of the soul,[482]
it is possible, according to Herbart, to say something about the soul in as far
as — transcending the realm of immutable being — it comes to 'appearance'
in the order of change. What it is possible for us to know about the soul and
thus can become the object of psychological knowledge, are the *ways in
which the soul-real reacts to (disturbing) influences from other, dissimilar
reals*, that is to say, the dynamics of disturbance and self-preservation of the
soul-reals, because (only in the specific case of the soul) *we know* — in fact
this is a hypothesis put forward by Herbart! — that ideas must be considered
as belonging to the class of 'self-preservations' of the soul.[483]

In this way, with the conception of a theory of the disturbance and
self-preservation of the soul-reals, the conceptual foundation is laid in rational
psychology for an empirical psychology which is concerned with further
working out the laws of the mechanism of ideas. This is the psychology in
which, amongst other things, the ego concept of common sense receives its
'scientific' translation or, to put it another way, is reduced to an in itself
changeable product of a complicated dynamics of ideas.

We will leave our discussion of the ego concept and related concepts in
Herbart's psychology at this point. I have restricted my summary of this
psychology to its most characteristic features, not so much because the
details of Herbart's 'empirical' psychology would not be relevant (since in
my view they are) to an *exhaustive* comparison of Herbart and Griesinger
(which it is *not* my intention to give), but because this restriction is adequate
for the proposed judgement of the significance of the ego in Herbart's (ego)
psychology in comparison with the role which this ego plays in Griesinger's
psychology.

3.7.2. *From Mathematical Psychology (Herbart) to Medical Psychology (Griesinger)*

It is, however, impossible to carry this comparison through satisfactorily if
one has not clearly established in what sense Griesinger's work is linked to
Herbart's psychology. How must this connection be viewed if it has been
established that Griesinger was solely or primarily interested in Herbart's

mechanistic psychology as a psychological theory which lent itself to being interpreted as an extrapolation of the physiological reflex model?

It very quickly becomes apparent to the reader of Griesinger's *Die Pathologie und Therapie der psychischen Krankheiten* that what emerges from it of significance as the source of his psychology occurs (almost entirely) in a few paragraphs in *Physiopathologische Vorbemerkungen über das Seelenleben.*[484] This fact is, it seems to me, significant. Speaking of Griesinger's naturalistic ontology we have already remarked that psychology, as Griesinger understands it, has as its object the mental side of the 'organic' (in the wider sense) being of man. In itself this means no more than that psychology, *as well as* physiology, forms part of a comprehensive doctrine of nature, and this is entirely consonant with Herbart's views on this point. Griesinger differs from Herbart, however, in that whereas for Herbart psychology's credibility as a science was based on the fact that mathematics could be applied to it, Griesinger evidently found this criterion inadequate. Griesinger's (cerebral) physiological 'prejudice' – a prejudice in favour of *empirical* (cerebral) physiology which was to be defended from the 'standpoint of experience' against the old 'philosophical' physiology – meant that, to him, psychology could only be considered as a true science in so far as it could be based on (cerebral) physiology, or at least offered the prospect of such a foundation. For this reason he develops the psychological issue with an eye constantly on the possibility of a connection with (the results of) empirical cerebral physiology (and anatomy). For this reason, too, he (unlike Herbart) does not treat psychology as an independent scientific discipline, but presents psychology in a 'physiopathological' context. Without the possibility of this connection, psychological theory would have to remain idle speculation, devoid of scientific interest.

One must not try to explain Griesinger's methodological option (which differed from Herbart's) simply and solely on the grounds that physiology in his time was 'more advanced' than it had been in Herbart's day.[485] It has undoubtedly primarily to do with the perspective of his research: it was because of his *medical orientation* that psychology, in this case Herbart's, could only be of interest to Griesinger within the perspective of the mind–body relationship.

All this makes it quite clear that an investigation into the traces of any *philosophical* influence which Herbart may have had on Griesinger should begin by considering Herbart's vision of the relationship of mind and body. We shall see that no more than a glance at Herbart's treatment of this problem is enough to enable us to deny, with good reason, that Griesinger can be described as a Herbartian on the issue of the (metaphysical) interpretation of the mind—body relationship.

3.7.2.1. Herbart's Interpretation of the Mind—Body Relationship. The factor which strikes one first of all in this connection is that the problem of the mind—body relationship, which occupies a central position in Griesinger's interest, is of peripheral significance in Herbart's psychology.[486]

It can certainly be stated that what Herbart found (relatively) the least interesting facet of psychology was the medical aspect, unlike, for example, Lotze, whose major psychological work was *Medicinische Psychologie oder Physiologie der Seele.* Jülicher is therefore right when he observes that Herbart's psychology is expressly based on the normal, healthy human being, and that Herbart himself, as a non-physician and philosopher, refused to look for psychological regularities by starting from the extraordinary, abnormal or psychopathological.[487]

To Herbart, the relationship between mind and body, as the relationship between a mind-real and a body-real, is no more than a special case of a relationship between two reals. He can therefore see no reason whatsoever for according its treatment any privileged status. In the preceding discussion Herbart has already exhaustively examined all the essential factors in the conception of reals which can 'disturb' one another and which react to such disturbances with 'acts of self-preservation', and the additional pages on the "link between mind and body" do not essentially contribute to this. I am referring here to the hypothesis mentioned earlier,[488] with which Herbart constructed the bridge between the world of (immutable) reals, i.e. the world of being, and the world of appearance and change.[489]

Mind and body, although strictly separated by Herbart in the sense of psychophysical parallelism,[490] have on another level a causal relationship (*Causalverhältnis*) which moreover, according to Herbart, "[is] no more difficult than that between any other beings".[491] The explanation that he gives is of particular interest in terms of the comparison with Griesinger, because he illustrates this causal relationship with the example of the causing of a muscle contraction by the will, that is to say, the causal relationship of mind and body is discussed on the physiological level.

Herbart's view is as follows. According to the (metaphysical) "fundamental theory of disturbance and self-preservation", the disturbance between two beings is reciprocal, that is to say, causally speaking there is an interaction. Each of the two beings reacts to the disturbance brought about by the other being with its *own* system of self-preservation. In other words, the systems of self-preservation called into being by each of the reciprocal disturbances by no means have to be of the same kind. The only similarity which must exist between these different systems of self-preservation is that — as acts of self-preservation — they are internal states of beings which are preserving themselves.

In the case in point, in which the will brings about a muscle contraction, it is clear that cause and effect (*Bewirktes*) are heterogeneous. We are concerned here, that is, with a heterogeneity of the internal states of different beings, i.e. with the heterogeneity of two systems of self-preservation.[492] When two internal states of two beings which are associated with each other, as two heterogeneous systems of self-preservation, correspond with each other in the context of the reciprocally effected disturbance, this is still not enough to explain how the will can bring about something like a muscle contraction. What is needed is some kind of medium through which these heterogeneous systems of self-preservation can make contact with each other. This mediating function, according to Herbart, is performed by the nerves.[493]

The way in which Herbart conceives the causal relationship between will and muscle contraction can be summarised as follows. If it is true that (1) the relationship (connection, *'Verbindung'*) between mind and body is, metaphysically speaking, the relationship between two beings or reals, which (2) only have a causal relationship on the level of the respective systems of self-preservation with which they react to each other's disturbing influence, this means that, given (3) the reduction of the mind—body problem to the question of the (nature of the) causal relationship between will and muscle contraction, the primary issue here must be the *specification* of these systems of self-preservation, without which there can be no question of a causal influence in this case. In the case in point, three *sorts* of systems of self-preservation must be assumed. These are the systems which are individually characteristic of the beings (reals) of the mind, the nerve(s) and the muscle(s). (The process of) causal influence thus occurs here between the (numerically but also qualitatively) differing systems of self-preservation which lie between the mind and the muscle. (Causal influences *within* one and the same dimension, for example that of the nerve, i.e. in the relationship of *similar* systems of self-preservation, and causal influence over more dimensions, for example from mind to nerve, or from nerve to muscle, i.e. in the relationship of *dissimilar* systems of self-preservation, are equally possible.) The causal influence of one system on the next system of self-preservation in the series which runs from will to muscle comes about through a *change* (however slight) in the state of self-preservation of one of the (innumerable) beings which lie at the root of the (equally innumerable) systems of self-preservation. This demonstrates that the (causally) affected system of self-preservation *conforms to* the (changed) state of the affecting system of self-preservation.[494]

In all this (as elsewhere in Herbart's work) the teleological viewpoint otherwise remains subordinate. Even that act of will which is the starting point of the chain reaction in the example we have quoted is, in his mechanistic view, not the absolute starting-point of a subsequent operation or action, but is understood as that system of self-preservation with which the soul (the soul-real) *reacts* to (the threat of) disturbance. When Herbart nevertheless from time to time appears to acknowledge the right to exist of a teleological viewpoint, this is invariably in order to express his surprise about the structure of the creation as it was brought about by the divine Creator. The way in which the mind makes the body subservient to itself or, more specifically, the way in which the soul subjugates the nervous system, is difficult to understand unless one presumes the existence of a divine creative plan.[495]

The significance which must be attributed to this justification of a teleological view is not altogether clear, but one would not be too wide of the mark if one stated that, to Herbart, the teleological viewpoint only played a part in an extrascientific context ('scientific' in a sense of 'science' which comprises science in the narrowest sense and philosophy), or, more accurately, in the context of an aesthetic and/or religious viewpoint.[496]

If this is correct, then we have here a not unimportant point of difference with Lotze, for whom the teleological approach had a legitimate place in philosophy (metaphysics), and also with Griesinger who, although his formulation seems to be reminiscent of Herbartian literature,[497] appears prepared to grant the teleological approach a (modest, provisional) place in *science*.

This will, I hope, have made it clear that when Herbart talks about the *'Causalverhältnis'* of soul and body, this can only refer to a body–soul relationship on the level of the disturbances and acts of self-preservation of the two reals. This is to say, it refers to a relationship which can only be conceived as a relationship between two — dissimilar — systems of self-preservation, of which one can be described as the 'affecting' and the other as the 'affected' system. The distinction between affecting and being affected is only one of aspect: each changing system is the affected system in relation to the one which precedes it in the series, and the affecting system in relation to the one which follows it.

Griesinger proves to have made a total abstraction from Herbart's fundamental metaphysical plan. He never talks of reals or beings, nor about systems of self-preservation, nor about the relationship between disturbance and self-preservation in the sense of Herbart's metaphysics. We should not see in this simply a sort of 'positivistic' reservation of judgement, a sign of his anti-metaphysical convictions. The point is that it is also impossible to describe Griesinger the 'metaphysicist' as a Herbartian. In defining the mind—body relationship Herbart defends a psychophysical parallelism, and Griesinger an identity theory monism. This in itself is enough to make it implausible that Griesinger would have been able to allow himself to be inspired in more than a superficial sense by Herbart's view of the causal relationship of mind and body.[498]

Against this background, Griesinger's so-called dependence on Herbart's psychology appears in a different light, for if it is true that Griesinger's assimilation of the fundamental concepts of Herbartian psychology takes place in a different philosophical framework from Herbart's, then it is only to be expected that the — unmistakable, albeit 'superficial' — kinship on the level of the psychological concept formation will prove, on more searching analysis, not to be supported by agreement on philosophical fundamentals. That this is indeed the case has already become clear in the discussion of the positions of the two men on the question of the mind—body relationship. This can be further borne out by rounding off the comparison of the ego concept in Herbart and Griesinger with an explanation of the way in which this ego concept functions in the context of Griesinger's psychology.

3.7.3. *The Ego in Griesinger's Psychology. Mechanism or (Principle of) Teleology?*

The points of agreement between the psychology of Griesinger and Herbart are not difficult to find. For both of them, psychology is the psychology of the consciousness, both adhere to the idea of the unity of the soul, both prove to opt for an intellectualistic variant of ego psychology, in which ideas are the vehicle of the 'empirical' ego. Griesinger also adopts Herbart's notion of an 'inner life-history' conceived as a mechanism of ideas, and follows his lead in that typically intellectualistic reduction of the mental activities of thinking, will, feeling, to modifications of the life of ideas. And, by no means least important, the Herbartian conception of an unremitting dynamics of mutually supporting (fusing) and mutually suppressing (contrasting) ideas which occupy the central stage of the consciousness also found a place in Griesinger's psychology. No wonder that posterity has emphasised Griesinger's dependency upon Herbart.

Nevertheless, there are differences which, however unobtrusive they may appear at first glance, prove to be important. Thus, for example, one can see a difference in the treatment of the dynamics of ideas in the consciousness. In Herbart the mathematical-quantitative viewpoint clearly prevails: the conflict of antagonistic ideas is actually a problem of force-relationship which can, in principle, be solved along arithmetical lines. The fact that the mechanistic conception of the process of consciousness harmonises in one way or another with the common experience of psychological conflict is not in itself denied, but it has no fundamental significance for the justification of this conception; for Herbart mechanistic psychology means a *victory* over the standpoint of common sense.

For Griesinger, in contrast, one gets the impression, the tie with everyday (self)-experience is preserved, notwithstanding his mechanistic language. In Griesinger, it is true, the 'standpoint of experience' as the standpoint of empirical science (in the sense of "mechanical natural science") does in fact have a *relationship of tension* with the starting-point of pre-scientific, everyday experience, but it does not imply — at least *de facto* — its total suppression.

It is more particularly the consideration of Griesinger's ego psychology [499] which forces us to emphasise the difference with Herbart outlined above. As a result of the progressive linking of ideas, says Griesinger, large, increasingly cohesive idea masses form in the course of our lives. Their singular characteristic (*Eigentümlichkeit*) varies from (human) individual to individual, and is not only determined by the particular content of the ideas evoked by sensory perception and external experiences, but also by the habitual relationships to the passions and the will, and by the inhibiting or stimulating influences which emanate from the organism as a whole and which have become permanent. Even the child, according to Griesinger, gains from his still relatively simple idea masses a total impression which he will, as soon as the material (of ideas) is sufficiently developed and strengthened, denote with an abstract expression, *the ego*. The genesis of the ego, as it comes about in the growing child, is still understood and expressed in sound Herbartian terms. The ego is the product of the mechanism of ideas which is active in the various mental processes. The description of the process through which the (old, already existing) ego assimilates new idea masses and through which it is enriched and strengthened — what is meant here is, of course, Herbart's 'apperception' — also seems, at least at first sight, entirely in line with the Herbartian example.

The bounds of Herbartian orthodoxy are, however, clearly overstepped when Griesinger describes the relationship of the apperception mass (of the ego already mentioned) to the new, as yet unassimilated mass of ideas as the opposition of an *I* to a *you*.[500] It is true that this is meant as an analogy or comparison, but the analogy is not an arbitrary one; in other words, it is an analogy *cum fundamento in re*. The conflict it describes between two or more souls which dwell within man is namely, according to Griesinger, *psychological reality* for every thinking human being. This is not in the sense of (psychological) reality, as it is conceived in terms of mechanistic psychology, but psychological reality as we know it from inner experience. The *I–you* comparison points to a change of perspective in Griesinger's viewpoint: a change towards the perspective of the psychology of content, towards the point of view of internal description.

Indeed, it is not only the conflict, but also the (manner of) solving the conflict which is presented, as it were, 'from the inside out'. The suggestion which this provokes is that of the existence of a 'deeper' ego-subject, which brings about the integration of new idea masses (ego-components, if you like) into the old ego, and as such functions as a principle of teleology. I do not propose to pursue the question of the extent to which this suggestion has a real foundation. The presumption of an ego-subject understood in this way – more in line with the ideas of Leibniz and Lotze than with Herbart's – would in any case accord with the idea of a "task of self-education" in the sense which Griesinger clearly links with it here.[501] The fact that Griesinger, in this description of the (psychological) ego, in which he exchanges the external-mechanistic perspective for the psychology of content perspective of 'pre-scientific' self-experience, abandons, as it were, his Herbartian role, is not an isolated instance in his work nor is it, in my view, pure coincidence. As far as the first point is concerned, the duality of the (Herbartian) mechanistic view versus the psychological (teleological) view is a *structural* characteristic of Griesinger's treatment of the forms of mental disturbance.[502] And, as far as the second point is concerned, if we take into account the fact that (1) Griesinger defended not only the programme of a mechanistic neuropsychiatry but also the indispensability of the standpoint of everyday experience, and that (2) the limits of Herbartian mechanism are nowhere more perceptible, the distance from everyday experience is nowhere greater than in the mechanistic treatment of the ego-subject, then it can come as no surprise that it is on precisely this point in Griesinger's psychology that the divergence from Herbart's view becomes manifest.

The fact that the breaking through of the mechanistic perspective which we have described is, as we have just observed, not simply incidental or coincidental, but structural, underlines all the more that in any case *for Griesinger* there could be no question of abolishing the standpoint of pre-scientific experience (i.e. experience not yet raised to the level of 'mechanical science'). His view of this was, as I have already explained,[503] that as long as the natural scientific researcher is still a human being, that is to say, has not yet attained the level of the "angel ... [who] explains all to us", the pre-scientific (and thus the psychology [of content]) perspective on the reality of the life of the soul remains indispensable. Without this standpoint the necessary point of departure for this systematic breaching of the mechanistic view, as we find it realised *de facto* in Griesinger's justification of the forms of mental disorder, would indeed be missing.

It is precisely the *combination* of this pre-scientific, if you like vulgar psychological, standpoint with the standpoint of identity-theory monism which makes the fundamental philosophical figure of Griesinger's psychiatry so paradoxical. (The same is, of course, true of the related duality of mechanistic and teleological-psychological views.)[504] This paradox should not, however, unduly alarm the historian of science. Certainly it cannot be denied that the *consistency* of a position which implies the simultaneous defence of (1) the identity theory thesis and of (2) the thesis of an epistemological dualism, which in a manner of speaking divides the cognitive access to reality over two perspectives, is a serious problem. This problem is, however, systematic-philosophical in nature and as such is not under discussion here.

Should my interpretation of Griesinger's philosophical position arouse doubts, because one considers such a position *philosophically* untenable or improbable, I sould like to reply by pointing to the philosophy of Herbert Feigl, one of the most outstanding contemporary representatives of the so-called identity theory, whose position indeed shows a striking similarity to the philosophical pattern which we perceive in the background of Griesinger's psychiatry.[505]

3.8. GRIESINGER'S RELATIONSHIP TO INSTITUTIONAL PSYCHIATRY

It will certainly come as something of a surprise to anyone familiar with the current picture of Griesinger as the great innovator in psychiatry, whose psychiatric conception meant the break with anthropologically-oriented institutional psychiatry, to find Griesinger discussed as a man "in whom progress and tradition were not in conflict, but strove for reconciliation".[506]

Nevertheless, there is a certain amount of truth in this. In order to be able to see this, however, it is necessary to define what is the new element and what the traditional in Griesinger's work in a way which has not so far been tackled in the historiography of science literature.[507]

The reformation of nineteenth-century medicine, to which Griesinger (with Rokitansky and Virchow) made decisive contributions,[508] was, as we have seen, part of that comprehensive development in nineteenth-century thinking in which the orientation towards the mechanistic ideal of science became determinative for the self-conceptions of, amongst other things, physiology, psychology, medicine in general and, therefore, also for psychiatry in particular. Like Rokitansky in pathological anatomy and Virchow in general pathology, Griesinger expressed the mechanistic point of view specifically in the field of the pathology of mental disorders.

What was new in Griesinger's psychiatry was indeed the result of a reorientation towards the mechanistic ideal of science. It is this reorientation which we are referring to when we speak of the "natural scientific self-conception of psychiatry" and it is through this natural scientific self-conception that we distinguish Griesinger's psychiatry from the anthropologically-oriented institutional psychiatry. Viewed in this way, it seems at the very least misleading to speak of Griesinger's psychiatry as (a form of) *anthropological* psychiatry, as, for example, Gerhart Zeller believes he can do.[509] The philosophical discontinuity which comes to light in the comparison of the philosophical frameworks of institutional psychiatry and Griesinger-style natural scientific psychiatry cannot simply be argued away.

Nevertheless, the historian of science will not hesitate to speak in this case of a continuity in the (clinical) psychiatric tradition, because – despite all the differences in philosophical presuppositions – an unmistakable connection between the two forms of psychiatry continued to exist. That this is so is connected with the fact touched on earlier (in our discussion of the work of L. Snell)[510] that the backbone of this psychiatric tradition is formed by the results of (clinical) description. Psychiatric experience could be transmitted in this way without losing substance in the change of philosophical frameworks – indeed, "Hypotheses, theories, aetiopathogenic conceptions come and go, but clinical observation remains the pedestal on which psychiatry rests" (H. Ey). Thus anything which was of lasting significance in pre-Griesinger French and German psychiatry did not have to be lost in the transition to natural scientific psychiatry.

Moreover, the assimilation of insights from anthropologically-oriented institutional psychiatry into natural scientific university psychiatry which

Griesinger created was made easier by the fact that Griesinger's framework was in accord, *in some aspects*, with the anthropological orientation in psychiatry. While it does not do to describe Griesinger's psychiatry as anthropological psychiatry (as Gerhart Zeller does), the consideration of these aspects of his theoretical framework make it in any event plausible to ascribe an "anthropological" character, in a somewhat loose sense of the word, to this psychiatry. I am referring here to three characteristic facets of Griesinger's psychiatry: (1) the recognition of the (viewpoint of the) individual, inner life-history; (2) the unitarian view, i.e. the view geared to man as a psychophysical unity; and (3) the commitment revealed in word and deed to a certain ideal of humanity (if you will, an ideal of existence), which was also characteristic of (the representatives) of institutional psychiatry.

All three aspects are fundamental to institutional psychiatry, so that the thesis that Griesinger, in spite of his avowed (natural scientific) naturalism, continued to be bound to essential presuppositions of institutional psychology is naturally defended through a more detailed elucidation of these three viewpoints. Although all three are closely related, in the following discussion of Griesinger's relationship to Albert Zeller, the man who really taught him psychiatry, the accent will lie on the second and third points, while in the concluding reflections on Binswanger's relationship to Griesinger's psychiatry we shall look more closely at the first (and to a lesser extent the third) point.

3.8.1. *Griesinger and Zeller*

Griesinger received his actual education as a psychiatrist during the relatively short period (1840–42) when he was employed as an assistant doctor in the institution in Winnenthal, which had been run by Albert Zeller[511] since its opening in 1834. The period in Winnenthal must have been of enormous importance to Griesinger's development and to his education as a psychiatrist. Not only was the foundation of his psychiatric knowledge and practical psychiatric experience laid at this time, but we may also assume that in these two years spent in daily association with Zeller he was able to clarify his attitude of institutional psychiatry and crystallise his own psychiatric conception. His publications in the years which followed (1842–45), which we have already discussed, bear witness to this.

In their obituaries of Griesinger, both his friend C. Wunderlich and C. Westphal, who succeeded Griesinger in Berlin, emphasised the good personal relationship which existed between Zeller and Griesinger, and Griesinger

himself expressed his gratitude to Zeller in the foreword to the first edition
of his *Pathologie und Therapie der psychischen Krankheiten* (1845).[512]
Westphal's description is of especial interest to us, particularly where he
observes that "the friendly and warm relationship with Zeller and his family,
right up until the end . . . [was] all the more remarkable because the two men
held totally different opinions, particularly in the field of religion".[513]

Indeed, a comparative reading of the works of Griesinger and Zeller
reveals that one of the most striking differences between these two psychiatric
authors is that whereas to Zeller the psychiatric task had a pronounced
religious (Christian) motivation, and was also expressly understood as a duty
of Christian brotherly love, there is no trace of specific Christian-religious
interest in Griesinger's work. The humanitarian ethos which Griesinger,
according to his contemporaries, so unequivocally demonstrated in the
practice of his work and his relationship with the sick, apparently derived
from another source, and was not expressed thematically in his written work.

The fact that a lifelong friendship could exist between the two men in
the face of such a deep-seated difference in self-interpretation must surely
be accepted as not the most negligible proof of the thesis that Griesinger
remained linked with the tradition of institutional psychiatry in the humanitarian motivation of his psychiatry. There is indeed nothing in what we know
of Griesinger's practical psychiatric work and his (theoretical) conceptions
concerning it which would lend support to Binswanger's assertion that
"depersonalisation of humanness"[514] became a fact in Griesinger's natural
scientific conception of psychiatry.[515] On the contrary, the belief in the
irreplaceable value of the individual personality, the authentic interest in
his individual fellow man which, as we shall see, was so clearly marked in
Zeller, were, although perhaps not expressly reflected, factors which inspired
Griesinger's psychiatric thinking and activity.

As we have remarked, it is not only this motivation, but also the *unitarian*
traits in Griesinger's psychiatric theory which lead us to speak of an 'anthropological' aspect of his work. In this context, certain suggestions which
Gerhart Zeller[516] makes concerning the medical-psychiatric school in which
Griesinger (*and* Albert Zeller) should be considered to belong, deserve our
special attention. This author places Zeller and Griesinger in the context of
what he calls the "old Tübingen school of psychiatry". This was a school
of psychiatry founded by the eminent clinician Johann Heinrich Ferdinand
Autenrieth (1772–1835). Its fundamental principles – (1) the so-called
'*Einheitspsychose*' (unitarian psychosis) theory, (2) mental illness is the
result of a *combination* of psychic and physical factors, and (3) a fundamental

segregation of the psychiatric hospital and the custodial mental institution
– can be seen in Autenrieth's work as well as that of Zeller and Griesinger.

The general pathological conception, within which the unitarian psychosis
theory can be understood as a specific pathological consequence, is said
to be[517] a conception developed by Autenrieth, following on from the
humoral pathology of Sydenham and Franz de le Boë (Franciscus Sylvius).
According to this, there is a relationship between various diseases and the
possibility exists of the transition from one disease to another (*transmotio
morborum*). Through Autenrieth, A. Zeller became familiar with the theory
of J. C. Reil, who was convinced, among other things, that almost all forms
of madness ('*Verrücktheiten*') were accompanied by a sort of melancholy.[518]
It was, however, not until his confrontation with the Belgian psychiatrist
Guislain (1797–1860) that Zeller developed his own psychopathological
system, i.e. his interpretation of the idea of unitarian psychosis. (Zeller
contributed a foreword and some supplementary material for the German
translation of Guislain's *Traitée sur les phrenopathies, ou doctrine nouvelle
des maladies mentales* (1838), which was published by Christian Wunderlich.)[519] What Guislain's theory amounted to was that (1) mental disorders
were the result of an impairment of mental sensibility brought about by
abnormal stimuli, and (2) melancholia was the 'fundamental alteration'
in every form of mental disturbance and the beginning of every type of
psychosis. On the basis of Guislain's thesis, Zeller's theory of unitarian
psychosis postulated that melancholia was the basic form and the start of
every psychic disorder and that the main forms of psychic disorder – (1)
Schwermut (melancholia), (2) *Tollheit* (mania), (3) *Verrücktheit* (paranoia),
and (4) *Blödsinn* (amentia) – are *stages of a single (psychic) disease*. The
melancholic stage can, but does not necessarily have to, change into other
forms of disease. It has its own laws of development and progress. If there
is no improvement in this melancholia it will change into mania (*Tollheit*),
and if the disease is not arrested *Verrücktheit* and *Blödsinn* will follow,
in that order.[520]

It is not difficult to see that the solution which Zeller's theory of unitarian
psychosis offered to the problem of nosological systematics was one which
could easily be reconciled with the aetiological conception contained in the
second principle of the 'old Tübingen school'. According to this, mental
diseases only occur when the psychic-reactive disharmony is accompanied
by a physical disease, i.e. when there is a combination of psychic and somatic
causes of disease. The (tacit) presupposition here is that what is diseased in
mental illness (whatever its form) is always the psychophysical individual,

that is to say, the individual human person as a unity of soul and body — a presupposition with which we have already become sufficiently familiar in our discussion of somaticist psychiatry.[521]

As far as the third principle of the 'old Tübingen school' is concerned — the fundamental separation of psychiatric hospital and custodial mental institution — this can be understood as a practical psychiatric consequence, which follows from a consequence of the unitarian psychosis theory. The first of the four forms of mental disease are distinguished from the last by their better prognoses, that is to say, they are 'curable' as opposed to 'incurable'. Accordingly, a distinction must be made in the organisation of psychiatric nursing between the treatment of 'curable' cases (in the psychiatric hospital) and the care of 'incurable' cases (in the custodial mental institution).

Although the information (argument is too strong a word) put forward by G. Zeller (1968) is not sufficient to confirm the probability of the assumption that Zeller was indebted to Autenrieth for his ideas,[522] there can be no doubt that Griesinger learned about the idea of unitarian psychosis as such through Zeller.[523] This does not, however, alter the fact that this conception was altered to a certain extent in the framework of his mechanistic theory, and one certainly cannot rule out the possibility that similar ideas, such as we encounter in Broussais' pathology, which Griesinger admired, directed his own psychopathological reflections towards the idea of a unitarian disease of the mind.[524] In addition, one should not lose sight of the fact that the philosophers who were important to Griesinger's theory, Herbart and Lotze, were both, given the basic assumptions of their theories, bound to reject the notion of a system of separate syndromes and to sympathise with the idea of a unitarian psychosis.[525]

3.8.2. Zeller's Position in Somaticist Institutional Psychiatry

"What God has joined together, let no man put asunder", says Zeller at the beginning of an important article written in 1838, in which he formulates the fundamental principles of his psychiatry.[526] These words are usually quoted to back up the notion that (1) the defence of a "psycho-somatic" standpoint is the main characteristic of Zeller's psychiatry, so that (2) he is distinguished from both the psychicists and the somaticists. I believe, however, that this interpretation is wrong. I should therefore like to defend the theses (1) that Zeller's "psychosomatics" (Bodamer) can be understood as a variant of somaticist psychiatry, i.e. that Zeller — in terms of the history of psychiatry — must be classed as a somaticist, and (2) that the factor which sets him furthest apart from his fellow-somaticists can be found in the religious basis underlying his psychiatric conceptions.

Although the religious element is not entirely absent from the work of any of the leading institutional psychiatrists, for none of them was the Christian-religious inspiration so crucial as it was for Zeller. This religious element becomes increasingly evident, particularly in his later work. In view of this strong religious tendency, one would be inclined to assume that Zeller's sympathies would in the first place lie with Heinroth's psychicist psychiatry, but although it is known[527] that the first psychiatric works he studied were those of Heinroth, Jacobi's (somaticist) views gained a decisive influence on his thinking. When Zeller wrote his article in 1838 psychicism in psychiatry, although past its peak, was nevertheless not yet dead — Heinroth lived until 1843 and remained productive until the end — and this undoubtedly motivated the young institution director to express unequivocally his connection with somaticist principles, which he had put to the test in the first four years of practice in his institution in Winnenthal. The emphasis therefore lies on the fundamental somaticist position, and more particularly on the second principle of 'the old Tübingen school' — the principle which states that psychic disease is the result of a combination of psychic and physical causes.[528] It is entirely unfounded to ascribe to Zeller a position outside somaticism and psychicism on the basis of his defence of this aetiological principle,[529] because this aetiological conception is wholly in accordance with, indeed presupposes, the fundamental somaticist thesis of the unity of body and soul.

The immediate background and source of inspiration in Zeller's case will not, I assume, have been Aristotelian psychology. What Zeller's writings suggest, in this respect, points rather in the direction of a romantically-tinged Spinozism — a belief in the (eternal) life which is individualised in each of us, that is to say, in the psycho-somatic totality of our being. This does not, however, detract from the fact that the 'psycho-somatic' concept which Zeller defended fits without difficulty into the somaticist framework and, indeed, was considered by him to be somaticist psychiatry. (More on this later.) Strictly speaking, this is not even surprising when one realises that (and to what extent) all somaticist psychiatry was 'psycho-somatic'.

What distinguishes Zeller most clearly *within* the somaticist position from someone like Jacobi is that where, in Jacobi, a dualistic accent is dominant — identified by us as the separation of the 'higher soul' or mind and the

'lower soul' (i.e. the soul which forms an entity with the body) – in Zeller
a monistic tendency prevails, a tendency to try to establish the unity of
body, soul *and* mind or spirit.[530] This is connected with the fact that Zeller's
attention – much more than could be true of Jacobi – is directed towards
the significance of man's spiritual side for the understanding and cure of
psychic disease. If mind (spirit), soul and body are one, then it is reasonable
to assume that (1) consideration of someone's (earlier) spiritual life can
throw light on the genesis of the (present) psychic disturbance and, therefore,
(2) care of the spiritual well-being of the patient must form an integral part
of total psychiatric care. Zeller's work illustrates both these points.

Thus we see that Zeller expresses the opinion that, regardless of the
question of whether a psychic disorder has been triggered by the body or
the soul, the heart or the head, "the behaviour of the soul itself, in particular
that of its higher part, the mind, [is] of the utmost importance".[531] In contrast to the mind, "which should realise and develop itself freely, according
to rational laws and in the knowledge of a divine order of things", the soul
(in the narrow sense) has only relative independence. It is the task of the
mind "to work itself up to real freedom and to develop the higher, universal
character of humanity, [starting from] eternally changing fluctuations and
moods of the life of the soul, and from natural tendencies, opinions and views.
The latter cannot have any value of their own and in themselves, as long as
they have not yet been impregnated by the higher element and they can,
depending on circumstances, just as easily turn to Good, as fall prey to
Evil. It is precisely this development that gives every single human being his
growing singularity and his higher individuality."[532]

The way in which this spiritual life has *de facto* developed in man in its
relation to the natural[533] life, i.e. the relationship of character and tempera-
ment, the habitual degree of dependence or independence of the mind with
respect to the states of the soul and to those of the (living) body is, therefore,
according to Zeller's express conviction, one of the most important things
in the investigation into the genesis of disorders of the soul – something
which, in his view, is all too often forgotten in the case history of the patient,
and which is of the utmost importance for the prognosis.[534]

It will be clear that if Zeller was critical of the psychiatry of his day on
this point, this criticism went together with his philosophical-psychological
(anthropological) presuppositions concerning the relationship of mind and
soul (in the narrow sense). They made it possible, indeed obligatory, to
take seriously the actual dependence (of mind and soul in the narrow sense)
in the psychiatric sense, whereas the dualistic prejudice of contemporary

somaticists like Jacobi predisposed them systematically to leave this dimension of research out of consideration and to limit quite stringently the psychiatric sphere of work to the area of the bodily soul. Even Jacobi, however, could not avoid recognising the real significance of ingrained habits for the life of the soul. He was nevertheless bound, by virtue of his own principles, to deny the physician the authority to occupy himself with this aspect of the soul (in the wider sense); the physician has to do only with the formal aspect of the soul, and not with its aspect of freedom. [535]

Although Zeller, on the grounds of his more monistic conception, had to draw the bounds of psychiatric competency differently from the more dualistic Jacobi, it cannot be denied that Zeller remains on the whole within the presuppositions of somaticism, convinced as he is "that the soul is everywhere in its bodiliness, as is life, and everywhere in its entirety regardless of all multiplicity of phenomena". [536] What becomes diseased in the case of psychic disorder is the soul, the soul can *actually* become diseased, but because the soul in question here forms an entity with the body, a disease of the soul is always (and necessarily) a disease of the soul in its bodiliness. [537] In this context Zeller uses a formulation which expresses the 'romantic' Spinozist origins and background of his (variant of) somaticism: "In every disturbance of the soul life itself falls ill, only in a specific way, in a particular part of its being". [538]

If we may assume that 'soul' here is always meant in that wider sense which includes the 'mind' (as the 'higher part' of the soul), then we must recognise that the standpoint formulated by Zeller himself implies criticism not only of psychicism but also of the somaticist position. The abandonment of the strict dualism of bodily soul and non-bodily soul (i.e. mind or 'higher soul') in favour of a position that implies recognition of the incarnate nature of the lower *and* higher soul (i.e. mind) brings Zeller in opposition to Jacobi. The differences with Jacobi, however, should not be overrated. First, it is clear that after all Zeller's psychiatry had most affinity with somaticist psychiatry and was understood as such by him. Second, the fact that in Zeller's work, too, something like a distinction between incarnate and non-incarnate soul (or as Zeller would probably say: finite and eternal life) was assumed, suggests that − religiously speaking − his position was not so far from that of Jacobi and his followers as would appear at first sight. Indeed, Zeller himself was firmly convinced that his somaticism could be reconciled with the Christian dogma of the immortality of the soul.

The extent to which Zeller was conscious of the difference between his conception and that of Jacobi, and of the possible criticism which his monistic interpretation of the body—soul—mind relationship could have provoked among the more dualistically-minded somaticists is proved, in my view, by a very characteristic passage from his 1838 article. "In a sense we can say that the soul cannot only fall ill, it can even, and has to, die. In no way does this theory lead to the acceptance of a spiritual death [read: as the Jacobi-style somaticists would like to throw in my face], but to an organic eternity, i.e. one developing and forming itself according to an inner law of life. For if the soul had nothing to go through, to develop, to suffer and to do during dying, death would be of absolutely no avail. Death is only the last metamorphosis of the bodily and spiritual life."[540] To put it another way, death is always organic death, but in the process of dying – in all its bodily and spiritual complexity – Eternal Life is at work and achieves in the dying man – of course, for that individual man alone – its last and highest manifestation.

It cannot be denied that Zeller's variant of somaticism is in tune with certain value-accents in Heinroth-style psychicist psychiatry, in its emphasis on the care for the spiritual well-being of the patient and its recognition of the significance of the relationship of the soul (in the narrow sense) and the mind in understanding the genesis and the prognosis of psychic disorder and in the treatment of disturbed people. However, the self-explanatory anti-psychicist pronouncements which occur in both his earlier and his later work make it absolutely clear that he did not want his own psychiatry to be seen as psychicist.[541]

However, describing Zeller's psychiatric position as (a variant of) somaticism, does not, in my opinion, sufficiently illuminate the source, the heart of his psychiatric interest. 'Somaticism' primarily describes a theoretical option in the discussion about the soul—body relationship. The emphasis of Zeller's psychiatric interest lay, however, not so much on theory as on psychiatric practice.[542] I mean this not in the sense that Zeller published relatively few scientific contributions (although this is indeed the case), but rather in the sense that the self-conception we find expressed in his work is essentially a self-conception of psychiatric practice. Psychiatric thinking ('theory') is of significance *in so far as it serves* psychiatric activity ('practice'), and the ultimate purpose of psychiatric practice is to fulfil a Christian ideal of life.

The fact that Zeller, as a psychiatrist, believed that his Christian convictions about life were not an extra-scientific appendage, not a dead letter, but the source of his inspiration in his psychiatric thinking and doing, becomes strikingly apparent in a letter he wrote to his father in 1842:

My concept of Christianity does not exclude anything really human. All my knowledge
and understanding find in it the keystone, as does the deepest and innermost passion of
my heart. From this source, I can only draw consolation and assistance for other people
in my profession of spiritual guide (Seelsorger), for all the pain that mischief and evil
has brought upon the lives of those entrusted to my care.[543]

To Zeller, man is essentially *homo religiosus*, and this means that if man
is fundamentally affected by mental disease the (possible) curative effect
which emanates from religion is important. The mentally ill patient can be
distracted in many different ways, but the deepest ground of his suffering is
not touched by such superficial distractions and only superficial psychic
wounds are healed: "only religion [can] alleviate and cure the most profound
pain of life". There are still many people, says Zeller, who simply do not
understand what point there is in preaching to the mentally ill:

because they do not understand how the mentally diseased lead a double life, a healthy
and a diseased one, a conscious and a dream life. Nor do they understand that there is
a difference between the mind and the soul of man, the latter being based in quite a
different way from the former on physiological processes. This shows what the wonder-
fully created bodiliness by itself can do, whereas the specific freedom of thought and
deed consists of what the body cannot accomplish, even if every really spiritual deed
involves in fact a spiritual as well as a bodily process.[544]

3.8.3. *Zeller and Griesinger: Opposition and Unity*

We may safely assume that among those who will have viewed Zeller's religious
zeal *in psychiatricis* with a certain degree of scepticism was his own right-hand
man in the Winnenthal institution, the assistant physician Wilhelm Griesinger.
This makes the question of Zeller's position with regard to Griesinger's
psychiatric views all the more interesting. One would expect to find something about this in Zeller's article which appeared in 1844, because, among
other things, it reports on the two years when Griesinger was staying in
Winnenthal.[545] Griesinger is not mentioned by name in the article, but it
is difficult to interpret some passages in it as anything but the expression of
an 'internal debate' with his younger colleague, or at least with aspects of
the mechanistic psychiatry which Griesinger defended and which Zeller saw
as a threat to the true fundamental principles of psychiatry.
    Thus Zeller criticises[546] the "new purely mechanistic and atomistic
view of psychic life", because in the consideration of psychic life it (1)
leaves the spirit aside and restricts itself to the soul (in the narrow sense)

and (2) comprehends the (remaining) life of the soul "only in an extremely one-sided way, and therefore does not explain anything, because it has failed to understand that the simplest sensation already presupposes a psychic unity, a being that feels, one that imagines and becomes conscious of an outer world, according to inborn laws". Of the greatest significance, in his view, is research into the 'physiological mechanism' – this makes us aware that much of what we put down to free activity of the soul must be ascribed to the bodily ego, and that the life of most people remains "a purely organic and animal act" which only has the character of reason because "all human nature is designed for rationality and most people are reasonable without wanting to be, indeed, even against their will". But – Zeller seems to be suggesting in this not entirely clear text – the physiological approach has its limits. It enables us to trace the, as it were, automated 'natural reason' of the living human organism (i.e. man in so far as he is such an organism) and to explain it as 'mechanism', but there always remains the aspect of reason which resists a mechanistic interpretation. Strangely enough, it is precisely where one would expect this "highest force in us, our actual self" to be entirely abolished, namely in mental diseases, that it remains clearly recognisable.[547]

The second point on which Zeller criticises the mechanists – the extremely one-sided view of the life of the soul which leaves no room for something like the unity of the life of the soul and is therefore incapable of explaining the most elementary phenomena of the consciousness, such as perception, for example – is more extensively dealt with elsewhere.[548]

The context is that of a criticism of Herbartian intellectualism (while it is not specifically mentioned it is certainly what is meant), which, as we have already seen, understood the psychic functions of thinking, will and feeling as modifications of the life of ideas. Nowadays we would say that what Zeller is criticising there is the *non-phenomenological*, not to say *anti-phenomenological reductionism* which is contained in the mechanistic psychology of ideas.[549] Is the mind's struggle with passion, with hate, love, pride, greed, lust, self-indulgence, sorrow, doubt, no more than a simple "static play (i.e. mechanism) of ideas"? In Zeller's view this is a ridiculous suggestion which entirely disregards the evident conflictory character of the life of our soul, without which "the whole psychic nature of man cannot be understood".[550]

Such passages[551] can be seen quite simply as a frontal attack on Griesinger, or at least as a criticism of those aspects of Griesinger's psychiatry which Zeller considered to be dangerous. In my view, this criticism indeed hits on a dubious, weak facet of Griesinger's work; but if our interpretation of Griesinger's relationship to Herbart is essentially correct, it is reasonable to assume that where Zeller's criticism appears to be directed against the consequences of an anti-phenomenological reductionism, this strikes at the final intentions of Herbart's psychology, rather than at Griesinger's. Compared with the metaphysicist that Herbart was, Griesinger was too much of an empiricist to be able to believe wholeheartedly in something like the *victory* over 'common sense' (the standpoint of everyday experience); too sceptical of the value of his own scientific theories not to be sensitive to criticism such as that expressed by Zeller; too 'phenomenological' not to identify with Zeller to a certain extent in this criticism. Thus we see how Griesinger, only a year before his death, faced with a choice between experience and maintaining the unity of his psychopathological systematics, opted for (clinical) experience.[552]

What, on the other hand, really divided Griesinger and Zeller was, as I said at the beginning of this discussion, a religious question, or perhaps it is more accurate to say a difference in (religious) philosophical self-conception. This difference must not, however, be exaggerated, because precisely on the level of philosophical presuppositions there existed a point of agreement which makes the excellent relationship between Griesinger and Zeller rather more understandable. The (naturalistic) monistic 'prejudice' of Griesinger's psychiatry[553] harmonised extremely well with the monistic tendency of that 'romantic' Spinozism which can be recognised in Zeller's work and which brought Zeller, to a certain extent, into opposition with Jacobi. The existence of this alliance against dualism does not, however, alter the fact that Griesinger's naturalism and Zeller's somaticism, with its Christian religious inspiration, were in the final analysis irreconcilable. (As, for that matter, similarly, there existed a latent relationship of tension *within* the thinking of Zeller himself between metaphysical presupposition and religious-ethical conviction, precisely because, and in so far as, the fundamental intuitions of Christian doctrine were not totally assimilable within the presuppositions of a 'romantic' Spinozist philosophy of life.)[554]

The undeniable discrepancy in self-conception between Zeller and Griesinger need not, as we have already observed, make us doubt that Griesinger remained linked in his basic humanitarian attitude to what was best in the tradition of institutional psychiatry. Nor need we doubt that his preference for a psychopathological unitarism (the idea of a unitarian psychosis) and the unitarian view of man connected with it precluded as sharp a break with anthropologically oriented psychiatry as posterity assumed to have existed between the psychiatry of Griesinger and the institutional psychiatry of his day. Finally, it can plausibly be argued that the third point (to be discussed hereafter), namely the recognition of the viewpoint of the individual inner history of life, meant a fundamental link with (the basic anthropological conviction of) institutional psychiatry.

When I, therefore, conclude this chapter by looking at Binswanger's relationship to (the tradition of) German institutional psychiatry in general and to Griesinger in particular, this is in the first place to throw into sharper relief Griesinger's links with this tradition which were postulated earlier. It is, however, also intended to draw attention to the fact that the tradition of clinical-psychiatric 'positivism', with its orientation around the individual, which linked Griesinger (in spite of everything) with the major representatives of institutional psychiatry, is essentailly the same as that which (*pace* Binswanger) brings Binswanger and Griesinger together, and thus links German psychiatry in the period we are studying with (related) forms of twentieth-century psychiatry.

3.9. BINSWANGER'S RELATION TO (THE TRADITION OF) INSTITUTIONAL PSYCHIATRY IN GENERAL AND TO GRIESINGER IN PARTICULAR

Ludwig Binswanger, known as the founder of so-called 'existential analytical' (*daseinsanalytische*) psychiatry, has left us a little-known historical study which is of particular interest to us because it throws some light on precisely that connection between nineteenth- and twentieth-century psychiatry which leads us to speak of the 'anthropological' *tradition* in psychiatry.[555]

The aim of the study is to reflect on the 'historical conditions and forces' which made possible the seventy-five-year history of the Bellevue institution in Kreuzlingen and which were realised there. As such, it in fact deals with a piece of family history, since the history of the Bellevue institution is the history of the life work of a dynasty of institutional psychiatrists, which starts with the founder, Ludwig Binswanger (1820–80)[556] and continues with his son, Robert Binswanger (1850–1910), who in his turn, was succeeded as director of the institution by his son, Ludwig Binswanger (1881–1966), the author of the study in question.[557]

Ludwig Binswanger 'the elder' was – like Griesinger, with whom he remained friends until Griesinger's death – a pupil of Zeller, a man "whom every psychiatrist bearing the name of Binswanger has admired, and always will admire".[558] What linked psychiatrists like Zeller, Jacobi and the elder Binswanger with one another was their common involvement in an ideal of humanity which they tried to realise in their existence as psychiatrists; humanity understood as "not simply a programme and a method, but a form of interhuman relationship".[559] The form of humanity of these doctors was based on a comprehensive personal *Bildung* (education) of the individual or of small groups brought together by friendship. It was characteristic of this *Bildung* (according to our author) "that it withstood and maintained the tension between the religious, or the philosophical, view of life and natural scientific thinking, without mixing the two or allowing the violation of one by the other".[560]

Speaking of the anthropological conception of his grandfather's psychiatry, Binswanger points out that "in this entire view of man, the *historical* factor, the historical development of his examination of the world of 'objects' is emphasised so strongly", and that we can already see here a concept which has now come into prominence again. This is the concept of the life *history* as a concept of the historical continuity of the present man with the much earlier man, including the stressing of the *uniqueness* of the *course of each individual's life*. All this is, of course, coupled with the demand for *individualisation* in the conception and treatment of the sick, in the sense of a *"loving* concern for their inward and outward fate in life and for the problem of how to bring them back, out of their 'subjective' insanity, to a 'normal' *distance* towards the world of 'objects' ".[561]

It cannot be denied that the description Binswanger gives here is partly determined by the interests and prejudices which governed his own psychiatric conception. In other words, the motives and insights which he recognised in nineteenth-century institutional psychiatry had (at least in part) been instrumental in determining the figure of 'existential analytical' psychiatry which he himself developed.[562] His retrospective look is, however, by no means a subjective projection of his own ideas on to the past history of psychiatry: it touches on an essential aspect of nineteenth-century institutional psychiatry and thereby gives substance to our thesis of a psychiatric tradition linking together the (anthropologically-oriented) psychiatry of the nineteenth and twentieth centuries.

In order to arrive at a precise definition of Binswanger's relationship to the psychiatry of Zeller and Griesinger, it is important to see that (and in what sense) one can speak of continuity, and this also implies seeing that (and in what sense) this continuity was more far-reaching than Binswanger was, or could be, aware of.

The concept of the "inner life history" is of central significance in this context. In his *Lebensfunktion und innere Lebensgeschichte* (1928), Binswanger himself put forward this concept as a fundamental category of psychiatric and psychological thinking. The inner (otherwise mental) life history (which is always the life history of a person) is described by Binswanger as the unique historical sequence of the contents of experience of the individual, mental person (person, conceived as the origin or centre of all experience).[563] What Binswanger is trying to bring out in this lecture is in fact the *categorial* difference which is thought to exist between the concepts of the function of life and inner (or mental) life history. (Because of this, his study can be said to be concerned with the investigation of the conceptual foundations of psychology [or, more broadly, human sciences] and to have as such both philosophical significance and scientific relevance.)

The starting-point is the important differentiation, introduced by Bonhoeffer in 1911, (within the class of psychogenic [in the broader sense] diseases or diseased conditions) between disease of a purely functional nature (psychogenic in the narrow sense) and so-called hysterical conditions (in which a psychological content factor seems to determine the direction of the will). Bonhoeffer's purely clinical distinction conceals, however, a distinction which is fundamental to the whole of psychiatry and psychology, namely that "between the mental (or physical–mental) functioning of the organism and its disorder on the one hand, and the sequence of the contents of mental experiences on the other".[564] The category of the life function is applicable to the psychophysical organism, the biological–psychological complex of functions, that of the inner life history to the soul, in so far as it is centred in the individual, mental person.

Psychiatry's object of investigation — and here we can see the philosophical (anthropological) implications of Binswanger's study — includes both areas: the corporeal–psychic organism (within which an abstract dualism of body and soul is no longer[565] assumed) *and* the human being or (mental) person.

Binswanger's categorial division (which implies a methodological dualism of functional and life-historical psychological-hermeneutic viewpoints or methods of thinking[566]) in fact repeated, as the reader will already have observed, the ontological scheme of nineteenth-century somatic psychiatry, however new its effect may have been in the fundamental psychiatric (psychological) discussion at the time. He was himself, however, apparently not conscious of this connection, although the philosophical-historical tendency of his study makes it reasonable to assume that he must have had some idea of the fact that the categorial division he was defending presupposed an

anthropological conception which united Aristotelian and Christian-philo-
sophical motifs.[567]

Binswanger was undoubtedly right when he said that the psychiatry of
the (somaticist) institutional psychiatrists like Jacobi, Zeller and the elder
Binswanger emphasised the individual 'inner life history' of the sick human
being (person).[568] The idea that the sick person entrusted to their psychiatric
care is (also) a spiritual person, the subject of mental life, was not only
never contested by institutional psychiatrists but was usually assumed to be
self-evident, and, after all we have said about the somaticists and psychicists,
it needs no more detailed explanation. It is, however, important to realise –
and on this point Binswanger's view must be corrected – that the recognition
of something like the individual, inner life history is not what *distinguishes*
Zeller, Jacobi, etc. from Griesinger, but is, on the contrary, precisely what
links them! This is not really surprising when one considers that Herbart's
psychology contains a theory of the individual development of personality
– the basis of his pedagogy – a theory which explains the *'factum'* of the
'inner life history'.[569] Griesinger built on this.[570]

In other words, it is not the possible failure to appreciate the existence of
something like an 'inner life history' which must therefore be seen as the
feature which distinguishes Griesinger from the above-mentioned institutional
psychiatrists, but the fact that the Herbartian view he adopted implied a
mechanistic conceptualisation of this inner life history, whereas anthropological institutional psychiatry (at least implicitly) preserved the character
of intentionality of this mental occurrence and (therefore) also adhered to
the idea of a spiritual person as the "origin or centre of all experience"
(Binswanger). The fact that, as we demonstrated earlier, this ego-subject,
which was conceived as the spiritual person in the tradition to which the
psychicists and somaticists belonged, is also presupposed as the centre or
origin of the 'inner life history' in Griesinger's psychiatry serves further to
relativise the difference between Griesinger and institutional psychiatry. It
is precisely this presupposition which – despite appearances to the contrary
– also links him with Binswanger.

What is established is that Binswanger in any event failed to recognise
this point of continuity with Griesinger's psychiatry, and that, despite his
undisguised admiration for his illustrious predecessor, he ultimately deplored
Griesinger's 'constitution of clinical psychiatry'[571] as no less than the greatest sin in psychiatry because, in his opinion, it brought about the *"depersonalisation* of humanness".[572]

This interpretation set the seal on a double separation – that between

Griesinger and Zeller on the one hand, and that between Binswanger and Griesinger on the other – a separation which has stood in the way of the understanding that (and in what sense) Zeller, Griesinger and Binswanger were linked to one another in the context of one and the same tradition of clinical psychiatry.

CHAPTER 4

SCHOPENHAUER, ROKITANSKY AND LANGE:
TOWARDS AN EXPLICIT PHILOSOPHICAL JUSTIFICATION
OF GERMAN 'MATERIALISM' (FROM ABOUT 1840)

In our background study in Chapter 2 we devoted considerable attention to
the philosophical and scientific influences which worked together in the
genesis and growth of the mechanistic ideal of science and the mechanistic
self-conception of, amongst other things, physiology, psychology and medicine
(including psychiatry).

Following on from this, in Chapter 3 we were able to provide an interpretation of the philosophical presuppositions which governed the thinking
of Wilhelm Griesinger. We were thus able not only to demonstrate the intrinsic
unity of his thinking, but also to show that it can plausibly be said that,
on the issue of his self-conception, Griesinger was 'a product of his time'.
His 'materialism', like that of the leading natural scientists in Germany from
the end of the eighteen-thirties onwards, was not to be understood in a
metaphysical sense or as *Weltanschauung*, but in a methodological sense.

In this chapter I should like to pursue and round off the reflections
on the theme of the self-conception of that German 'materialism', on the
understanding that the emphasis will no longer primarily lie with science
(in this case psychiatry), but with philosophy. The central issue here is
not the historical identification of that 'materialistic' self-conception as
being of a particular nature or as characteristic of one or another scientist
who called himself a 'materialist'. Instead, we are concerned with the question of the role played by some philosophers in the process of crystallising
this methodological materialistic self-conception, which is present in the
work of Griesinger, among others, and which was to achieve its explicit
philosophical expression and justification in the philosophy of F. A. Lange.
In this context, Schopenhauer and Lange represent the beginning and end
of a development in which nineteenth-century German philosophy and
the physiology of the brain and senses had a fruitful interaction. Between
these two, chronologically and systematically, there is the imposing figure
of Rokitansky − (still) linked on the one hand with Schopenhauer, but
on the other already pointing ahead to Lange. Rokitansky, it is true, was
neither a philosopher nor a physiologist (of the brain or senses); he was, in
fact, a (philosophically-trained) pathological anatomist who, because of
his leading role in the history of medicine, cannot be omitted from any

156

discussion or clarification of the relationship between nineteenth-century philosophy and medicine.

4.1. SCHOPENHAUER AND PHYSIOLOGY

The very fact that the results of physiological research made such an important contribution to Schopenhauer's philosophy (particularly his epistemology) may tempt one to assume that Schopenhauer's philosophy was attractive to the physiologists or primarily physiology-oriented scientists of his day, and that it was of *direct* significance for the determination and articulation of the self-conception of the physiology-oriented (natural) sciences of his time. On closer examination, however, this presupposition does not hold up. In order to understand how this comes about, it is necessary to look more closely at Schopenhauer's relationship to the physiology of the period. The line of research thus proposed is, however, a complex one, because it contains at least two questions: the question of the significance of physiology for Schopenhauer's philosophy (epistemology), and that of the significance of Schopenhauer's (physiologically-based) epistemology for the determination and explicit articulation of (early) nineteenth-century physiology.

By way of introduction, I should like briefly to discuss the place occupied by physiology, according to Schopenhauer, in the scientific system, and the relationship between the systematics he suggested and the doctrine of the four-fold significance of the "*Satz vom zureichenden Grunde*" (the principle of sufficient reason), so central to his philosophy.[573]

4.1.1. *The Position of Physiology in Schopenhauer's Classification of the Sciences*

Assuming (with Kant) that (1) knowledge is always knowledge of the *phenomenal* world (the world of 'appearance' or, to use Schopenhauer's terminology, 'Idea'), and that (2) the world of ideas (or better [but less used]: presentations, *Vorstellungen*) which we know in this way is *ordered*, in the sense that we can speak of a system of ideas, Schopenhauer postulates the principle of sufficient reason as a universally-valid principle of ordering. It is the *a priori* assumption that everything has a reason which gives a rational basis for our always and everywhere asking 'why?', and he is therefore able to describe the principle of sufficient reason as the foundation of all science.[574] Within the phenomenal world, on which we are brought to bear as knowing subject and which is explored in the process of our asking why, that is to say, in sustained, consistent application of the principle of sufficient reason, various regions of being, object-areas, can be distinguished. Within each object-area there rules a particular type of causality which is valid only for that object-area and not outside it. This rules out the univocality of the principle of sufficient reason: a why-question about objects which are considered to belong to inorganic nature cannot be transferred, without a change of meaning, to the investigation which takes place in the realm of organic nature or the realm of human activity, etc.. Ontological differentiation therefore demands a differentiation of the meaning of the why-question. The general formula, borrowed from Wolff,[575] thus reveals itself as "the common expression of a number of a priori insights".[576] The one a priori of the principle in question can, as it were, be resolved into four different a prioris which correspond to a similar number of areas of being distinguished by Schopenhauer. Each of these four a prioris specifies a direction of scientific why-questions, and can thus be understood as regulating[577] a specific group of sciences. In this sense, the differentiation of the four a prioris in scientific cognitive activity distinguished by Schopenhauer also served as the foundation of his philosophy of science systematics.

For our purposes — answering the question concerning physiology's place in Schopenhauer's classification of the sciences — it is not necessary to go into all the details of this "highest classification of the sciences".[578] Suffice it, therefore, to say that in Schopenhauer's scheme physiology is defined as an empirical science (or, science a posteriori), and is thereby more specifically included in that subgroup of empirical sciences to which the causality category of the stimulus is specific.[579] To say that physiology is an empirical science is to say that the variant of the principle of sufficient reason which is of regulatory significance to it is the variant in which 'reason' is understood to mean the 'reason of becoming', i.e. 'the law of causality' ('principium rationis fiendi'). What can meaningfully rank as 'cause' is, however, not the same for the various empirical sciences, and a distinction must therefore be made between three different modes of causality, each of which specifies a separate meaning of 'reason of becoming', namely as (mechanical) 'cause', as 'stimulus', and as 'motive'. The division of the class of empirical sciences into three subgroups follows the differentiation of causality modes, and when, in the further detailing of the classification scheme, physiology proves to be classed in the subgroup of empirical sciences of the kind of 'stimulus'—type causality, this simply means that physiology is a life science which, as such, occupies an intermediate position between natural sciences (in the narrow sense) and what we should perhaps more properly speak of as sciences of man in a specified sense of the term.[580]

The fact that Schopenhauer classifies physiology in this way, in the system of sciences, is of relevance in terms of the history of science. It draws attention to the fact that the physiology he is talking about was not (yet) that strictly mechanistic physiology which predominated after the middle of the century in the so-called physics and chemistry school of physiology (Müller's pupils, Ludwig and his followers), but the physiology of the period which preceded it; to be precise, it proves to be the French (and to a certain extent the English) physiology which was prominent in Europe before Müller's 'assumption of power' in 1830.

The fact that this was the case is understandable not only for chronological reasons,[581] but is also, and primarily, based on material grounds. It could only have been the (vitalist-oriented) French physiology (Cabanis, Bichat) which could achieve fundamental significance for a philosophy like Schopenhauer's voluntaristic metaphysics. Janet is therefore not incorrect when (in relation to Cabanis and Bichat) he speaks of "the French origins of Schopenhauer's philosophy", because, as he has demonstrated, the two fundamental theses of Schopenhauer's doctrine – the principle of the Will as the ultimate ground of all (natural) phenomena, and the opposition of Will and intellect – were anticipated in the work of Cabanis and Bichat, respectively.[582]

After this warning about the relevant history of science, we are better prepared to look in more detail at the significance of physiology in Schopenhauer's philosophy.

4.1.2. The Role of Physiology in Schopenhauer's Philosophy

As we have more or less indicated in the preceding section, what made (chiefly French) physiology of interest to Schopenhauer is in the first place the fact that *through the object of its research* – the realm of organic nature, of the phenomena of life – it was particularly relevant to his metaphysics of the Will. In the second place it is based on the fact that – in the meaning conceived by Schopenhauer – it lent itself extremely well to an empirical-scientific foundation of (an, in a sense, 'reconstructed' form of) Kant's aprioristic theory.

As far as the first point – metaphysical relevance – is concerned, it has to be admitted that it is not *a priori* understandable why, judged from the metaphysical viewpoint, physiology (or, more broadly, the life sciences) should have to occupy an exceptional position. After all, the phenomenal world, the world of ideas (or presentations) is in its entirety, and thus in all the regions of being distinguished earlier (if you will, on all levels of objectivity), a manifestation of the in itself unknowable Will 'in itself'. The phenomenal world — which Schopenhauer summed up, as we have seen, in the term 'matter' — is the place where the Will becomes visible.[583] In other words, it is "the bare visibility of the Will or the bond between the world as Will and the world as Idea".[584]

It will be clear that in this metaphysical interpretation *all* phenomena, whether they belong to inorganic nature, organic nature or to the realm of human endeavour, are *ipso facto* the expression, the 'objectivisation' of the Will. The only distinction which can still be made within the class of objectivisations of the Will is one of levels, of degrees of objectivisation of the Will. The Will has its lowest degree of objectivisation in unconscious inorganic nature, its highest degree in the (conscious) activity of man. This means that some phenomena, more than others, make the in itself unknowable Will 'visible' to us, and for this reason deserve our special interest.

The guarantee that what we encounter as real in 'outward' directed knowledge can be meaningfully interpreted as an objectivisation of the Will, is provided by the self-experience we have of our own bodies; that is to say that in this self-experience, in one way or another, we have direct 'knowledge' of the Will, because what becomes 'experience' to us is the identity of the acts of our bodies with the Will, which attains consciousness in the acts of the body, on the highest level of objectivisation which it can achieve, and in the highest attainable degree of 'visibility'. This should not be misunderstood. In the strict sense of the word 'knowledge', the Will in its 'in itself-ness' is not knowable for us — no physiological, anatomical or other knowledge of physical phenomena, however comprehensive, enables us to overstep the bounds of the phenomenal world. Knowledge in the strict sense is by definition "*outwardly* directed knowledge, transmitted by the senses and enacted in the intellect and, as well as time, also has *space* as its form. In this knowledge the two are most intimately united by the intellectual function of causality, and this is how it becomes *Anschauung*",[585] i.e. knowledge of the phenomenal world. This knowledge can never be the origin or legal ground of metaphysical interpretation.

This fundamental limitation of the possibilities of our knowledge is, however, compensated for by the fact that, *from the inside*, namely in the consciousness of self, we have access to and awareness of the real nature of things, i.e. the Will. Because we can only speak of 'knowledge' in the strict sense of the word when we are dealing with (the product of) synthesis of the forms of time and space and the category of causality, there can no longer be any question of (real) knowledge in the case of the Will, because it is true of the thing-in-itself or the Will that, in its self-awareness, "it sheds *one* of its phenomenal forms, space, keeping only the other, time".[586]

This means that the Will cannot be perceived as the enduring substratum of its activities (*Regungen*), cannot be seen as a 'persisting substance', but can only be 'known' in the sequence of its acts, movements and states for as long as they last, directly, it is true, but not in the way of *Anschauung*. "The knowledge of the Will, in the consciousness of the self, is consequently not an *Anschauung* of the latter, but a direct awareness of its successive activities".[587]

What we thus "become cognisant of" in self-awareness is, as we pointed out initially, *the identity of the body and the Will*, namely in the separate activities of the two. What is known in the awareness of self as a direct, real act of will, manifests itself at the same time as a movement of the body. Every human being, according to Schopenhauer,

sees his momentary decisions of the will, caused by equally momentary motives, immediately represented as faithfully by as many actions of his body . . . as the latter itself [is represented] in its shadow; for the unprejudiced the insight simply arises from this, *that his body is only the outward appearance of his will, i.e. the form and way in which his will presents itself in his intuitive (anschauendes) intellect; or his will itself in the form of idea (presentation).*[588]

It is this metaphysical truth concerning the identity of body and will, that is, the thesis according to which the body (metaphysically speaking) is the objectivisation of the Will or, conversely, the Will objectivises (thus: manifests) itself in the body, which explains Schopenhauer's interest in the life sciences, especially physiology. Because if the living, animal organism or body is defined as the place where and the form in which the Will becomes 'visible' to us, then it is only logical to consider the scientific knowledge of the body as being, in principle, metaphysically relevant or, in other words, open to metaphysical interpretation.

The connection which Schopenhauer here assumes to exist between metaphysics and science (in general) is the connection of two different, but mutually complementary (theoretical) perspectives of reality: one and the same 'thing' is approached by science 'from the outside' and by metaphysics 'from the inside', in other words, reveals itself in the 'objective view' in science, or in the 'subjective view' in metaphysics. The relationship between 'subjective view' and 'objective view', between metaphysics and science, is thus to be understood in the sense that the scientific, empirical view is in itself inadequate to satisfy our theoretical needs and must therefore be supplemented by a metaphysical interpretation.[589]

It is in the concentration on the problem of our cognitive faculties, specifically the 'intellect', that this idea of the 'subjective view' and 'objective view' becomes important for the theory of knowledge. Therefore, the central (metaphysical) thesis in the doctrine of the "objectivisation of the Will in the animal organism" reads: "what in the consciousness of one's self, i.e. subjectively, is the intellect, presents itself in the consciousness of other things, i.e. objectively, as the brain".[590]

This thesis formulates the foundation of Schopenhauer's interpretation of the problem of the theory of knowledge, the metaphysical starting-point which is assumed in his efforts to put Kant's epistemology on an empirical-scientific, i.e. (cerebral) physiological footing, and thereby to free it of an essential shortcoming: philosophical reflections on (the structure of) our intellect which, like Kant's aprioristic theory, limit themselves to the 'subjective view' of the intellect and neglect the 'objective', i.e. physiological, view are one-sided and thus inadequate. They leave "an unbridgeable gulf between our philosophical and our physiological knowledge which makes it impossible for us ever to find satisfaction again".[591] The gulf between philosophy (metaphysics) and science which Schopenhauer criticises has, of course, its root in Kant's metaphysical presupposition of an absolute dualism of the phenomenal and the noumenal world. What in Kant is unbridgeably separated, Schopenhauer however manages in a sense to link together in his metaphysics through the thesis that the thing-in-itself which is the Will *objectivises itself*, that is, becomes (for us) Idea. The distinction between the phenomenal and the noumenal world is not thereby abandoned; the (metaphysical) thesis of the absolute separation of the two worlds is merely relativised. This relativisation makes it possible to conceive the relationship between philosophy (metaphysics) and science in the sense Schopenhauer has in mind, namely as a connection in the sense of a mutual dependency and complementariness of the 'objective view' and the 'subjective view'.

Given Schopenhauer's identification of the knowledge of the objectivisations of the Will with scientific knowledge, one understands why Schopenhauer believed that, in contrast with the one-sided 'subjectivism' of Kantian epistemology, the insights gained in the 'objective' perspective of physiology were important; in other words, why Kant's aprioristic theory must be augmented and supported with physiological arguments. In his transcendental-philosophical reflections Kant had accounted only for the 'inside' of the intellect. It was now necessary also to look at the intellect 'from the outside', that is, to make room for the 'objective view of the intellect'.

This 'objective', physiological way of looking at things, which is sometimes also referred to as genetic[592] (as opposed to transcendental), is concerned with the intellect in so far as it is itself Idea (presentation) or, more precisely, a physiological phenomenon. That is to say, it is concerned with the intellect as the brain or, more specifically, with thought or perceptions (*Anschauungen*) as "nothing but ... the physiological function of a viscus, of the brain". Or, to formulate it in yet another way, this 'objective' way of looking at things allows us

to state that the entire world of objects, unlimited as it is in space and time and so unfathomable in its perfection, is actually only a certain movement or affection of the pulp in the skull.[593]

Summing up, we can thus state that the significance of (mainly French) physiology for Schopenhauer's system lay in the fact that it made it possible to give substance to what he called the 'objective view' of the intellect.

The problematic consequence of this 'objective' conception of the intellect − the fact that it implied the antinomy, that the intellect must be regarded at one and the same time as *prius* and as *posterius* of this world −[594] was only later to be realised in sharp definition when in the revival of Kant's transcendental-philosophical intention the concern for a neat separation of the logical and the empirical point of view in the theory of knowledge gained the upper hand.[595]

In the period we are studying this did not yet play any part. The principal effect of Schopenhauer's express recognition of the relevance of sensory and cerebral physiological research to epistemology, one may assume, lay in the fact that it provided a philosophical basis and a stimulus for later developments, both in the (scientific) physiology of the senses (Müller, Helmholtz) and in epistemology (which to a greater or lesser extent relied on this sensory and cerebral physiology), such as that which was developed before the Neo-Kantianism of Cohen and others (Helmholtz; Lange).

Because of these later developments which it made possible, and also to make it clear why Schopenhauer's theory of knowledge did not (and could not) determine − or at least only to a very limited extent − the self-conception of the physiological (medical) German 'materialists' of the eighteen-forties,

we shall now look in somewhat more detail at some aspects of Schopenhauer's
theory of knowledge.

4.2. SOME ASPECTS OF SCHOPENHAUER'S THEORY
OF KNOWLEDGE

Indeed, if the conviction that the intellect is the *prius* of the world (of
phenomena, ideas or presentations) is what links Schopenhauer to Kant, then
the idea that the intellect is at the same time also the *posterius* of the world,
with its related interest in the (sensory and cerebral) physiological approach
in epistemology, is what divides them.

Although it is certainly true, as Cassirer observes,[596] that to Schopenhauer
it was not primarily Kant's epistemology, but sensory and cerebral physiology
which formed the true access to the critique of knowledge, it is preferable on
expositional grounds to begin with Schopenhauer's relationship to Kant (and
Kant's epistemology).

In brief: what Schopenhauer tries to bring out in relation to Kant's epistemology in his philosophical work of 1813, and presupposes (three years
later) in his − primarily physiological − writings[597] as a theoretical basis, is
a certain conception of what '*Anschauung*' (perception) is. His theory is that
Anschauung is *ipso facto* 'intellectual' *Anschauung*, that is, "that it is mainly
the work of the *intellect*, which by means of its characteristic form of causality and the pure sensibility, i.e. time and space underlying it, is first to create
and produce this outer world of objects from the pure material of some
sensations in the sense organs".[598] In other words, our everyday empirical
Anschauung is an "intellectual *Anschauung*",[599] and that, in Schopenhauer's
view, is the only true meaning which this much misused philosophical concept can have.

With the thesis of the "intellective character (*Intellektualität*)[600] of
empirical *Anschauung*", however, Schopenhauer not only takes a stand
against philosophers like Schelling, Fichte, etc., who reserved the concept
of intellectual *Anschauung* for a mode of thinking which is a kind of supersensory intuition, but also (and this is more important here) against Kant.
Kant's mistake is that he "either did not see that empirical *Anschauung* is
mediated by the law of causality, of which we are aware, prior to any experience, or wilfully bypassed it because it did not suit his purpose".[601]
Where the relationship between causality and *Anschauung* is discussed, this
is only in the context of things-in-themselves as the causes, transcending

experience, of phenomena; the genesis of *Anschauung* itself (for which something appears) is not discussed. This is connected with the fact that "to Kant, perception (*Wahrnehmung*) is something totally direct, occurring without any assistance from the causal connection and, consequently, from the intellect: he virtually identifies it with sensation".[602]

In other words, Schopenhauer accuses Kant of failing to distinguish between (subjective, immediate) sensation ('*Empfindung*') that is to say, "a process in the organism itself, but as such restricted to the region beneath the skin [and] therefore by itself not capable of containing anything beyond it, i.e. outside us"[603] and objective perception ('empirical *Anschauung*'). Alternatively, to formulate it in yet another way, Schopenhauer reproaches Kant for his blindness to the fact that the category of causality is already operative on the level of what Kant conceives, without distinction, as sensation/perception, and what Schopenhauer calls 'empirical *Anschauung*'. Precisely because Kant works on the assumption that the law of causality only exists and is possible on the level of thought ("in reflection, i.e. in an abstract, conceptually clear understanding") and has absolutely no idea that its application already takes place on the *pre-reflective* level, he fails to explain the genesis of empirical *Anschauung*: "as far as he is concerned it is a miracle, simply a matter of the senses, thus coinciding with sensation".[604]

The consequence of Schopenhauer's thesis of the 'intellective character' of *Anschauung* for the theory of knowledge was that it nullified the Kantian dichotomy of active and passive cognitive faculties. That is to say, the distinction which Kant made between the *a priori* forms of sensibility (time and space) on the one hand, and the (*a priori*) categories of the intellect on the other, as a distinction between (the *a priori* forms of) our receptive cognitive faculties and our spontaneous, active cognitive faculties, respectively, was relativised by Schopenhauer's emphasis of the active (thus 'intellectual') element in the exercise of our sensory faculties. Given his formula, perception is, after all, something which implies the activity of the intellect while, on the other hand, the realm of sensibility as being "restricted to the region beneath the skin" is a matter of organic reaction.

That which does not belong to the realm of our subjective, organic sensations, but is operative in the ordering of the 'material' of sensation, i.e. the *a priori* forms of time, space and causality, is *ipso facto* an ordering method of the (active) cognitive faculty of the intellect. The result is that in Schopenhauer's work time and space acquire the significance of categories (forms of intellect). If one adds to this the fact that Schopenhauer reduced Kant's table of twelve categories to the single category of causality, it becomes clear

that Schopenhauer changed Kant's theory of knowledge in a fairly drastic
fashion.

However, it also becomes clear that — and how — this modification of
the Kantian theory was able to define the perspective of Schopenhauer's
physiological approach, as he conceived it in his — physiological — reconstruction of Goethe's theory of colour. The general problem which had first to be
solved — before the seeing, the perception of different colours could be
explained — was how (objective) perception was possible at all. In other
words, the problem which first had to be solved was the question of how
the transition from a state of organic affection, from subjective, 'internal'
sensation to a state of the perception of objective 'things' outside ourselves,
could be conceived. Since objectivity, according to Schopenhauer, could
only come into being in the function of our intellect, more specifically, in
the application of the causality category to the material of our sensations,[605]
the general physiological question at issue could really only be understood
as the question of the (sensory) physiological mechanism through which the
postulated activity of the intellect in perception is assumed to operate.

I shall not go into the details of this physiological—optical 'projection'
mechanism, which enables us to build up an objective (perceptual) world
from subjective sensations.[606] What is important for us is to establish
that, in general, Schopenhauer's physiological interpretation of Kantian
apriorism — (the activity of) the intellect is, as we have seen, a function of
the cerebral nervous system — although implying a fairly arbitrary narrowing
of the apriority concept of the Kantian critique to knowledge,[607] became
important in two ways: on the scientific side as an anticipation of the physiological view which was later to achieve major significance in the monumental
researches of Müller and Helmholtz; on the philosophical side specifically
as a preparation for that figure of physiologically-oriented critique of knowledge which we encounter in the work of F. A. Lange and which we shall
discuss later.

It must, however, be borne in mind, as far as the first point is concerned,
that the sense in which it is permissible to speak of the significance of
Schopenhauer's theory for the sensory physiology of Müller and Helmholtz
is related not so much to the details of physiological theory-formation as
to the creation of the conditions for theory-formation of this kind. To put it
more precisely, Schopenhauer's contribution to the subsequent development
of sensory physiology lay in the fact that he helped to make his (German)
contemporaries familiar with the idea that a *scientific-physiological* approach
to the theory of perception was, after all, meaningful. He had here not only

to overcome a philosophical (Kantian) prejudice but also to demonstrate the viability of a specifically *physiological* theory of perception which was not only reconcilable with the principles of the Kantian critique of knowledge but was also in agreement with the general insights of physiology (which, as we have seen, was mainly French at that time).

While the work of physiologists like Cabanis and Bichat — primarily important because of the inspiration it provided for Schopenhauer's metaphysics —[608] and the Swiss physiologist A. von Haller, whom we have not previously mentioned, and whose 'irritability' physiology gave a scientific foundation to the causality category of the stimulus as specific to phenomena in the realm of organic nature,[609] was of major significance, it was nevertheless primarily the ideas about the perception of colour put forward by Goethe in his *Farbenlehre* (theory of colour) which, although they could not themselves he called 'physiological', provided the impetus for Schopenhauer's physiological theory about the perception of colours. It is in his treatise *Über das Sehn und die Farben* (1816, 1854²) ("On Vision and Colour"), that is, in his confrontation with and appropriation of Goethe's *Farbenlehre*, that Schopenhauer's own — physiologically-oriented — epistemological conception was crystallised,[610] and it is therefore correct, in terms of the history of philosophy, to lay equal emphasis on his relationship to Kant and his relationship to Goethe when putting Schopenhauer's theory of knowledge in context.[611]

In retrospect, Goethe's *Zur Farbenlehre* (1810) and Schopenhauer's *Über das Sehn und die Farben* (1816), despite all the differences between them,[612] display an important similarity which can help to explain why, with the rise and spread of strictly mechanistic scientific experimental research methods in physiology and psychology, as advocated by Helmholtz and his followers, neither was of much lasting significance for the developments in the area of perceptual psychology and physiology which took place after about 1850. Both of them account for the phenomena of colour in terms of a polarity model and thereby reveal themselves as bound to pay tribute to precisely that 'romantic' philosophy of nature which the Helmholtz generation of scientific researchers had *de facto* (and not only in intent) left behind them.[613]

We have thus (at least in part) already answered the question of why Schopenhauer's theory of knowledge (in particular his theory of visual perception) could not be (or at least only to a very limited extent) a determining factor in the rise of the self-conception of the German physiological (medical) 'materialists' of the eighteen-forties. Schopenhauer's physiology, with its vitalistic overtones, could no longer be seen as 'up to date' by the younger generation of progressive physiologists and doctors who made themselves heard in the eighteen-forties and -fifties.

The fact that these young men drew so little on Schopenhauer's work did not, however, rest solely on the fact that the natural scientific literature on which Schopenhauer's work was based was for the most part considered to be out of date. The reason must also be sought in the fact that the explicit *metaphysical* intention of Schopenhauer's philosophical work could no longer count on gaining approval, because of the predominantly anti-metaphysical, 'positivist' leanings of the younger generation of scientists. In this context, we must remember that what Schopenhauer criticised as 'naturalism' corresponded to the position in (natural) scientific thinking current around the middle of the century, of which someone like Griesinger, among others, was representative.

There seems, however, to be one important exception in the history of nineteenth-century medicine to the state of affairs we have described. This was Rokitansky, the founder of pathological anatomy and the spiritual father of the so-called 'second Viennese school of medicine'. Rokitansky, it is said, was an admirer and adherent of Schopenhauer's theory.

To what extent this is indeed the case I shall explain, if only briefly, hereafter. We shall see (1) that on the issue of his self-conception Rokitansky did *not* occupy any exceptional position in respect of the methodological self-conception of the natural science of his time, and (2) that he can be considered as a witness and exponent of an idealistic naturalism which — closer to Kant than to Schopenhauer, but with ties to both — [614] points ahead to Lange.

4.3. ROKITANSKY AS AN EXPONENT OF IDEALISTIC NATURALISM

Together with Rudolf Virchow and Wilhelm Griesinger, Karl Freiherr von Rokitansky must be numbered among the great reformers of nineteenth-century medicine in the German-speaking world. Rokitansky is the oldest of this trio: born in 1804 (Griesinger, 1817; Virchow, 1821) he would seem rather to belong to the generation of J. Müller (b. 1801) and J. L. Schönlein (b. 1793). Virchow was to acknowledge both these men as his teachers.[615] In terms of content, too, there is certainly good reason to emphasise this chronological difference: in the position between old and new which Rokitansky occupies there is a striking similarity with that of Müller. Both preferred morphological-anatomical work, and it can be said of both that they took this method of working to its utmost limits and had an eye to future developments in natural scientific physiology. As Lesky remarks, Rokitansky was *de facto* a macromorphopathologist, but nevertheless "considered as the primary task of pathological anatomy the elevation of pathology to a physiological pathology".[616] We were able to ascertain a parallel state of affairs in our discussion of Müller's relationship to the physics and chemistry school of research, which was developed by his pupils and close colleagues.[617]

If, nevertheless, Rokitansky is primarily remembered in the history of science as the pioneer of the mechanistic view, while in Müller's case his ('underground') link with the fundamental vitalistic attitude is stressed, this difference of accent is determined to a significant degree by the energetic defence of the mechanistic viewpoint which we encounter in the often more philosophical treatises which Rokitansky wrote towards the end of his life,[618] and which has no parallel in Müller's work.

It is specifically these later essays which give an insight into his 'philosophy' and lead historians of medicine to talk about the (considerable) influence of Schopenhauer's *Weltanschauung* on Rokitansky.[619] In this context, special reference is usually made to his lecture *Die Solidarität alles Thierlebens* (1869), delivered at the 'Imperial Academy of Sciences', as the most complete expression of a pessimism which, culminating in the Byronic "sorrow is knowledge", provides documentary evidence of an unmistakable kinship with Schopenhauer's feelings about life. As far as the question of the figure of the self-conception of Rokitansky's medicine (and Schopenhauer's possible influence on it) is concerned, however, the publications of 1859, 1862, and 1867, referred to earlier, and particularly the essay *Der selbständige Werth des Wissens* of 1867, which has been inadequately dealt with in the (in any case very scanty) literature, are of greater importance.

In 1859 Rokitansky found "that medicine, as far as intention and method are concerned, has entered the ranks of the natural sciences" because "in its research it has rid itself of any assumption of or appeal to vital forces, which differ from the known natural forces, and has taken the path of a strictly physical consideration of the processes in the organism". The research principle of the so-called 'physical school' – i.e. "that something unknown cannot be explained by another unknown thing, and that within the phenomenal world only a physical understanding of the processes and their stimuli is possible" – has shown itself to be viable and tenable, and it gains more support every day (through the results of research).[620] It is therefore beyond dispute that only this (natural scientific) method "as a method of research which matches the nature of the phenomenon, [can] lead to an increasingly more profound insight into the organism and into the vital processes", but the belief that this method is the only true method in (natural) scientific research is accompanied by the conviction that there exist specific ('characteristic') differences between inorganic and organic matter and between lower and higher forms of organic life (differences which, according to the terminology used by Rokitansky, make it necessary to speak of the causality which operates on these last two levels – shades of Schopenhauer – in terms of stimulus (and reaction) and motives).[621]

The idea of a physicalistic reductionism is absent here, in fact, it is expressly denied:

indeed, this physical research would lead to explanations of the utmost importance, controlled by an inexorable causality. But it would certainly never solve the mystery of life completely and satisfy man's inborn metaphysical need that goes beyond its limits, even if it were to create a science, exact in all its details.[622]

The philosophical background to this limitation of the possibilities of natural scientific research is explained in *Der selbständige Werth des Wissens* (1867). In this we see how the distinction made by Schopenhauer between the objective and subjective view of the intellect recurs (hardly by coincidence in an exposition of the function of visual perception), namely in the contrasting of the 'physiological' (also called 'realistic') standpoint and the 'idealistic standpoint'.[623]

In this lecture, Rokitansky considers the visual mechanism through which we perceive things as being outside ourselves, and comes to the conclusion "that the observable world around us is essentially a creation of the personality, that it is functions of the organs through which things can form themselves as things outside us, as things of a certain quality, form, of a certain size and power".[624] But that is not all, "in causality [we exercise] a form of activity, inherent in our organs, by means of which we unite the different things and the changes they undergo, one with the other".[625]

This is said from the viewpoint of physiological research, which is concerned with providing clarification of the mechanisms which play a part in the perception of things outside ourselves. This physiological viewpoint has, however, to be augmented by a consideration which interprets the physiological mechanisms in question as (the expression of) the subjective *a priori* form of our observation.[626] This subject-dependency of our observation, Rokitansky seems to want to go on to say, cannot be adequately understood as long as it is conceived solely from the physiological standpoint

(Schopenhauer would say: in the "objective view of the intellect"). However, there are major obstacles which stand in the way of the recognition of the necessity to go outside the physiological perspective (namely in the direction of the so-called 'idealistic' standpoint): "for we always fall back on the realistic, physiological standpoint, according to which things do exist in fact by and for themselves and in addition have to be known".[627] Even Kant was not able to free himself entirely of this realism, the realism of the natural attitude, Husserl would say, but it is reasonable to assume that if Kant

had held and followed his course unswervingly, he would have come to the idealism of Schopenhauer, since he would have eliminated those 'given things' through the insight that it is only by perception (*Anschauung*) that all and everything can be given.[628]

It is the idealism of Schopenhauer, i.e. an "idealism of the most decisive form", purified of every "realistic element", an idealism for which the object is subject-dependent Idea,[629] which Rokitansky acknowledged and which — at least nominally — is put forward as the philosophical position underlying his qualified defence and criticism of materialism and naturalism.

This last point in particular — Rokitansky's position with regard to materialism and naturalism — is of great importance for the exact definition of his relationship to Schopenhauer, and to his contemporary 'colleagues' in natural scientific biology, physiology and medicine.

In brief: there is for Rokitansky a meaning of 'materialism' which he is prepared to defend: a materialism which is regarded as irreconcilable with metaphysical materialism of Vogt, Büchner, and Moleschott, but which, on the contrary, is thought to be reconcilable with idealism (in the Schopenhauerian sense). It is this meaning of 'materialism' he is referring to when, in the introduction to his 1867 lecture, he states that, although he is not a materialist (by which he means in the sense of Büchner and his followers), in the following investigation he "still [adopts] a materialistic, or if you prefer, organicistic, physiological standpoint".[630] 'Materialism' here means belief in natural scientific *method*, and a 'materialist' in this usage, which we find in the work of virtually all the leading natural scientists (in Rokitansky as well as Griesinger, Virchow, Schleiden, etc.), is someone who "pursues materialistic studies",[631] i.e. someone who carries out *natural scientific* research and is consequently bound to natural scientific method.

In all cases, therefore, this 'materialism' is a methodological materialism which remains distinct from the *Weltanschauung*-oriented materialism of Vogt, Büchner, and Moleschott, and which thus dissociated itself, more or less expressly, from the application of natural scientific method outside the realm of the phenomenal world. This Kantian motif certainly also played a part in Rokitansky's thinking and, in precisely the degree to which this was the case, it had to alienate him from Schopenhauer's metaphysical intention. This is borne out by the fact that Schopenhauer's actual metaphysics of the Will, *as metaphysics*, is, as it were, ignored in Rokitansky's considerations. The idealism which he *de facto* defends, despite his declared following of Schopenhauer, is therefore not so much the *metaphysical* idealism of Schopenhauer, for which he mistakenly takes it, but rather *transcendental* idealism in the Kantian sense, which he considers to have been superseded by Schopenhauer.

That this is the case seems to me, without a shadow of doubt, to be implicit in the way in which Rokitansky describes idealism. Idealism, to him, is the philosophical position which says

that things, in addition to being phenomenon and idea (presentation), have to be something else too; that besides the appearing, perceptible being, [i.e.] the relative ideal being which is conditioned by the knowing subject, there necessarily exists another real being – a reality which is something unperceivable, a thing-in-itself, that stands *outside all relation to the knowing subject*. Idealism consequently postulates *something transcendental* (*ein Transzendentes*) which has come to a perceivable expression within us and in the world around us, [that is, something] divorced from all perceptual knowledge"[632] (my italics).

In brief, idealism (according to this description) is nothing more than the philosophical conviction *that* there is a 'metaphysical something', but it refrains from making any pronouncements about *what* that metaphysical something is. This is, as we have observed, transcendental idealism in the Kantian sense, rather than Schopenhauerian metaphysical idealism.

Defined in this way, idealism can be reconciled without more ado to (the demands of) natural scientific research into the "phenomenal world of objects as a material realm, bounded by absolutely immanent laws" but "opposes it (i.e. empirical research), however, *whenever it wants to form itself into a realistic, materialistic view of life*, by showing that it failed to turn to account the data, collected in its own field, from a higher level"[633] (my italics).

Schopenhauer's idealism, Rokitansky had earlier remarked,[634] was only seldom understood correctly, chiefly because of the aversion it aroused, "which has its roots in the immanent realism of everyday life". Generally speaking, however, idealism was accepted sooner than materialism, primarily because "it is not materialism, and because it acknowledges something metaphysical".[635] This recognition that there is 'a metaphysical something', he continues, does not yet, however, confer the right to identify that idealism as a form of spiritualism. It is, rather, the case that this idealism exists *alongside* materialism "because the immaterial substance assigned to matter changes nothing of the real conception of things — and naturalism is either pure pragmatic materialism or it thinks up something to add to matter, an essence which itself must not be matter but which acts according to the scheme of matter. There can only be materialism or idealism".[636]

This passage, which is crucial to our interpretation, can only be understood if we see that the materialism which is explained at the beginning as existing *alongside* idealism, i.e. can be reconciled with it, is not the same as the materialism which is placed in (exclusive) opposition to idealism in the last sentence. The former materialism is, indeed, the materialism which Rokitansky is prepared to endorse, i.e. what I have previously described as methodological materialism. It is the materialism which he describes in the passage quoted above as that variant of naturalism which postulates "something to add to matter ... which acts according to the scheme of matter" (the idealistic variant of naturalism, if you will) and which he distinguishes from a pragmatic variant of naturalism ('materialism') in which one can recognise without difficulty the figure of the 'positivistic' self-conception (Helmholtz, Du Bois-Reymond, etc.). The latter materialism, in contrast, is the materialism which he abhors from the bottom of his heart, and rejects[637] because it does not acknowledge a '*meta*-physical something'.

This has certainly not exhausted the theme of 'Rokitansky and Schopenhauer'. It is, however, possible, summarising, to formulate a (provisional) conclusion. Rokitansky's relationship to Schopenhauer is more complex than is generally supposed (at least in his *Der selbständige Werth des Wissens*). On the one hand, it is much closer than has been realised until now (particularly in the justification of the physiological view of visual perception), on the other hand, it is not as close as he himself would have us believe. What links him with Schopenhauer is the *starting-point* of the physiological view (Schopenhauer's 'objective view of the intellect'); what however separates him from Schopenhauer is that, whereas Schopenhauer consciously oversteps the bounds of the physiological perspective and embraces idealistic *metaphysics*, Rokitansky, more cautious, does not *de facto* go much further than a *transcendental* idealism.

This accords with the fact that, whereas to the metaphysician (which is what Schopenhauer was), *every* form of materialism (and therefore that 'halfway materialism' which he criticised as '(pure) naturalism') was unacceptable, Rokitansky defended the figure of methodological materialism as being compatible with the insights of transcendental idealism and with opting for an idealistic *Weltanschauung*.

If, following Rokitansky's suggestion, we distinguish this 'materialism' as an idealistic variant of naturalism from (by definition non-idealistic) pragmatic naturalism, we can say that what Rokitansky owes to Schopenhauer's philosophy in the formation and articulation of his (natural scientific) self-conception amounts to not much more than the fact that it provided the conceptual means for conceiving what Lotze, in his way, had tried to make clear: the compatibility of the demands of the natural scientific thinking of his time with the metaphysical needs of the heart — methodological materialism or an idealistic variant of naturalism.

As far as I know, Rokitansky was the only noted (natural) scientist of the eighteen-forties who drew his inspiration on this point from Schopenhauer. For those who sympathised with the idealistic interpretation of natural scientific naturalism, Lotze, the younger man, had found the convincing formula, while those scientists for whom Lotze's metaphysics was too much of a good thing were more inclined to seek support in the orientation around Kantian criticism in one variant or another (Schleiden, Virchow),[638] or defended a sort of pragmatic naturalism which was linked with more or less pronounced forms of (metaphysical) agnosticism (Helmholtz, Du Bois-Reymond).

Considering all this, it is indeed difficult to come to any conclusion other than that, taken as a whole, Schopenhauer's significance for the formation of the self-conception of natural scientific physiology and medicine in the eighteen-forties was extremely minor. When it comes to the point, even Rokitansky was too much a product of his time (which is to say, a defender of methodological materialism) for it to be possible to describe him as an adherent of Schopenhauer's metaphysical idealism in more than the sense of his *Weltanschauung*.

The two variants of (natural scientific) naturalism mentioned by Rokitansky, the idealistic and the pragmatic variants, describe the two basic figures which characterised the self-conception of the so-called 'materialists' from the eighteen-forties onwards. The idealistic motive (the concern with 'saving' that which transcended science, the *meta*-physical) and the pragmatic (agnostic) motive (the concern with a non-metaphysical justification of the natural scientific undertaking), which for a long time remained in a relationship of unresolved tension, were ultimately to be reconciled in the thinking of F. A. Lange. In philosophical terms, Lange's work can thus with justice be regarded as the leading justification of the methodological materialistic self-conception of the generation of natural scientific researchers who, after the collapse of the German idealism of Fichte, Hegel, and Schelling, and before the triumph of Kantian reflection in the neo-Kantianism of Cohen and Natorp and the rise of Haeckel's naturalism inspired by Darwin's theory of evolution, sought the emancipation of the preceding philosophy of nature and the formation of the natural scientific self-conception, in particular in the life sciences, psychology, and medicine.

What makes Rokitansky of interest to us in this context is that — de facto closer to Kant than Schopenhauer as far as his self-conception is concerned and, as it were, the historical hyphen which links Schopenhauer and Lange — he points ahead in his thinking to that synthesis of methodological materialism and idealism which was to be realised in Lange's philosophy.

4.4. F. A. LANGE (1828–75), PHILOSOPHER OF METHODOLOGICAL MATERIALISM

In a letter dated 1858 Lange wrote:

A critique of psychology, which would prove the major part of this 'science' to be idle chatter and self-deception, and which would, in essence, follow Kant's Critique of Pure Reason as a second big step [forward], that is the book I would most like to write.[639]

Lange's wish was not to be fulfilled: the book was never written. Nevertheless, all the viewpoints which are essential to such a critique of psychology can be found in the second part of his Geschichte des Materialismus (1875) in the chapters "Brain and Soul", "Natural Scientific Psychology" and "The Physiology of the Sense-organs and the World as Idea (Vorstellung)".[640]

What makes Lange of particular significance for our (historical) study is that he found the philosophical formula for the self-conception of the majority of the natural science oriented scientists in Germany in the third quarter of the nineteenth century. This is the period in philosophy during which efforts were made, in a reaction against uncritical materialism and the equally uncritical theologising idealism (more properly, spiritualism) which had been opposed to each other, particularly in the materialism conflict (Wagner versus Vogt, etc.) which we discussed earlier, to achieve an epistemological reflection which was linked to a renewed interest in and an intensive occupation with the Kantian critique of knowledge. We have seen that even where the concern for an empirical-scientific basis predominated in epistemology, as in Helmholtz's case, Kant's philosophy was a source of inspiration.

The period we are now discussing was, however, also important for a different reason: it was the period that also witnessed the rise of the so-called classical psychology of the consciousness. We see in it how the (metaphysical) "fight about the soul" found its natural sequence in the defence of an autonomous *scientific* psychology (i.e. a psychology free from metaphysics, and distinct from psychophysics).

Epistemological reflection and the pursuit of an independent (empirical) scientific psychology are very closely connected, and it is therefore no coincidence that Lange, whose thinking so clearly interpreted the reigning self-conception of the (natural) science of his time, has earned a place in the annals of both nineteenth-century philosophy *and* psychology. Viewed in this light, it is certainly remarkable that, whereas Lange from the outset acquired a permanent – if modest – place in the historiography of nineteenth-century German philosophy,[641] the recognition of his role in the history of psychology did not come until more than ninety years after his death. The credit for repairing this omission in the historiography of nineteenth-century psychology belongs to L. J. Pongratz. He showed that Lange was one of the fathers of modern psychology who, in his theses, had "already anticipated the essential points in the programme of objective psychology".[642]

4.4.1. *Lange's Relationship to Psychology*

A description of Lange's position in psychology means, in the first place, a description of his relationship with Kant's views on the possibility (or impossibility) of (natural) scientific psychology, and in the second place a definition of his position with regard to the psychology of Herbart and the so-called Herbartians.

Kant, we have established,[643] occupies an exceptional position in the history of scientific psychology. On the one hand, with the postulate of the determinedness of all phenomena, and thus of psychic phenomena, his work laid the foundations for a scientific psychology in accordance with the mechanistic ideal of science, that is to say, for the later, so-called 'natural scientific psychology'. On the other hand, he declared that psychology as a (natural) science was impossible because, he said, psychic phenomena are given one-dimensionally (in time as a succession) and can therefore not be determined mathematically.

Herbart, however, as we have seen, conceived in his statics and dynamics of ideas a psychology in which precisely that application of mathematics which Kant said was impossible in psychology took place on a grand scale, thereby starting a development in psychology which, flying, as it were, in the face of Kant, demonstrated that the mechanistic conception of science was viable in this field of research too.

The abstract character of Herbart's psychology could not be satisfactory in the long run. Lotze's medical psychology meant a shift of emphasis in the psychological approach in the direction of the problem of psychophysics, through which the path to an empirical, namely physiological foundation was opened up. However, Lotze (in view of the limited development of the empirical physiology of his time) also could not travel far along this road, so that looking back, and from the viewpoint of empirical-scientific psychology, the distance between Herbart and Lotze does not seem to be great. Both appear as representatives of a practice of psychology which is still too speculative to pass for 'real' empirical science.

This, combined with the fact that it was Herbart, and not Lotze, who created a school of thought in Germany with his psychology — I am thinking here of 'Herbartians' like Drobisch, Waitz, Lazarus, Steinthal, etc. —[644] make it understandable that Lange, interested as he was in (the defence of) natural scientific (i.e. physiologically-based) psychology, 'purified' of all metaphysics, should in the first instance direct his attack against that conception of psychology which was defended by the Herbartian school — so-called 'mathematical psychology'.[645]

Herbart himself, as the author of this mathematical psychology, is not only offensive to Lange, he is also a source of amazement:

It is a remarkable monument to the philosophical turmoil in Germany that so subtle a thinker as Herbart, a man of admirable critical acuteness and great mathematical skill, could have hit upon such a bizarre idea as that of *finding by speculation the principle of the statics and mechanics of ideas*. It is still more striking that so enlightened a mind, with a genuinely philosophical approach to practical life, could lose itself in the laborious and thankless task of working out a whole system of mental statics and mechanics from this principle, without having any empirical guarantee of its accuracy.[646]

Like Lotze before him,[647] Lange takes exception to Herbart's definition of the absolute, singular, immutable 'reals' which, when it comes to the point, nevertheless interact with one another: reals which, although considered in themselves as devoid of ideas, nevertheless react to one another with an act of self-preservation (that is, an idea). The efforts of the Herbartian school to establish a natural scientific psychology were thereby burdened from the outset with a highly problematical, metaphysical mortgage, and attempts like those made by Waitz — whom Dilthey later described, probably not without

justice, as the first advocate of an 'explanatory psychology'[648] — to make
this legacy acceptable by reinterpretation must be regarded as having failed,
and bear witness to the disastrous influence of metaphysical prejudices. Thus
Waitz tried to salvage Herbart's disputed fundamental thesis by making a
distinction between *dispositions* to a state and actual *states* (of the soul). The
soul can only be disposed towards states, because the assumption of (actual)
states of the soul would cause the absolute unity of the soul to be lost. One
must, however, be a metaphysician to close one's eyes to the fact that a
disposition towards a state is also a state, that self-preservation in the face
of the *threat* of an influence is impossible without an *actual* influence,
however slight.

The metaphysician does not see this. His dialectic has brought him to the edge of the
precipice; he has brought out, turned over and rejected every idea a hundred times,
until at last he reaches the point where he absolutely *has* to know something. So he shuts
his eyes and boldly takes the plunge — from the heights of the keenest criticism into the
most common confusion of word and concept.[649]

Even the relativisation contained in Waitz's interpretation of mathematical
psychology as a *hypothesis* about the nature of the soul is by no means
enough for Lange. A hypothesis like this, or even a hypothesis about the
existence of the soul, can be of no use "as long as we still have so little
accurate knowledge of the *particular phenomena* which are the first things
to be considered in any exact investigation".[650]

The Kantian inspiration of this criticism is clearly recognisable. The
object of (scientific) psychology can only consist of the *phenomena* of the
soul, and because this is so one should not hesitate (according to Lange) to
assume a *psychology without soul*.[651]

The slogan of a psychology without the soul symbolises a turning-point
in the history of psychology's conception of itself. What this means can only
be clarified by looking more closely at the position of Lange's thinking,
especially as far as his ideas about psychology are concerned.

In general, Lange's position is defined by criticism on two fronts: on the
one hand, against speculative philosophy (or, more generally, metaphysics)
and its pretension that "speculative knowledge is higher and more credible
than empirical knowledge, to which it is related simply as a higher to a lower
stage"[652] and, on the other, against the excesses of unbridled, introspective
psychological practice.

For the moment the first point is of minor importance. In order to prevent
misunderstandings I should simply like to point out that Lange's criticism of

everything which bears any resemblance to speculation must be understood as the result of his 'critical' starting-point. He is by no means concerned with a total rejection of philosophical reflection, but with a denial of those forms of it which, in their pretensions to knowledge, overstep the limits drawn by Kant's critical philosophy. There remains, therefore, room for philosophical reflection, albeit only in the sense of critical philosophy.[653]

In contrast, the second point — the criticism of introspective psychology — is important in this context, because this criticism forms the prelude to Lange's development of the concept of psychological method and the definition of the concept of ('real') scientific psychology which it entails.

Introspective psychology should be understood to mean a form of empirical psychology which bases its results on self-observation. The example of such a form of psychology which Lange quotes as a warning to us is Fortlage's *System der Psychologie als empirischer Wissenschaft aus der Beobachtung des inneren Sinns* (1855), in Lange's view eloquent testimony of a disastrous development in psychology for which Kant, paradoxically, had laid the foundations by allotting a separate area to the 'inner sense' in his *Critique of Pure Reason*. Paradoxical indeed, for it was none other than Kant who had been extremely critical of — and in fact had rejected — the introspective approach because it was constantly exposed to the danger of the "projection of arbitrary experiences in the ostensible field of observation of the inner sense" (Lange) and had therefore primarily based his own empirical psychology *de facto* on external observation.

Anticipating the following discussion, one could state that Lange's contribution was that he explained the *methodological* meaning of Kant's orientation around observation from outside (*Fremdbeobachtung*); in other words, in formulating a psychological method he worked out (in greater detail) what was contained as a suggestion in Kant's work.

The problem of psychology is the same on this point as that which confronts the other sciences, namely the problem of how the influence of the subjectivity of the researcher can be neutralised in an investigation or, to formulate it positively, the problem of objectivity. The question is thus whether a method can be found for psychology which performs the same services for psychology as the rules of natural scientific method perform for the natural sciences and, if so, what this method is.

The answer Lange gives us is contained in the formulation of what he calls the *somatic method*. (This method might also be called "materialistic", he remarks, "were it not that this term also includes a reference to the basis of the world-view [of materialism] in its entirety which is totally out of place here".) The method requires "that in psychological research we should as far as possible keep to the physical processes which are indissolubly and by law connected with the psychical phenomena".[654]

There is one interpretation of this methodological rule which is immediately obvious because it is supported by two (earlier) formulations which represent the highest maxim of this method as the requirement "to indicate for each mental activity corresponding processes in matter . . . [to indicate] the physical mechanism of sensation and of thought".[655] In this interpretation the somatic method is the method of a form of psychology which is essentially (natural-scientific) psychology based on (cerebral and sensory) physiology. More precisely, it is psychology such as that which was, and is, practised in the context of the *psychophysical* approach. Lotze, Fechner and Helmholtz had led the way in this field and had − each in his own way − given substance to the idea of a natural scientific psychology *within* the perspective of the psychophysical approach.

The formulation and explanation of the concept of the somatic method which Lange gives in his last work, *Geschichte des Materialismus*, Volume II (1875), suggest, however, that the restriction to the psychophysical perspective is not essential. We see that Lange conceives a natural scientific psychology which, transcending the psychophysical perspective, broadens the psychological approach to include the "investigation of human action and speech, and generally of all manifestations of life, in so far as an inference can possibly be drawn from them as to the nature and character of man",[656] and argues that psychological research should also encompass animal psychology, the systematic study of newborn babies, and linguistics (as part of ethnological psychology).[657]

The fact that this shift in the object of psychology is coupled with an almost imperceptible change in the meaning of the word 'somatic' (or 'physical') and a shift of emphasis in the tenor of the somatic method which, incidentally, will prove to be of far-reaching significance, can be deduced from the comment which Lange makes in the introduction to his concept of the somatic method. He emphasises here that the methodological requirement he has formulated does not prejudice in any way at all the question of whether the psychic phenomena studied can or cannot be reduced to material occurrences. But, he says, the alternative (whether or not they can be reduced to material occurrences) is false, "because empirically ascertained facts, and even 'empirical laws', have their own rights, quite independently of their reduction to the ultimate grounds of phenomena".[658] Psychological research, experiment and theory-formation are perfectly possible without involving our brains or nerves in our considerations. On the other hand, it would also be nonsensical to declare the whole of neurophysiology to be inadequate "because it has not yet been reduced to the level of the mechanics of atoms, which must, in the final analysis, lie at the root of the explanation of all natural phenomena".[659] In other words, the application of the somatic method to psychology in no way necessarily implies a commitment to materialism (psychophysical materialism) or to the idea that the mental occurrence does not have a physiological basis. To put it another way, the somatic method, in *this* interpretation, is in itself neutral as far as the reducibility of mental phenomena is concerned. The change in the tenor of the somatic method from the first interpretation to the second interpretation is, in short, a shift of accent from the requirement of reducibility to the requirement of 'visibility'. That is to say, Lange is concerned with formulating a method which helps to realise something like objective psychological *observation*. The important question for the psychologist is "whether an observation is such that it can also be made by others at the same time or later, or whether it eludes any such supervision and confirmation".[660] Psychology must be based on observation which is capable of being verified intersubjectively.

It will be clear that in this explanation of the 'somatic method' the word 'somatic' no longer refers to (neuro)physiological or other mechanisms which (*ex hypothesi*) lie at the root of mental phenomena, but now in principle relates to the total range of expressions of life in which the mental life of man becomes visible in a way accessible to, and capable of being verified by, any observer. The term 'somatic' (as I interpret it) thus refers in this case to that class of phenomena which are interpreted as *objectivisations* of human life, and anyone who, as a psychologist, commits himself to the application of the 'somatic method' proceeds on the assumption that psychological investigation must begin where human life (the life of the soul) has become 'objective', i.e. 'somatic' in Lange's terms.[661]

An argument in favour of this interpretation of 'somatic' can also be found, in my view, in the greater intelligibility it gives to Lange's criticism of introspective psychology (i.e. that based on 'self-observation') with which he prepares the way for his concept of method. What Lange most dislikes about this introspective psychology is essentially its *dualistic* assumption which, in the analysis of mental phenomena such as sensation, perception and thought, traps one into making a radical separation of inside and outside, inner and outer, form and matter. Just as the aesthetic value (significance) of a drawing can (in thinking) be *distinguished* from the lines of the drawing, but the two cannot be *separated* from each other (in the perception of the

drawing), so form and matter can be conceived *in abstracto* as being separate
from each other in the mental phenomena of perception, thought, ideas,
etc., but are not separate in the phenomena themselves.[662]

The comparison with (the perception of) a drawing contains a suggestion
about what mental phenomena are, structurally speaking. They are in principle phenomena with a structure of expression, i.e. phenomena which as
expression point to a subjectivity which objectivises itself in it, acquires
'body', becomes 'somatic'. The implication of this for the definition of
psychological method is evident. If mental phenomena are *ipso facto* 'somatic'
phenomena (in the sense indicated above), then the method appropriate to
psychology is that method which dictates that we limit ourselves in psychological investigation to 'somatic' phenomena. Purely introspective psychology
is then of course unacceptable because it erroneously proceeds on the assumption that mental phenomena exist outside their objectivisations. In the
final analysis it therefore builds on quicksand (thus we may interpret Lange),
despite all the subtlety it displays in the refinement of its ideas, because
instead of starting from the concrete manifestations of our mental life, it
proceeds on the basis of a number of 'bodiless' abstractions.

For the psychologist who bases his research on observation, the question
of whether his observation must be internal or external is thus irrelevant
– this is an abstract alternative; the only question is (I repeat) "whether
the observation is such that it can also be made by others at the same time or
later, or whether it eludes any such supervision and confirmation". Psychological observation which is of any use to us is – and how can it be
otherwise – observation of the first kind and thus, of course, concerned with
'somatically' given mental phenomena; and what Lange recommends as his
'somatic method' is, according to this interpretation, really nothing but the
expression of the methodological intention in psychological research to admit
only *this* kind (in fact the only 'real' kind) of psychological observation.

The shift of emphasis of the somatic method from the factor of reducibility
to that of observability reflects a development in psychology itself which,
stepping outside the limits of psychophysical investigation, ventured into
areas of research (I am thinking particularly of the ethnological psychology
brought into being by Lazarus and Steinthal) where the idea of a possible
reduction to (neuro)physiological mechanisms had to appear as premature,
not to say illusory or pointless. Empirical psychology proves to be possible
even if it cannot meet the requirement of reduction to physiology.

If, in conclusion, we ask what place Lange's work must be allotted in
the history of the self-conception of psychology, I should like first of all

to point out that the slogan of a *psychology without the soul* with which Lange, undaunted, made his stand against the traditionally-hallowed authority of metaphysics in psychology, represents a milestone in this history. It made it clear, in an unequivocal fashion, that the ties which had of old bound psychology to metaphysics (as was still the case with Schopenhauer, Herbart and Lotze) and which had been seriously eroded by positivistic and materialistic criticism from the middle of the century onwards, were now irrevocably severed. Since philosophy (metaphysics) had had to give way as the authority which determined the self-conception of the sciences to the methodology of natural scientific thinking, the even more recent natural scientific psychology, which had made its voice heard during the last quarter of the nineteenth century, was also bound to attempt to anchor its self-conception in the concept of natural scientific method.

The consequences of this were diverse. Where psychology's orientation towards the natural scientific example ultimately itself again served the metaphysical intention, as in the case of the inductive metaphysics of Lotze and Fechner, there was no essential opposition between natural science (natural scientific psychology) and metaphysics. It was a different matter in the case of the (naive) materialists, who believed that they could, and must, put natural science *in the place of* metaphysics and thus themselves 'backslid' into metaphysics, and different again in the case of the positivists, who sought to defend a metaphysics-free natural science *in opposition to* metaphysics, without being able to prove convincingly that the sought-after emancipation from metaphysics had already succeeded.

When Lange, in accordance with the spirit of his age, conceives the problem of psychology as the problem of its method and in this connection develops his concept of the somatic or materialistic method as the method of a psychology which also has applications outside the narrower realm of psychophysical research, in the strict sense,[663] in the study of all expressions of human life, this implies a dissociation or, if you will, emancipation from the 'nothing but' reductionism of the psychophysical materialism of the psychophysicists[664] ("mental phenomena *are* nothing but physical occurrences"). This does not, however, alter the fact that Lange remained true to his natural scientific concept of psychology in the conviction that mental phenomena, if they were not *reducible* to the phsyical, were nevertheless *based on* neurophysiological (or physical) mechanisms. Because of this he was also able to find points of contact with Wundt's physiological psychology (on a psychophysical *parallelistic* basis).[665]

From the point of view of the establishment of (natural scientific) psychology as an *independent*, experimental scientific discipline, Lange's concept of the 'somatic' method can therefore be understood as the expression of a twofold development: on the one hand as the index of an emancipation from the metaphysical position which is abandoned (psychophysical materialism), and on the other as evidence of psychology's detachment from physiology.[666]

4.4.2. *Lange's Philosophical Position*

We have already briefly touched on the fact that Lange, in his rejection of speculative philosophy or, more generally, metaphysics, upheld a 'critical' standpoint. In his case, there is no question of a point-blank positivistic or materialistic denial of the meaning and possibility of philosophical knowledge, but of a tempering of the claims to knowledge to the level of a philosophical reflection on the possibilities and limits of our knowledge, in the spirit of the Kantian critique of knowledge.

We shall now go on to discuss how Lange's conception of the somatic or materialistic method in psychology is linked to what is sometimes called his *methodological materialism*, as well as the connection between that methodological materialism and his 'standpoint of the ideal' on the one hand, and his interpretation of Kant's epistemology on the other.

4.4.2.1. *The 'Standpoint of the Ideal'*.

The problem which underlies Lange's *Geschichte des Materialismus* is in fact the problem of the reconciliation of religion and natural science. We have seen how, in the eighteen-fifties, the so-called materialism conflict broke out, in which the supporters of natural scientific materialism on the one side, and the representatives of theologising 'idealism', or more accurately spiritualism,[667] on the other, were locked in bitter conflict. To the more critical minds among the (natural) scientists and philosophers of the Helmholtz—Lange generation it was clear that an epistemological reflection was necessary here, if one really took seriously the 'scientific-ness' of science and philosophy. In the recognition of this need, the endeavour to 'return to Kant' took root.

Lange, too, was involved in this movement, but it must be borne in mind that in his case the return to Kant's philosophy served the endeavour to formulate a philosophical standpoint from which it would be possible, in a manner of speaking, to save (the meaning of) religion, without detracting from the viewpoint of natural science. We are concerned here with what Lange calls the 'standpoint of the ideal' and the way in which, in his view, the conflict between materialism and religion (theology) should be brought to a 'peaceful conclusion' — described by him as "the elevation of religion to the realm of the ideal".[668]

The reconciliation he is aiming at implies, however, not only the 'elevation of religion', but at the same time that reinterpretation of the 'materialistic' position which we have previously called methodological materialism. Because both issues — Lange's relationship to materialism and to religion — are very closely connected in what he calls the 'standpoint of the ideal', it is advisable to introduce the discussion of Lange's attitude to materialism with some observations about his philosophical standpoint.

This philosophical standpoint is — most characteristically — not embodied in a systematic philosophical work, but is contained in a historical-philosophical work, his *Geschichte des Materialismus*, which we have mentioned earlier. This is not pure coincidence: one could argue that Lange's rendering of the history of materialism as the history of the defence and criticism of materialism from its origins in early Greek philosophy up to his own time is one long, sustained *persuasive argument* for the 'standpoint of the ideal', that is to say, for the standpoint of transcendental, critical idealism in the interpretation given by Lange, which we shall go on to explain.[669] Lange's critical historiography should therefore, in my view, be regarded as the way in which he *de facto* philosophises, the way in which he gives substance to his philosophical mission,[670] the ultimate sense of this way of philosophising being the realisation of an ideal of humanity, as Lange himself conceived it in his 'standpoint of the ideal'.

Underlying Lange's philosophical conception is the time-honoured — metaphysical — figure of thought of the two worlds, which he distinguishes as 'the world of reality' and 'the world of the ideal'. Through our senses and our intellect we have access to (the world of) reality, through our emotional life (feelings) we have a relationship to (the world of) the ideal. Measured against the yardstick which is valid for the 'world of reality', the 'world of the ideal' is unreal but, conversely, it is also true that without the 'world of the ideal' our lives would be considerably impoverished in *spiritual* terms. "We are not created simply to know, but also to write poetry and to build."[671] Science is not enough to sustain and nourish our spiritual lives: man is not only senses and intellect, he is also feelings, and for this reason he does not only want to know, he also wants to be spiritually enriched, i.e. to elevate himself above the 'world of reality' to the 'world of the ideal' which he himself, as it were, brings about through great cultural creations in the fields of religion, art and philosophy (besides science).[672]

This 'ideal world', created by man himself, is in fact nothing but the highest manifestation of the same function of creative, 'poetising' synthesis which is already active in the ordering of the material of sensation and thence leads to the causally ordered world of things, that is to say, to the constituting of the 'facts of science', from where it climbs to the constituting of science itself.[673] The difference between the two forms of this one synthesising activity is that whereas "[the creation of the ideal] deals quite freely with the materials . . . synthesis in the province of science has only the freedom of its origin in the creative [*dichtende*] mind of man". The latter form of synthesis is, namely, tied to the task "of establishing the utmost possible harmony among the necessary factors of knowledge, which are independent of our will".[674]

From the lowest to the highest forms of this synthesis, that is, from the level of the most elementary cognitive organisation, on which the individual still seems entirely bound to the 'basis of the species', to the highest level of synthesis in the 'creative exercise of poetry', the act of synthesis is directed towards establishing unity, harmony, perfect form.[675]

Where, as here, the poetic imagination represents the highest level of human creativity, it can come as no surprise that it is not a philosopher, but a poet (specifically, not Kant, but Schiller)[676] who is Lange's shining example. To Lange, Schiller's significance lies in the fact that he makes the intelligible world of Kant's philosophy, which, according to Kant himself, can only be conceived and not observed, *visible* in his poetry – thereby following in the footsteps of Plato, "who, in contradiction to his own dialectic, produced his highest creations when he made the supersensory become sensory in his myths".[677]

It was Schiller, as the 'poet of freedom', who openly dared to transfer freedom into the 'realm of dreams' and the 'realm of shadows', "for in his hands dreams and shadows were raised to the ideal. What was wavering became a fixed pole, what was fleeting became a godlike form, the play of caprice became an everlasting law, when he confronted life with the ideal. Whatever good is contained in religion and morality cannot be more purely or forcibly expressed than it is in that immortal hymn, which closes with the ascension of the tortured son of God. Here we find embodied the flight from the limits of the senses into the intelligible world. We follow the god who 'flaming, parts from man', and now the roles of dream and truth are reversed – the heavy *dream-picture* of *life* sinks, and sinks, and sinks."[678]

Thus, at last, it is Schiller who points the way to the only correct interpretation of the intelligible world – understood by Plato as the world of ideas and by Kant as the noumenal world — or what Lange calls the 'world of the ideal'.[679]

The significance of the world of the ideal, conceived as the 'world of poetry' (in a wide sense of the word "poetry"), may not therefore be measured against the yardstick of (theoretical) truth — this yardstick is only valid in respect of our knowledge of the 'world of reality'. The truth pertaining to the world of the ideal is poetic, 'pictorial' ('symbolic') truth, i.e. a truth which makes itself true by and in the (aesthetic) educational effect which it has on us; in other words, by the inspiring effect which it exerts on us and which motivates us — by its guidance — to seek the 'elevation to the ideal'. This is not what is true in the scientific sense, but "only what endures when measured by the standards of poetic purity and greatness can claim . . . to serve as *instruction in the ideal*".[680] To serve this goal — 'instruction in the ideal' — is the ultimate purpose of all poetry, and the 'truth' of all poetry is measured by the degree to which it contributes to this goal.

The full consequences of Lange's reinterpretation of Kant's intelligible world only become clear, however, when we realise that, to Lange, the word 'poetry' embraces not only poetry itself, but also all creative work in the fields of art, religion and metaphysics. From the viewpoint of their function of bringing man to spiritual elevation, poetry, metaphysics, art, religion are simply different forms (Cassirer would call them 'symbolic' forms) in which the function of 'poetising' synthesis achieves expression.

The formal coordination of what I shall call, for the sake of convenience, 'symbolic forms', which results from relating them to the single goal of 'elevation to the ideal' or — as is already implicit in this — the conception of metaphysics, art, religion, etc. *as* forms of 'poetry', *as* ways of manifesting the one 'poetising' function, has as its corollary a far-reaching abstraction from the differences in content. This accords with the fact that in Lange's work the emphasis is placed on the formal aspect, a point which is particularly striking, for example, in his explanation of what he sees as the 'core' of all religion, and one which makes clear the way in which he understands the salvation of religion in the face of all (natural) scientific materialistic criticism. The heart of religion he believes, lies in "overcoming all fanaticism and superstition by conscious elevation above reality and by the definitive renunciation of the falsification of reality through myth, which can, of course, render no service to knowledge".[681]

If one takes this definition seriously, there need exist no fundamental difference between the most refined forms of religious imagination and the "gullibility of the uneducated masses", and Schiller's generalisation of the

Christian doctrine of redemption as the idea of aesthetic redemption in his 'realm of shadows' can thus appear to us to be a high point of religious thought: it is, after all, essentially the same 'physic processes' [!] which are called up here by the various products of the human imagination.[682]

In the face of the standpoint of dogma and orthodoxy, it must be borne in mind that the core of religion is not concealed in one dogmatic 'truth' or another, but is a matter of our emotional life. It is, after all, no coincidence that 'truly devout minds' have always attached exceptional value to inward experience as a proof of faith.

Many of these believers . . . know very well in theory that the same emotional processes [!] are also found to have the same result and the same force of conviction in connection with entirely different articles of faith — indeed, among the adherents of entirely foreign religions. They do not as a rule recognise the opposition to these or the equivocal nature of evidence which supports contradictory ideas equally well, since it is rather the common oppositon of all belief against disbelief which arouses their feelings. Does this not make it clear that the essence of the matter lies *in the form of the spiritual process, and not in the logical and historical content* of any particular view of doctrine?[683] (My italics.)

Religion can survive and has a future, even in this 'enlightened' age, so long as there are people who are susceptible to the rousing influence of the religious 'poetic muse', so long as there are people who, attempting to live up to the standard of true humanity, endeavour to go beyond the 'world of reality' and raise themselves to the 'world of the ideal'. This, in short, is the conception which lies behind Lange's attempt to save religion and to reconcile religion and natural science.

If we try to draw up a provisional balance-sheet of the implications and consequences of Lange's rescue attempt (in so far as we have discussed it above), we must bring out the following points. The 'elevation of religion to the realm of the ideal' for which Lange pleads, implies an interpretation of the meaning of religious 'truth' as a mode of 'poetic' truth, namely as that mode which is (or has become) active in the medium of the products of religious 'poetic thinking'. On this point there is no *essential* difference between religion and other forms of 'poetising', creative synthesis, such as (the 'conceptual poetry' of) metaphysical speculation and art. Differentiation between religion, metaphysics, art, and poetry, according to the criterion of *content* has become entirely subordinate, indeed, even *within* the category of, for instance, religion, a differentiation according to differences in content (religious truths) can no longer play a significant part. Schiller's poem about the ascension of Hercules is classified without hesitation as religious,[684] and the differences which exist in this respect between different religions is considered as being of less importance than that which binds believers together against unbelievers.[685] Thus, we can clearly see that Lange's interpretation of religion, metaphysics, art, etc., expresses a pronouncedly *formalising* view. Viewed according to their objective aspects, religion, metaphysics, art, etc. are *forms* of 'poetic art', according to their subjective aspects they are *forms* of spiritual life.

This formalism (which some may criticise as the levelling-down of differences in content, while others prefer to see it as the anticipation or foundation of psychological typologies in the style of Spranger's *Lebensformen* or Jaspers' *Psychologie der Weltanschauungen*) is rooted in the fact that the philosophical problem, as Lange saw it, essentially consisted of the question of the fundamental purpose and possibility of human life as *spiritual* life. The alternative here was not between one definition of the content of spiritual life or another, but between a position which — as in (metaphysical) materialism — excluded the possibility and meaning of (aspirations to the) 'elevation to the ideal', because it limited itself to and absolutised the 'world of reality', and a position — Lange's idealism — which, on the basis of the argument that the *form of spiritual life* was given with human nature, implied the fundamental possibility and meaningfulness of the elevation to the ideal.

Another point that must be made in this context is that Lange's interpretation of religion as a form of poetic thought (*Dichtung*) implies a rejection of a strictly agnostic standpoint with respect to religion.[686] Religious imagination has an innate, poetic, 'pictorial' or symbolic truth, which proves itself through the elevating effect it exercises. It is, however, clear that the élan to which this 'poetic thought' can arouse us is linked to the 'pictorial' presentation in which the ideal becomes 'visible' to us. This form of knowableness is the condition on which religious imagination can become active in us; in other words, the appellative, evocative function of religious 'poetic thought' cannot be divorced from its cognitive-theoretical significance.

Thus, we see that in Lange religion is 'saved' from the agnosticism which we encounter in his sceptical materialist contemporaries, albeit at the price of a relativisation of the essential differences between religious, metaphysical, aesthetic and ethical content and values, as they are revealed to us in the highest creative works of 'poetising' synthesis.

The defence of the 'standpoint of the ideal' implies, however, not only the 'salvation' of religion in the sense referred to above, but also a reinterpretation of the materialistic position. Without this reinterpretation there can be no question of a reconciliation of religion and science, such as Lange had in mind; just as, conversely, the ultimate purpose of Lange's attitude towards materialism remains obscure as long as one does not associate it with his interpretation of religion or understand both facets — the interpretation of religion and that of materialism — as the two essential steps which must lead to the sought-after synthesis.

4.4.2.2. *Methodological Materialism.* The central problem, which the materialistic undertaking in the course of the history of our thinking has come to grief over, is the problem of explaining the consciousness from material occurrences. However convincingly the dependence of the consciousness on material events is demonstrated, "the relation between *external movement* and *sensation* remains inconceivable and exhibits a contrast which becomes sharper the more light one sheds on it".[687]

As Hume has, however, demonstrated (according to Lange), this inexplicability is not restricted to the transition (causal relationship) from spatial movement to ideas and thoughts, but is true of every causal relationship; that is to say, in the final analysis, *all* natural occurrences are inexplicable (in the sense of a causal explanation which makes it clear that if A happens, B *must* follow) and this means the end of materialism as a philosophical principle. However, while materialism may well be untenable as a *philosophical* position, it can survive as a "maxim of detailed scientific research", and this view characterises the majority of present-day 'materialists'.

They are essentially sceptics; they no longer believe that matter, as it appears to our senses, contains the final solution of all mysteries of nature; but they behave basically as if this were the case and wait until they are compelled by the positive sciences to accept other assumptions.[688]

In other words (as I read this), the majority of 'materialists' (living in Lange's time) were *de facto* adherents of methodological materialism.

Although the defence of the 'materialistic maxim' — a general characteristic of nineteenth-century naturalism — proved to go hand in hand with various philosophical (metaphysical) assumptions,[689] Lange probably has the most influential variant of methodological materialism particularly in mind — the variant which Rokitansky described as pragmatic naturalism and which we have ourselves identified as the position of natural scientific *positivism*.

Lange (apparently) considers the methodological 'materialism' to which he is referring as incompatible with the "law of the inexplicability of all natural processes",[690] hallowed by Humist scepticism, which means the end of materialism as a philosophical principle. Lange's position is therefore essentially that he says that it is true that the sceptical 'law of inexplicability' must be endorsed, but that it is nevertheless meaningful in (the practice of) natural scientific research to behave *as if* these natural processes *are* susceptible to (causal) explanation; in other words, it is useful to seek a materialistic—mechanistic explanation of natural occurrences, even though in the final analysis any such explanation belongs to the realm of the impossible.

Why we should find this paradoxical recommendation meaningful is, it is true, not expressly argued, but it can be surmised from certain remarks he makes. The scepticism which Lange sees as characteristic of (the majority of) the 'materialists' of his time and which he is himself interested in defending, is not primarily scepticism with regard to materialism as dogmatic metaphysics, but a scepticism about the belief that our knowledge comes about through the fact that an outside world independent of our knowledge achieves representation in us (as passive receivers) — so-called naive realism. This is, in other words, the scepticism which is inherent in the position of epistemological subjectivism. To put it another way, 'our modern materialists' are 'essentially sceptics' because, and in so far as, they subscribe to epistemological subjectivism, no more and no less.

The epistemological subjectivism, which also has Lange's approval, stands outside and beyond naive realism and (radical) epistemological scepticism and is defended — in a way which is highly characteristic of Lange's physiologically 'coloured' critical philosophy — by reference to "the astonishing progress in [the] field of [the physiology of the organs of the senses]".[691]

To sum up briefly: the 'astonishing progress' in the modern physiology of the senses argues in favour of the assumption of an epistemological subjectivism which shows that our experience is independent of the structure ('organisation') of our cognitive apparatus. This (scientific) pragmatic argument goes some way towards meeting the sceptical doubt about the existence of the outside world, namely precisely as far as sceptical criticism is necessary to refute naive realism. The reference to the sensational successes of modern sensory physiology also conceals, however, an argument against radical epistemological scepticism. The explanation of this is that scepticism may be irrefutable, but the results of modern sensory physiology give us good reason to believe that the scientific undertaking is still meaningful. No more is needed to justify the recommendation to follow the materialistic 'maxim'.

It would, however, be wrong if the above discussion were to suggest that it was only *epistemological* considerations which played a part in Lange's embracing of the methodological materialistic position. It will be clear that, while Lange's fundamental problem concerned the reconciliation of natural science and religion, an important motive for the defence of the philosophical figure ('methodological materialism') which he chose lay in the fact that this made it possible to hold on to the meaningfulness and possibility of the natural science without thereby excluding in advance all meaning of religion. One can formulate this somewhat differently by saying that the standpoint of methodological materialism must have been attractive to Lange because the salvation of natural sciences which it brought about could be combined without difficulty with his interpretation of the religious world of ideas as a mode of 'poetic thought' in which man raises himself to the 'world of the ideal'.

From the 'standpoint of the ideal', any interpretation of materialism other than the methodological one is untenable. The defence of the standpoint thus necessarily implies a criticism of materialism, specifically where this position — to be consistent — jettisons the religious world of ideas not only as a system of dogmatic doctrines but also, as we have seen, in the interpretation put forward by Lange, in which religious ideas have a symbolic significance.

4.4.2.3. *Lange's Interpretation of Kant's Theory of Knowledge and the Impact of Schopenhauer's Philosophy on it.* Our proposed situating and description of Lange's thinking would be materially lacking if, in conclusion, we did not say something about his relationship to Kant and the way in which the (sensory and cerebral) physiology of Lange's day is important in his interpretation of Kant's theory. The situation is strongly reminiscent of that of Schopenhauer, and it is therefore the linking of Kantian epistemology and empirical physiology achieved in the idealistic standpoint, as Lange defended it, which makes it possible to grasp the closeness of his position to that of Schopenhauer.

As in the case of Schopenhauer's theory of knowledge, in Lange's case physiology also proves to provide the best access route. It is therefore best if we approach his interpretation of Kant by first throwing some light on the significance of (sensory) physiology for his own epistemological conceptions.

"The physiology of the organs of sense is developed or amended Kantianism, and Kant's system may, as it were, be regarded as a programme for modern discoveries in this field".[692] It is modern sensory physiology which, in Lange's view, has destroyed the naive belief of the materialists that what is real is what we perceive with our senses. As we have just said, according to Lange it is modern sensory physiology which makes (naive) materialism defenceless against sceptical criticism. Materialism can only still be defended as a methodological figure in combination with an epistemological position which does justice to the unmistakable dependence of our experience on the structure of our cognitive apparatus as this was argued, much earlier, on the philosophical level in Kant's theory of knowledge. On the issue of this 'subjectivistic' orientation concerning the problem of knowledge, this sensory physiology (Lange is probably thinking primarily of Helmholtz) was, so to speak, an extension of Kant's work. From the passage quoted above, it appears that Lange goes so far as to say that (modern) sensory physiology is essentially a further development and correction of Kant's theory of knowledge and that the significance of Kant's theory of knowledge actually lies in the fact that it provides the framework for the formulation (conception) of the questions which should be posed by (modern) sensory physiological research.

We thus see that Lange interprets Kant's critical philosophy in such a way that the relevance of sensory physiological research is already established for certain in advance. I am referring here to the central concept of (psychophysical) organisation, which Lange uses as the concept of the essence of the subject of knowledge, as the concept of the 'cause of synthesis *a priori*'.

It is not purely and simply a question of interpretation of Kant's theory when Lange introduces the concept of 'physical—psychic organisation'; it is also — as he was aware — a matter of criticism of Kant. This concept indicates how far and in what sense Lange rises above Kant, because the expression 'the physical—psychic organisation' attempts to show "that the physical organisation, as *phenomenon*, is at the same time the psychic one".[693] That is to say, the modification suggested by Lange "provides a very intelligible and easily conceived notion, instead of the scarcely comprehensible Kantian idea of transcendental presuppositions of experience".[694]

The difference with Kant thus lies in the fact that, whereas Kant substitutes for the thing-in-itself, which we cannot know, concepts or categories which are accessible to us, and speaks of these categories as if they were the *origin* of the *apriori*, Lange is prepared to go no further than to state that they are at most only its simple expression. "If we wish to denote the true cause of the *a priori*, we cannot speak of the 'thing-in-itself' at all, since the concept of cause does not extend to this. . . For the 'thing-in-itself' we must *substitute the phenomenon*".[695]

We can probably explain Lange's ideas by saying that the only way in which we can meaningfully speak about the (true) cause of the *a priori* is through the introduction of the term 'organisation', namely as an expression of something which is *phenomenally given*.

We must rather understand by the simple term organisation, or physical–psychic organisation, what to our external sense appears to be that part of the *physical organisation* which stands in the most immediate causal relationship to the psychic functions, while we may hypothetically assume that at the basis of this phenomenon there lies a purely spiritual relationship of the things-in-themselves, or even the activity of a spiritual substance.[696]

We thereby in fact take upon ourselves only the semblance of materialism, and escape the Scylla of a Platonic-idealistic hypostatisation of the categorial concept, and the Charybdis of a purely tautological interpretation of Kant's critique of reason ("synthesis *a priori* has as its cause synthesis *a priori*").[697]

The psychophysical 'organisation', as Lange conceives it, is thus not a thing-in-itself (as the materialists believed) but 'appearance'; as such, however, it is nevertheless conceived as something which precedes and conditions experience, that is, an *a priori* of experience[698] or, if you prefer, the sum total of the "*a priori* elements of thought".[699]

Lange's physiologistic interpretation of Kant's critical philosophy is coupled (how could it be otherwise?) with a relativisation of Kant's apriority theory. What Kant, according to Lange, had overlooked was that the method of transcendental reflection he used in the discovery of the *a priori* of experience was a method of *induction*, and that consequently the result of his work of discovery had not achieved apodictic certainty (necessity and universality), but only probability. The line between real and apparent necessary and universal knowledge is a fluid one, not *per se*, but for us. That is to say, the 'basic concepts and supreme principles' which Kant 'discovered' do not in principle provide absolutely valid knowledge beyond any doubt. The consciousness of the universality and necessity of a thesis can indeed deceive; in the course of the empirical process every judgement which is considered to be evident can reveal itself as pseudo-evident.[700] On the other hand, however, even if it were to prove that a series of evident truths in the Kantian theory did not withstand the test of experience, i.e. were to be unmasked as pseudo-evident, there need still be no doubt whatsoever about the existence of *a priori* fundamental concepts and principles in our minds, towards which experience is 'psychologically constrained' to direct itself.[701]

The problem then is, of course, to find out which instances of evidence out of the class of all evident judgements mediate true and which untrue *a priori* knowledge. To state that we can, after all, have *a priori* knowledge (true and/or untrue), i.e. necessary knowledge, which "is given before each particular experience and which is therefore immediately apparent in the first experience, without the mediation of induction",[702] is another way of saying that our knowledge is based on our *physical–psychic organisation*. Our physical–psychic organisation predisposes us, as it were, to the possession of (true and/or untrue) evidence (in this case evidence relating to the structure ['organisation'] of the human cognitive apparatus). What distinguishes the real evidence from the pseudo-evidence is, however, that only the former can maintain itself in the face of *experience*: anything which holds up in experience thereby proves its truth. Lange can therefore also say that in *tracing* and *testing* this sort of evidence, we are thrown back on 'the usual means of science' (i.e. empirical science), and that we can only make *probable* pronouncements about it, "whether or not the concepts and ways of thinking, which we now have to accept as *true* without any proof, originate in the abiding nature of man; whether, in other words, they are the true, basic concepts of all human knowledge or will eventually turn out to be *a priori* mistakes [an expression originally used by John Stuart Mill]".[703] Thus, more particularly, experience will show us how much of Kant's philosophical theory will remain standing. As far as Lange's opinion on the matter is concerned, it must be added that the experience gained in modern sensory physiological research has already provided partial support for or even contributed to a revision of the Kantian insights in question.

I shall not go into all the details of Lange's reinterpretation of Kant's transcendental apriorism, but will restrict myself to Lange's interpretation of the apriority of the concept of causality. Generally speaking (according to Lange) it is not the categories which are available to experience "but only such arrangements by which the influences of the outer world are immediately united and ordered in accordance with the rules of these concepts".[704] Perhaps, he goes on to speculate, it will be possible in the future to find the basis of the concept of causality in the mechanism of reflex action and sympathetic excitation: "then we would have translated Kant's pure reason into physiology and in this way made it more vivid".

He is fully aware of the fact that in these views he goes beyond Kant. Where Kant states that the concept of causality is a "basic concept of pure reason" and, as such, lies at the root of our experience, Lange states: "The concept of causality *is rooted in our organisation and is prior to all experience, according to its nature*".[705] The fact that Lange was able to believe that, with this physiological interpretation of Kant's critique of knowledge, he had not fallen into the trap of the materialism he was criticising is based on the fact — as I have already pointed out — that this 'physical–psychic organisation' was understood by him as 'appearance' (and not as thing-in-itself). With this, he placed himself in the company of Schopenhauer, where the latter safeguarded his cerebral physiological interpretation of Kant's theory of knowledge against the criticism of materialism by conceiving the human brain as the objectivisation of the Will, thus, as Idea. However convinced Lange himself may have been that his own standpoint was essentially that of Kant's critical philosophy it seems to me that on closer examination his position betrays a greater kinship to Schopenhauer than to Kant. Although the physiological interpretation of our *a priori* forms of knowledge, which we have discussed above, points clearly enough in this direction, I should like in conclusion to look briefly at some of the ideas in Lange's idealism, which lend themselves particularly well to the confirmation of this relationship to Schopenhauer's philosophy.

In our discussion of Lange's 'standpoint of the ideal' [706] we pointed to the central significance of the concept of ('poetising') synthesis. The activity of synthesis is operative from the highest level of speculative 'conceptual poetic thought', art, poetry, religion, to the 'lowest' regions of the level of cognitive organisation, i.e. to the most elementary processes of thought and sensation. Without this concept of synthesis even the most elementary fact of sensation remains totally incomprehensible, and even if this synthesis is itself an inexplicable mystery to us, the fact that there is something like a synthesising activity operating even in the 'lowest' phenomena of consciousness, such as sensation, is one which can no longer seriously be called into question since the convincing sensory physiological confirmation of Kant's epistemological subjectivism.

At first sight, this modern sensory physiology appears to play into the hands of materialism, because it demonstrates the basic mechanism of our sensations and thereby, in fact, also demonstrates the applicability of the mechanistic viewpoint in the explanation of the most elementary phenomena of consciousness; or, formulated somewhat more carefully, because there seems to be nothing in the (sensory physiological) facts it presents which prevents us from assuming "that all these effects of the constellation of simple sensations rest upon mechanical conditions which we may be able to discover when physiology has progressed far enough". [707]

Belief in the mechanistic explicability of the phenomena of sensation studied by sensory physiology, the belief that physical mechanisms can be discovered which explain the activity of our consciousness, does not necessarily imply the acceptance of materialism (in the sense in which this belief contains the 'in-itself-ness' of matter). To see this, it is necessary, as Lange has it, "to stride away right through the consequences of this materialism", [708] that is to say, to take note "that the same mechanism which thus produces all our sensations *also produces our idea of matter* ... Matter in general may just as well be merely a product of my organisation – must, in fact, be so – as colour or as any modification of colour produced by the phenomena of contrast".[709] This also explains why he can talk about our 'organisation' in a way which leaves undecided whether physical or psychic (mental) organisation is being referred to, "for every physical organisation, even if I can demonstrate it under a microscope or with a scalpel, is still only my idea, and cannot differ in its nature from what I call mental".[710]

In other words, let us assume (with materialism) that there exists a physical mechanism in the human body which produces the operations of our intellect and senses – but then we are immediately confronted with the questions: what is the body, matter, the physical? And the answer which modern physiology, like philosophy, must give is "that these are all only our ideas; necessary ideas, ideas which follow the laws of nature, but all the same not the thing itself".[711]

Thus the materialistic view proves, when its consequences are thought through, to turn into a consistently idealistic view. The ontological dualism on the basis of which we assume a gulf between the physical and the mental sides of our being is thereby destroyed.

We must not attribute certain functions of our being to a physical nature and others to a spiritual nature; but we are within our rights if we presuppose physical conditions for not only the outer world as it appears to us but *also the organs* with which we apprehend it, as mere images of what really exists. The eye, with which we believe we see, is itself only a product of our ideas; and when we find that our visual images are produced by the structure of the eye, we must never forget that the eye too with its arrangements, the optic nerve, together with the brain, and all the structures which we may yet find there as causes of thought, are only ideas, which indeed form a self-coherent world, yet a world which points to something beyond itself.[712]

The similarity with Schopenhauer's epistemological conceptions, as we have previously explained them, is indeed too striking to need extensive elucidation. The idea that the 'world of reality', i.e. the world of phenomena as it is given to our senses, is no more than a complex of ideas, that this world of ideas is the 'product' of our (psychophysical) organisation, are notions which display more than a superficial agreement with some of Schopenhauer's ideas. The fact that Lange, nevertheless (and not entirely unjustly) is usually reckoned to be a Neo-Kantian is probably becuase he takes seriously the critical idea of the fundamental unknowableness of the thing-in-itself, whereas Schopenhauer, as it were, opens the door to metaphysics (again) through the assurance that in any case the Will in its 'in-itself-ness' is accessible to us (namely through 'inner experience'). Lange is even quite emphatic on this point — "the transcendental basis of our organisation remains . . . as unknown to us as the things which influence it. We always have only the product of the two [by which he means the phenomenal world as the world of sensorily given ideas] before us".[713] It is therefore only to be expected that anyone who wishes to emphasise the critical elements of Lange's philosophy will want to stress his affinity to Kant, while anyone who considers the physiological or anthropological nature of this critical philosophy to be its most striking feature will want to give the greatest weight to the influence of Schopenhauer.

4.5. CONCLUSION

The question of Lange's primary dependence — on Kant or on Schopenhauer — is not an academic one, because it is precisely this ambiguity, this Janus face displayed by Lange's philosophy, which explains the way in which his work was received by his contemporaries. Despite the tremendous esteem which Lange's work inspired in his younger colleague in Marburg, Herman Cohen, it was precisely his physiological interpretation of Kant's apriorism which met with opposition from Cohen. The idea that the fundamental principles and concepts of (natural scientific) research, matter, atoms, the forces and the principles of mechanics, were rooted in the physical–psychic organisation and had their existence there; the idea that matter, its forces and laws, are nothing but the products of our minds, i.e. ideas, were notions which Cohen was, he said, unable "to accept as an adequate expression of Kantian *apriorism*".[714] The endeavour to "solve the mystery of science with the magic word 'organisation'" seemed to him to be a late, feeble reminiscence of the old theory of innate ideas, that is to say, as a form of nativism which — in part — contributed to "the delay in the complete and final break-through of the transcendental method". In this context he also expressed the suspicion that, for example, Lotze's dismissive attitude towards 'modern epistemological works' (by which Cohen means the epistemological works of the rising Neo-Kantian school, of which he was himself to be one of the leading figures) and the misjudgement of its peculiar character which appears from the fact that Lotze describes it as 'psychological dissection' . . . "could only be understood, however, in terms of the impression which Lange's exposition of Neo-Kantianism had made on him".[715]

The suspicion of psychologism (physiologism) was extremely obvious — certainly in the camp of the Neo-Kantians like Cohen and Natorp. The idea of a 'physical—psychic organisation' as the ultimate *basis* of the principles of reason seemed to be nothing less than a misjudgement of the 'pure' methodological meaning of the transcendental question,[716] and it was precisely this which took Lange (as he was well aware)[717] beyond Kant.

The fact that Cohen rejected the formulation of the transcendental question as the question of the 'principles of human reason', where such a formulation was not offensive to Kant, does, however, support the assertion that the strictly logical interpretation of Kant by Cohen and his colleagues was just as one-sided as Lange's physiological explanation of Kant's critique of knowledge, albeit in a very different sense, and implied an alienation from Kant's intention.[718]

In the context of our study, this question is, however, of less direct importance than the question of the relationship between Lange's philosophy and (natural scientific) naturalism in the middle of the nineteenth century, and I should like, finally, to say a word or two about this.

One can describe Lange's philosophy as *transcendental naturalism*. As such it was distinct both from the transcendentalism of the Neo-Kantians in the narrower sense, and from dogmatic materialism,[719] and was able to provide 'living space' for the different variants of naturalism which knew themselves to be connected with each other, at least in the *formula* of methodological materialism. In this latter capacity it was also able — not without foundation — to conceive itself as a making explicit and a justification[720] of the self-conception which was in force in nineteenth-century natural sciences.

The motives which played a part in naturalism in the natural sciences from the eighteen-forties onwards, and which determined, amongst other things, the opposition between idealistically-oriented and pragmatic—agnostic naturalism, were reconciled in this transcendental naturalism. The defence of the methodological meaning of the 'materialistic hypothesis', as Lange demonstrated, is *compatible* with a defence of the values of spiritual life which had been an integral part of idealistic metaphysics from ancient times, without thereby itself again lapsing into (idealistic) metaphysics (as, for example, was the case with Lotze), and also without detracting in any way from the requirements of natural scientific method. Both standpoints, that of the 'world of reality' (i.e. in fact the natural scientific standpoint) and that of the 'world of the ideal' have a — relative — right to exist, and, provided they are properly understood, are not mutually exclusive.

The significance of Lange's philosophy for his 'materialistic' contemporaries, we may assume partly on the basis of what we have learned about the philosophical preferences of the leading natural scientists we have studied, lay in the fact that Lange made these contemporaries aware of something that they actually 'knew', but rarely dared to say out loud: that in their hearts they did not believe in (metaphysical) materialism, but that they were in reality (in one sense of the word or another) 'idealists'. In this sense it can be said of Lange's philosophy what Hegel, whom he reviled, said of philosophy in general: it was "its time, expressed in thought".

APPENDIX

MAIN LINES IN THE HISTORY OF PHILOSOPHY AND SCIENCE LEADING TO 'CLASSICAL' MEDICAL ANTHROPOLOGY AND ANTHROPOLOGICAL MEDICINE (PSYCHIATRY) IN GERMANY FROM ABOUT 1780 TO ABOUT 1820. A PHILOSOPHICAL AND HISTORICAL OUTLINE.

A1. THE SCOPE OF THIS OUTLINE

A question which to date has scarcely been asked, let alone answered, is that of the way in which the philosophy of the past and the 'modern' empirical study of man in the sixteenth to eighteenth centuries worked together in bringing about the eighteenth-century German anthropological tradition, and how, more particularly, in its turn this anthropolgical tradition (in its medical variant) was able to combine with motives from Aristotelian psychology and anthropology in constituting the most influential variant of nineteenth-century anthropological institutional psychiatry. An immense amount of research still needs to be carried out in this field before the historiographer can get down to work with any degree of confidence in the outcome of his undertaking.

For this reason — at the risk of labouring the point — I want to make the intention and pretensions of these historical notes quite clear. In short, their aim is to indicate some of the principal features of those developments in the history of philosophy and empirical human science which are of special relevance in the reconstruction of the history which led up to anthropological (predominantly medical-anthropological) and anthropologically-oriented thinking in the eighteenth and nineteenth centuries. I hope to make it clear, in the following observations, that the task confronting the historian of science is a comprehensive one.[1] The aim of these notes is not completeness (whatever that may be), rather it is to give an outline of the great questions which are still waiting for a historical answer, and this means that I have had to be selective in my treatment of the material. Anyone who regrets this selectivity is looking not only for something this outline does not aim to give, but also for something which it cannot give, when one considers that research in this field is at present still in its infancy.

201

A2. ARISTOTLE AND THE BEGINNINGS OF ANTHROPOLOGY

There are good reasons for beginning the clarification of the philosophical antecedents of 'modern' anthropological thinking by looking at the work of Aristotle. If Aristotle's philosophical 'psychology', particularly in its reception in Christian thought until well into the twentieth century, was a source of inspiration for *philosophical*-anthropological reflection (in the sense of metaphysical anthropology), the observations about man scattered throughout his biological writings laid the foundations of anthropology in the sense of empirical human science, which in this case means an empirical discipline concerned with describing physical man. Aristotle's view of man is thus, in fact, the cradle of later developments, both those in the history of the philosophy of man and those which form part of the history of the origins of so-called empirical human science.

In the Christian era, the developments which stemmed from the biological and psychological works of Aristotle were to find themselves in opposition. The scholastic reception of the Aristotelian theory of the soul (St. Thomas Aquinas) became a medium for the Christian (or, if you will, 'personalistic') conception of man and as such, many centuries later, its influence was still in evidence in the Neo-Thomism of the twentieth century, while the renaissance of empirical research into somatic man which, returning to Aristotle's biological anthropology, became a fact with the rise of 'modern' anatomy and physiology in the sixteenth century, demonstrates the viability of the naturalistic motive.

A more detailed consideration of Aristotle's 'anthropological' ideas and their treatment in the subsequent history of Aristotelian thought will serve to clarify this further.

A3. THE 'BIO-LOGICAL' VIEWPOINT IN ARISTOTLE'S ANTHROPOLOGY AND PSYCHOLOGY

In his biological writings, Aristotle refers to man as a living being (ζῷον, *zoön*) among living beings.[2] This means that his anthropology forms part of the study of living beings, 'zoology', which in turn is part of general biology. In this way it takes its place within the totality of his philosophy of nature. In the view of this philosophy of nature, man, as a living being, stands on the highest rung of a hierarchy of natural forms (the so-called 'ladder of being', *scala rerum*), which climbs from inanimate to animate beings. Man's precedence in the ordering of nature, the natural cosmos, is based on a 'psychic' quality, specific to man, which differentiates him from other living beings. Besides the psychic functions which he shares with other living beings, man is the only living creature with a soul capable of 'reasoning'.[3] But leaving this aside, man's physical constitution is enough to demonstrate that he is immensely superior to other living beings. He has the highest body temperature, and therefore proportionately more blood, and the largest brain.[4] It is only in man that we find, as a result of his warmer and more noble nature, the correct proportions of stature and the corresponding upright stance. The difference between right and left is also most clearly developed in man. His blood is the purest, and he therefore has the most sensitive touch, the most highly-developed faculty for sensory differentiation, and the keenest intellect. Mouth and windpipe, lips and tongue are used not only for the other functions, but also for speech ($\lambda\acute{o}\gamma o\varsigma$), which distinguishes man from all other living beings. In contrast to the animals, nature has endowed him not with one means of protection, but with innumerable ones which vary according to his need. His hand is $\check{o}\rho\gamma\alpha\nu o\nu$ $\acute{o}\rho\gamma\acute{a}\nu\omega\nu$ i.e., the tool of tools, so meaningfully constructed for the most diverse functions that it furnishes and replaces all other tools. In short, man is the first and most perfect of all creatures,[5] and because this is so, all other living beings are destined to be used by him, in accordance with the principle that the purpose of the less perfect always lies in the more perfect.

Aristotle's biological anthropology must not, however, be viewed in isolation from his psychology. Generally speaking, in Aristotle's conception, the body is *for the benefit of* the soul (which uses that body) and the soul is not real *without* that body (which it 'animates').[6] In fact, Aristotelian anthropology and psychology are united in a sort of philosophical *biology* or 'theory of life'.

The 'bio-logical' orientation of Aristotle's thinking becomes immediately apparent in the way in which he defines the object of psychology, the soul namely as a principle of life. Psychology (in the Aristotelian sense), in accordance with its general concept, is not *per se* concerned with the human soul, but with what constitutes the life not only of man but also of animals and plants. Thus in *De Anima*, the work in which Aristotle's ideas about the soul are most mature and, for later generations (particularly for scholasticism), most representative, the soul is defined as "the first entelechy (actuality) of an organic natural body which has life".[7] What makes this definition of the mind of such interest to us is not only or primarily that it makes the 'bio-logical' orientation of Aristotle's 'psychological' thinking obvious nor that by successfully applying the metaphysical principles of hylomorphism to the area of living, i.e. 'animated', nature it departs from the Platonic dualism of soul and body (the soul is the form of the body). Its importance lies rather in the fact that the *redefinition of the soul–body relationship* which it gives became the foundation and starting-point of a tradition of philosophical (i.e. metaphysical) anthropological thinking which, as we have observed, survived by way of the reception of Aristotelian psychology in the philosophical psychology of St. Thomas Aquinas until well into the twentieth century.

The relationship of soul and body as a relationship of form ($\epsilon \hat{\iota} \delta o\varsigma$, $\mu o\rho\phi\acute{\eta}$) and matter ($\mathring{\upsilon}\lambda\eta$), the soul as the actuality ($\grave{\epsilon}\nu\tau\epsilon\lambda\acute{\epsilon}\chi\epsilon\iota\alpha$) of the body, the (potentially living) body as the matter which is to be 'formed' by the soul – all this, viewed in the context of the Aristotelian conviction that 'reality' can truly be predicated only of $\pi\rho\tilde{\omega}\tau\alpha\iota$ $o\mathring{\upsilon}\sigma\acute{\iota}\alpha\iota$, the 'first substances', or individuals, was pushing in the direction of the one great 'anthropological' conclusion, which was recognised by St. Thomas Aquinas and which he made the centre of his psychology: man as a psychophysical totality, man as a unity of soul and body, that is greater than the sum of its parts, not an aggregate but, in Aristotle's words, $\sigma\acute{\upsilon}\nu o\lambda o\nu$.[8]

Aristotle's psychology, however, contained a stumbling-block which was to prove of far-reaching significance in the history of anthropological thought, particularly in the Christian era. This difficulty had nothing to do with the way Aristotle had defined the soul–body relationship in the *general* sense, but was concerned with an implication of his development of his psychological ideas which, on close analysis, was irreconcilable with his psychological postulate as laid down in the definition of the soul as a principle of life, quoted above. In more concrete terms, a problem arose when Aristotle, in working out what we would nowadays call a stratification of psychic functions (actually, vital functions), made a distinction between the psychic function which man has in common with other forms of organic life, namely the 'plant soul' (which is the nutritive or vegetative soul encompassing the functions of feeding, growth, and reproduction) and – one rung higher – the 'animal soul' or sensitive soul which, moreover, includes the functions of sense perception, desire, and local motion on the one hand, and, on the other, the 'higher' powers of the soul which are the distinguishing characteristic of man and stamp him as a rational animal, scientific thought ($\lambda\acute{o}\gamma o\varsigma$, $\nu o\tilde{\upsilon}\varsigma$ $\theta\epsilon\omega\rho\eta\tau\iota\kappa\grave{o}\varsigma$ or $\tau\grave{o}$ $\grave{\epsilon}\pi\iota\sigma\tau\eta\mu o\nu\iota\kappa\acute{o}\nu$) and deliberation ($\delta\iota\acute{a}\nu o\iota\alpha$ $\pi\rho\alpha\kappa\tau\iota\kappa\acute{\eta}$ or $\lambda o\gamma\iota\sigma\tau\iota\kappa\acute{o}\nu$).

The problem of the relationship between higher and lower forms of the soul, as it has become familiar, particularly as related to the cognitive faculties, lay in the fact that one of the powers of reason distinguished by Aristotle, which has become known as νοῦς ποιητικός,[9] could, according to the properties Aristotle ascribed to it, in no way be interpreted as a specification of the soul in the sense of a principle of life, or so it seemed at least to later commentators. In contrast to the νοῦς παθητικός – *intellectus passivus* – which in this theory of dual reason was considered to be the receptive, 'passive'[10] part of the mind, that is to say as something that, of (biological) human origin, comes into being and passes away in time, this νοῦς ποιητικός is defined as 'separable' (χωριστός) (with regard to the physical, individual existence of man), impassive, imperishable, eternally in action, etc..[11] If Aristotle's definition of the soul as a principle of life meant in the first instance a victory over the Platonic dualism of soul and body and – against Plato's theory of the separate parts of the soul – the promise of a unitarian conception of the soul, it was all placed in a different light by this description of active reason, which clearly seemed to create another dualism, no less radical than Plato's.

The importance of the problem of the unity of the Aristotelian conception of the soul which this raised was to be clearly expressed in the *Wirkungsgeschichte* of Aristotle's psychology. Aristotle's words left room for two interpretations which, in all common efforts to salvage the unity of the (human) soul, went in diametrically opposite directions. Depending on the question of what was considered more important – preserving the unity of an *imperishable* soul or preserving the unity of an *individual* soul – scholars opted for one or other of the possibilities of interpretation. That is to say, they either maintained that active reason was supraindividual (i.e. one and the same in all human beings) and immortal, and thus denied that individual man possessed his *own* substantial soul and enjoyed personal immortality – the so-called monopsychistic viewpoint which one finds in Averroës (1126–88), the Arab commentator on Aristotle – or, taking as their starting-point the soul as a principle of life, they defended the individuality of the soul, like the 'naturalists' of the Aristotelian school itself, such as Straton of Lampsacus (first half of the third century BC),[12] but were thereby forced to abandon the thesis of the immortality of the soul.

From the point of view of Christian thinking, neither of the two interpretations was able to find favour because, according to the Christian self-conception, man conceived as person was considered to be a psychic-subjective being which exists (and continues to exist) *in time and eternity* as essentially *the same individual*. It was therefore obvious that the Christian reception of Aristotle would be characterised on the one hand by the endeavour to ensure the ontological integrity (that is, the 'substantiality') of the (personal) individual[13] − the guarantee of his immortality − and, on the other, by the tendency to suppress the naturalistic viewpoint (in particular that of the Alexandrianistic interpretation of Aristotle) which was certainly no less offensive to the Christian self-conception than was Averroistic monopsychism, because its consequence was not only to deny the immortality of the soul, but also the transcendency of God.

Thus it was, in short, the naturalistic implications of the ancient (in this case, Aristotelian) view of man which − whether or not they were explicit − were bound to offend a form of anthropological thinking which was inspired by the insights of Christian revelational anthropology.

We therefore see how, prepared by related assumptions in the tradition of Neoplatonism, a development occurred in Christian thinking which, in accordance with Christian tenets (individual−personal immortality, the transcendency of God, etc.), allowed man to be understood as *more and something other than nature*. I refer here to the genesis of the (early) Christian conception of the person − in fact *the* conception of the person − which was unknown to antiquity and could only achieve validity within the framework of Christian thinking. Its first formal ontological definition occurs in Boethius,[14] the way having been prepared in the early Christian theology of the Trinity and incarnation.

That the (early) Christian conception (and definition) of the person had to go beyond the bounds of ancient ontology was inherent in the fact that Christian thought was concerned with comprehending the exceptional position of God, the angels and mankind *as opposed to* the order of nature to which the rest of creation belonged. Man, created by God in His own image, receives his identity by virtue of his eschatological calling, namely to represent God's acts on earth in order that he may participate in the perfect state.

Where, in Aristotle, man appears as the highest rung in the ascending series of natural forms, according to Christian self-conception, man is viewed in the perspective of a *supernatural* history of salvation. The thing that is called 'person' in this Christian thinking cannot, therefore, be entirely included in the 'natural' cosmos nor understood as a part of it. Seen in this light, it was quite logical that (early) Christian thought, in its endeavours to arrive at an acceptable ontological definition of man as person, should transcend the limits of ancient, essentially cosmological ontology.

The contrast between ancient and Christian thought which we have described was too great to be definitively overcome: the subsequent history of the concept of the person was to bear witness to the fact that a lasting harmonisation of ancient objectivism and Christian personalism was impossible. This can be seen even in the brilliant reconciliation of Christian theology and ancient (Aristotelian) philosophy brought about by St. Thomas Aquinas. The Christian point of view, that is to say the concern about the foundation of personal immortality, definitely put a strain on Aristotelian hylomorphism.[15]

Just as, given the Christian-anthropological assumptions, we can see that this departure from ancient 'naturalism' was inevitable, so we can understand its corollary — the suppression of everything in the philosophical tradition which could be explained as supporting a naturalistic conception of man. This explains the fact, which is perhaps somewhat surprising at first glance, that the rediscovery of the works of Aristotle and the tremendous concern with his philosophy, which is so characteristic of (Christian) scholastic thinking, did not lead to a revival of interest in his anthropology. What Aristotle had to say about man was, precisely because of the above-mentioned 'naturalistic' implications, not so much lacking in interest as downright dangerous, and thus it came about that, however willingly scholars acknowledged Aristotle as their mentor in logic and metaphysics, they took care not to follow his lead on the issue of anthropology.[16]

As long as the viewpoint of Christian scholastic theology dominated spiritual life in Western Europe, the impetus given by Aristotle to natural philosophy oriented anthropology was doomed to remain fruitless. The fact that at long last, in the sixteenth century, this began to play a significant part in the so-called Averroism of Padua, was only possible in a spiritual environment which, although not opposed, was at least indifferent to the Christian—theological—metaphysical tradition of scholasticism.

A4. THE FOUNDATION OF 'MODERN' ANTHROPOLOGY IN THE ITALIAN RENAISSANCE

A4.1. *The Beginning of the 'Renewal' in Christian Humanism and Platonism*

The fact that, after so many centuries of scholastic-theological domination, the natural philosophical viewpoint was able to assert itself in the consideration of man is a development which must be seen in the context of the wholesale change in orientation and feeling of life which is characteristic of the so-called Renaissance.

This change or, if you will, renewal in the cultural life of Western Europe was a result of the undermining of the mediaeval ordering of existence which, thanks to the domination of the one church, the feudal system and the agrarian economy, had managed for centuries to preserve a comparatively great spiritual, political and economic stability, but which finally found its socio-economic foundation eroded with the rise of the middle classes and of trade, and saw its spiritual basis disappear in the face of the growing scepticism regarding the established doctrinal authority of the church.

It is typical of the dawning Renaissance feeling about life that the scepticism about religious-metaphysical tradition was not primarily seen as a loss of meaning, but rather as a liberation, as the opening up of the individual's own creative potential. As J. Burckhardt rightly says, the (Italian) Renaissance can be called the era of the '*Entdeckung des Menschen*', or more precisely as the era in which man discovered himself and the world.[17] Scholars are fond of describing the spirit of this Renaissance in contrast to the spirit of scholastic, mediaeval thought and feeling in terms of a shift in emphasis in the system of values: a shifting of the accent from the hereafter to the here and now, from transcendency to immanency, from man's dependence to his creative powers.[18] In line with this general characterisation, albeit with an idealised exaggeration[19] of the contrast, the distinction between the mediaeval and the Renaissance world has been described in terms of two contrasting ideals of existence. On the one hand there is the mediaeval ideal of the saint, the monk, who tries, by constant directedness towards his heavenly destination, by ascetic simplicity and pious submission to the authority of the church, to give substance to the Christian commandments; on the other hand there is the typical Renaissance ideal of existence, which finds its fulfilment in the concern for worldly goods and which, in an occasionally even 'unchristian', heathen accentuation of life on earth, is expressed in the joy of living, the pleasure of discovery, boundless curiosity and a hungering after the individual riches of all life in the here and now.

Be that as it may, the discovery of the world in this period of Renaissance in Italy is in any event the discovery of man himself, man as the individual — personal creator of his own cultural world.[20] Man is a creature who has come from the hand of God like all creatures, but what distinguishes him from these other creatures is that the Creator has given him the gift of creation itself, so that it is true of man that he only attains his destination and fulfils his being in the exercise of this creative power. He is, in the words of the famous oration by Pico della Mirandola, called upon as a free and sovereign artist to model and mould himself into the shape he himself chooses.[21]

The fact that freedom and human dignity were central themes in Renaissance thinking serves to underline all the more that the Renaissance period was pre-eminently an 'anthropo-logical' age (in the sense of speculation about man), and also in what way this was so.[22] The removal of man from the centre of the cosmos, which was associated with Copernicus's heliocentric system (1543) and which put an end to the anthropocentric ideas of the mediaeval view of the world, was as it were compensated during this period by an unprecedented intensifying of the original Christian awareness, which had already existed in the Middle Ages, of the infinite value of the human person.

The Renaissance humanists' view of man, as it asserted itself in Petrarch ('the first modern man'), Bocaccio and other representatives of Christian Neoplatonism of the Florentine academy (Ficino, Pico della Mirandola), undoubtedly meant, with regard to the mediaeval view of man, an important shift in emphasis in the self-interpretation of man at the time, but it is certainly true, as far as the contribution of Christian humanism is concerned, that this did not really go beyond the bounds of the Christian self-conception. To put it more forcibly, whereas Burckhardt, in his concern to draw a clear line between 'Renaissance man' and 'mediaeval man', left obscure the factors which still linked the Renaissance with the Middle Ages, subsequent research, in a relativisation of the contrast created by Burckhardt, was to make it clear that the philosophy of the Quattrocento was still governed by scholastic theology, and to what extent this was the case.[23]

Thus the conflict which broke out during the early Italian Renaissance, first and foremost between the Neoplatonist and Aristotelian schools, is, all things considered, a conflict about the true Christian self-conception. The Neoplatonists reproached the humanistic Aristotelians for their naturalism and considered themselves, because of their directedness towards the transcendental, to be allied to (true) Christianity, while, conversely, the Aristotelians condemned as unchristian the pantheistic-monistic tendency which asserted itself in Neoplatonism.[24]

A4.2. *Aristotelian Naturalism and so-called Italian Natural Philosophy*

As far as the origins of 'modern' anthropology[25] are concerned, however, the opposition between Christian-theologically oriented thinking and so-called lay thinking is more important, because the formation of 'modern' anthropology is inseparably linked to the domination which the naturalistic viewpoint achieved and the suppression of Christian 'prejudice' which accompanied it. In particular, it was Aristotelian naturalism in seventeenth-century Italy which was important in creating the conditions for the genesis of 'modern' anthropology. The University of Padua was particularly significant in this respect; its celebrated medical faculty was generally considered the stronghold of Averroism.[26] The eventual revival of anthropological research in the spirit of Aristotle's empirical human biology, which heralded the start of 'modern' anthropology, is indeed due to this Paduan Averroism. In Gusdorf's words:

It is the Averroism of Padua, Aristotelianism of strict observance, which, putting Christian presupposition in parentheses, was to regain the objective inspiration indispensable to positive knowledge. Caesalpino, Vesalius and even William Harvey have links with the Paduan springtime. There and nowhere else, by virtue of [the cry] 'back to Aristotle', the preparations were made for modern anthropology's soaring flight.[27]

It is debatable whether Gusdorf's description of Pauduan Averroism as Aristotelianism 'of strict observance' is very felicitously expressed. It is well known that the question of the 'true' Aristotle was an issue that divided many Aristotle commentators from the thirteenth century onwards (cf. the controversy between Averroists and Alexandrinists), while an 'outsider' like the Christian Neoplatonist Ficino even stated that he believed that the — supreme — Aristotelians of his time "had also (i.e. as well as from Christian doctrine) apostatised from Aristotle himself".[28] It is, however, certainly true that Arab naturalism, in its study of Aristotle's work, generally permitted a carefree appropriation of precisely those texts of the Stagirite which Christian theology preferred to disregard. It is also indisputable that Arab medical literature (of the ninth to the thirteenth centuries), made accessible though Latin translations, which brought about a spiritual revival in Europe and was a major element in university lectures in the Middle Ages, was one of the factors which contributed to the flourishing of medical studies in Italy, particularly in the 'Averroistic' Padua of the sixteenth century.[29]

The significance of the return to the 'real' Aristotle (i.e. not interpreted or, if you will, censored along Christian theological lines) for the establishment of 'modern' empirical studies of man certainly did not lie entirely, or even primarily, in the renewal of interest in the *content* of Aristotelian biology and medicine. It was no less through the revaluing of the objective, given in experience, the positive attitude towards that which could be experienced in 'external' reality, that the Aristotelian example was able to give decisive momentum to the early anatomy and physiology of the sixteenth century. It served here, as it were, to cancel out the negative consequences of the turn towards the inwardness of the religious subject, which had come about with the Christian Platonism of St. Augustine, and to compensate for it by a renewed shift towards objective reality, in this case that of the human body.

The contribution made by the so-called Italian 'natural philosophers', either indivually or collectively, cannot be said to have been of comparable significance for the growth or practice of empirical studies of man.[30] If a few among them — I am thinking particularly of Cardano and Bruno — nevertheless earned a place for themselves in the history of sixteenth-century anthropology, it was either because (as in Cardano's case) they are known as the authors of 'anthropological' works[31] or because (as we see most clearly in Bruno's case) their theory of nature was, because of the monistic-pantheistic tendency it contained, significant for the subsequent history of 'anthropological' thinking (i.e. that concerned with speculations about man).

Bruno, in whom this tendency is strengthened by religious passion and radicalness, represents in this respect the peak of a development in the natural philosophical thinking of the Italian Renaissance, and if his name appears in the history of anthropology — and here Dilthey's studies of Renaissance anthropology[32] spring immediately to mind — this is undoubtedly justified by the fact that the vision of man which emerges from his theory of nature (i.e. the theory of the infinite, divine universe) represents an ideal of man which was to be a source of inspiration to later generations wherever an element of pantheistic religious thought determined man's conception of himself.[33]

None of this alters the fact that, as far as the establishment and extension of the so-called naturalistic variant of 'modern' anthropology are concerned, it was, first and last, not the speculative flights of the theories of the Italian 'natural philosophers', which were more often than not almost without empirical support, that appear to be of decisive significance, but rather the rediscovery of the "objective inspiration indispensable to positive knowledge" (Gusdorf), as we associate this with the non-Christian theological reception of Aristotle.

At the risk of labouring the point, we must emphasise the fact that Renaissance thinking was a true matrix of anthropological motives which encompassed not only the aforementioned medical-somatically oriented anthropology, but also displayed the beginnings of a form of empirical psychology which was psychologically-subjectively oriented, and that it gave vital impulses to a tradition of moral-anthropological reflection.[34]

Without in the least wishing to disparage the significance of these alternative schools of thought about man, it must nevertheless be pointed out that what was understood as anthropology from the sixteenth century onwards and especially in the seventeenth century, particularly in Germany, was, generally speaking, as far as its *area of study and sphere of interest* was concerned, a continuation of the somatic direction of research into human biology which had been started in the medical school at Padua, and was not an extension of the (practical) psychological and pedagogic-moralistic reflections of Christian humanism.

A5. ANTHROPOLOGY AS THE EMPIRICAL STUDY OF MAN IN THE PERIOD FROM ABOUT 1500 TO ABOUT 1660

If we accept Marquard's assumption that, in general, anthropology as it developed, in particular in the German School philosophy of the sixteenth to eighteenth centuries, constituted itself as a discipline which had dissociated itself from the scholastic-theological-metaphysical tradition, but which had not yet been assimilated by mathematical, experimental natural science,[35] it is obvious that in considering the period in the history of empirical studies of man between about 1500 and 1660 (which is in any case a period which lacks clear structure) the two complementary points of view concerning the relationship to philosophical tradition on the one hand and to 'modern' mathematical, experimental natural science on the other must be preserved as points of reference in an initial arrangement of the literature. I restrict myself perforce[36] to a sketchy discussion of the principal trends.

A5.1. *The Medical School of Thought (Anatomy, Physiology)*

Although the first anthropology to be called by that name – the *Anthropologium de hominis dignitate* by the Leipzig anatomist Magnus Hundt[37] – dates from 1501, the birth of 'modern' empirical studies of man must be said to have taken place in 1543, when Andreas Vesalius from Brussels (1514–64) published his great work *De humani corporis fabrica*. This work was the fruit of the lectures which he gave as professor of anatomy at Padua to a host of enthusiastic listeners, and it differed from its predecessors in its empirical descriptive orientation. Although Leonardo da Vinci had, albeit unsystematically, tackled the subject of descriptive anthropology forty years earlier in his anatomical drawings, which were intended to illustrate the internal structure of the body and were accompanied by short notes,[38]

Vesalius's lavishly-illustrated *Fabrica* was the first systematic university textbook of topographical anatomy in the style of the 'modern' empirical studies of man. It is thus a work that, in the judgement of some historians of science,[39] is not only credited with having laid the foundation of modern empirical scientific medicine, but must moreover be considered as the first, real scientific contribution of the New Age; indeed as a work whose importance is on a par with that of Copernicus's *De revolutionibus orbium coelestium*, which appeared in the same year.

The same trend towards the rehabilitation of sensory perception, which was expressed by Vesalius from the medical viewpoint (which at that time usually meant the medicine of Galen), also manifested itself in the work of Andreas Caesalpino from an Aristotelian biological angle.[40] Although he also wrote, amongst other things, about medicine, and later even became Pope Clement VIII's personal physician, his greatest achievements were in the field of botany and he has gone down in the history of biology as "the first botanist of the modern age to build the study of plants on a *scientific* basis".[41]

The reason for mentioning him here is the same as the reason for mentioning Vesalius and is also why, finally, we must discuss the English anatomist and physiologist William Harvey (1578–1657). In all three cases we see how, within the traditionally established bounds of Galenism (Vesalius) and Aristotelianism (Caesalpino and Harvey), an empirically-oriented research trend was pioneered, which in time was to lead not only to the overstepping of these bounds but even to their total abandonment and their replacement by the framework of modern natural scientific mechanism.[42]

Harvey, as the third in this series of founders of the empirical scientific approach to the phenomena of life, pupil of the renowned anatomist and embryologist Hieronymus Fabricius ab Aquapendente (1537–1619), Aristotelian and professor at Padua, owes his name to his work *Exercitatio anatomica de motu cordis et sanguinis in animalibus* (1628), which is of immense interest from the point of view of the history of science. This work, in which Harvey put forward a carefully thought-out theory concerning the circulation of the blood, supported by years of research, is generally considered to have laid the foundations of modern physiology and thus of modern scientific medicine.[43] Although his orientation was decidedly Aristotelian and he has gone down in history as a Neo-Aristotelian, indeed Aristotle's theory (of the blood circulation of the embryo) is thought to have contributed to his discovery of the circulation of the blood,[44] it is nevertheless those aspects of his work which point to a divergence from Aristotle's style of scientific practice and herald the advent of a quantitative-mechanistic conceptualisation in biology which have drawn the attention of historians of science to Harvey. Since the view of these historians of science was determined to a significant degree by later scientific development, in descriptions of Harvey's work the emphasis was automatically placed on that which *distinguished* Harvey as a physiologist (biologist) from Aristotle. Thus it could, for example, be said that, despite the traces of Aristotelian-Galenic tradition, Harvey's work introduced a new, mechanistic conception of biology which meant that he could be placed in the company of the astronomer-physicists Kepler and Galileo and the philosopher Descartes, that is to say in the company of those who, since about 1600, have given crucial impetus to modern scientific thinking and have led the way along the new (i.e. mechanistic) path.[45] However, the question of what makes Harvey's work specifically 'modern' in comparison with that of his predecessors − the innovative significance of his method of *proof* in physiology; the greater part played by *experiment*; the move towards the *quantitative* treatment of physiological phenomena contained in his theory of the circulation of the blood[46] − is of less importance in the context of our exposition than is the ascertainment of the fact that Harvey's thinking, for all its 'modernity', does not abandon the framework of Paduan Aristotelianism.[47] But, even by not taking this into account, Harvey can be described as an 'Aristotelian' if we realise that the Aristotelian orientation in biology was characterised of old by an emphasis on the indispensability of sensory perception. It is this empirical attitude − which is also typical of the Neogalenist Vesalius and the Neoaristotelian Caesalpino − which continues to link Harvey to the example of Aristotle, even when his own researches bring him into opposition with Aristotelian doctrine.[48]

A5.2. The 'Psychological' Variant in (Medical) Anthropology in Germany (Sixteenth to Seventeenth Centuries). Descartes as 'Troublemaker'

Besides the developments in the sixteenth and beginning of the seventeenth centuries in the field of empirical studies of man, as presented in the anatomical, biological and physiological work of Vesalius, Caesalpino and Harvey, mention must also be made of those works which had 'man' as the thematic focus of their speculation and which presented themselves under the title of anthropologia or psychologia (Melanchthon, Goclenius). It is difficult to say anything specific about this anthropological literature without more detailed research, but what is known about it permits one, it seems to me, to voice a conjecture about the extremes within which its different variants are situated. Thus, in an idealised exaggeration, I should like to contrast the not very philosophical, eclectic-encyclopaedic, *synoptic* anthropologies of this period with the more philosophical (which means in this period of German history with an Aristotelian slant), *synolistic* anthropologies.

The synolistic extreme is, so it seems, most closely approached among the (German) philosophers, or, more precisely, among those philosophers (theologians) who were representative of so-called Protestant scholasticism,[49] that is, the movement in German university philosophy in the sixteenth and seventeenth centuries which breathed new life into the tradition of Aristotelian philosophy and paved the way for the philosophy of Leibniz — I am referring here in particular to Philipp Melanchthon (1497–1560) and Rudolph Goclenius (1547–1628).

Melanchthon's psychology [*sic!*] appeared in 1540 under the title *Commentarius de anima;* in 1553 the title was changed to *Liber de anima.*[50] It was the first psychology written by a German since Albertus Magnus. Despite the Aristotelian example, it proved to deviate quite considerably from the psychology of the Stagirite, if only because it incorporated not only the ideas of Aristotle but also those of Galen and of contemporary physicians like Vesalius and Leonard Fuchs. What is presented here is, indeed, as has been observed,[51] not so much a work on psychology as one on anthropology. It is, however, not as strange as it may appear at first sight that an anthropological work could appear at this time under the title of psychology, since Aristotle, in his work *De anima*, had primarily contributed to the solution of the problem of the *relationship between body and soul* and thus anticipated the central 'anthropological' intuition of man as σύνολον, as we have known it since the Christian reception of Aristotle in St. Thomas Aquinas.[52]

Fifty years later, this 'anthropological' meaning suggested by the Christian reception of Aristotle was still determining what scholars understood by a 'theory of the soul'. Where Rudolph Goclenius is concerned, 'psychologia' is to be conceived as the 'theory of man', i.e. as anthropology; the theory of the soul is a part of this.[53] Since Goclenius was the first person to use the word psychology in the title of a book,[54] we may assume — strange as it may sound to modern psychologists — that the original meaning of 'psychology' had less to do with psychology in the modern sense of the word than with (philosophical) anthropology.

Why thinkers like Melanchthon and Goclenius did not call their psychology 'anthropology' cannot be said with any degree of certainty. The desire to underline their ties with Aristotle through their choice of a title coupled with their concern to avoid identification with the more eclectic-encyclopaedic (medical) anthropology after the style of Hundt[55] may have been motivating factors here. In any case, it is interesting to note that Otho Casmann, whose voluminous anthropology of 1594 and 1596 was to become representative of the natural philosophical-anthropological tradition in German School philosophy until the eighteenth century,[56] called his work *Psychologia anthropologica*,[57] which emphasises all the more the affiliation of this anthropology to the Aristotelian tradition and gives additional support to the thesis that what was called anthropology, and understood as such from the modern age onwards, was originally natural philosophical anthropology.

While in the case of some authors, like Melanchthon and Goclenius, the total anthropological vision betrays *comparatively* the strongest ties to the philosophical (i.e. primarily Aristotelian) model, it is generally true of the anthropological literature of this period that the philosophical orientation or inspiration is not very pronounced and the idea of a total view survives only in the form of an encyclopaedic-eclectic synopsis. So, in spite of more 'synolistic' works (that is, predominantly supported by the model of Aristotelian psychology) in the German 'psychological' literature of the sixteenth century, the general picture of the — chiefly medical — anthropology during the one hundred and sixty years which separate Magnus Hundt's *Anthropologium de hominis dignitate, natura et proprietatibus* (1501) from the anthropology of the Leiden anatomist Albert Kuyper, who represents the culmination of this period in anthropology,[58] displays unmistakably all the signs of a transitional period. What presents itself here as anthropology, whether or not it actually calls itself by this title — the work of Cardanus, Melanchthon, Meletius, Magirus, Casmann and others — although it expressly aims at man *in toto* is, in fact, based on the study of human anatomy and physiology, and demonstrates, despite all due respect to philosophical tradition, not so much a rational continuation of this tradition as the expression of a collector's mania which in time became completely indiscriminate.[59] We see here an encyclopaedic mind at work which originally — as in Hundt's case — absorbed all the nuances of the (medical) philosophical tradition which formed the basis of medical education at that time (Hippocrates, Pliny, Galen, Avicenna, Averroës, Albertus Magnus, St. Thomas Aquinas), but which, as the years passed, acquired independence and, with its express criticism of Galen and Aristotle (as in Kuyper), laid itself open to the results and ideas of the science of the modern age.[60] Despite all this, we must not lose sight of the fact that the works which were described as anthropology

and understood as being typically anthropological from the second half of
the seventeenth century onwards, particularly in German medical circles,
remained, irrespective of all their mutual differences, united on one point,
that is, in their common opposition to the "fractionating way of looking
at things"[61] initiated by Descartes.

It is well-known that Descartes' dualism (and objectivism) in defining the
mind—body relationship meant the abandonment of the synolistic conception
of man and that it forced the Aristotelian-scholastic anthropological tradition
into the background. Indeed, where Cartesianism became dominant, this
anthropological tradition was doomed to be lost from sight, just as, conversely, it was in precisely those areas where, at a later date, the central
intuition of this anthropological tradition survived, that the influence of
Cartesianism remained superficial and minimal.[62] But Descartes also appears
as the great 'troublemaker' from the point of view of the history of modern
anthropology, in the sense of a total science of man. This is connected with
the fact that, paradoxical as it may sound, with the categorical separation of
mind and body which characterises Descartes' philosophy and the limitation
of (scientific) anthropology to the corporeal part of that hybrid composite,
'man', anthropology was in a sense 'halved'. The object of anthropology was
no longer man in his concrete individual totality, but in the reduction to his
physical aspect.[63] It is true that because of this the somatically-oriented
anthropology which sprung from traditional medical thinking was confirmed,
but, at the same time, the implicit intentions and pretensions of a total
science of man were rigorously barred. From then on, a scientific study of
man was only possible as the science of 'half' of man — the physical half. We
must therefore say that if, partly under the influence of German Protestant
Aristotelianism, the seeds of anthropology as a scientific total view of man
(whether or not of Aristotelian, synolistic character) were preserved from
definitive destruction, this occurred in spite of the unmistakable influence
of the Cartesian viewpoint on German anthropology (until at least about
1750).[64]

To sum up, it must therefore be stated that the Cartesian redefinition of
the object of anthropology — anthropology as the physics (mechanics) of
the human body — meant a clear alienation from the holistic intention which
had of old characterised anthropological thinking.

Viewed in this light, it is no more than logical that when, after 1750,
German (medical) anthropology established itself as an independent discipline
and, moreover, as one concerned with the study of man as a psychophysical individual, this implied a *reaction* against the Cartesian influence in

anthropology and a return, conscious or otherwise, to the idea of anthropology as a total (if not primarily synolistic) view of man, which had been forced into the background by Cartesianism in the period up to about 1750.

Thus we see that, with the criticism of the reductive conception of anthropology inspired by Descartes, efforts to achieve a more philosophical approach to man gradually took hold. The term 'philosophical' did not imply inclusion in the existing philosophical systematics, but either was intended to describe the objective of the anthropological undertaking, or meant a return to the traditional anthropological problem of the mind—body relationship. While Kant's 'pragmatic' anthropology (1798), with its aim of serving not 'schoolbook knowledge' but 'knowledge of the world',[65] is the model for one of the variants of 'philosophical' anthropology in this period, the medically-oriented anthropology represented by Platner, among others, at this time, exemplifies the second major variant of anthropological thinking.[66]

A6. SUMMA IGNORANTIAE

In conclusion I should like to dwell for a moment on the question which motivated our excursion into the history that led up to 'classical' medical anthropology in Germany from about 1750 to about 1820, and into that of the anthropologically-oriented medicine (psychiatry) which subsequently came to the fore. This question was prompted by the ascertainment of a striking parallelism[67] between the philosophical-anthropological presuppositions of (a section of) nineteenth-century German institutional psychiatry on the one hand, and what we know of the philosophical conception of man (more precisely, the interpretation of the soul—body relationship) of the so-called Aristotelian-scholastic tradition on the other. The actual question is how it was possible for elements of the Aristotelian-scholastic tradition of philosophical psychology (anthropology) to achieve a new validity in German psychiatry at the beginning of the nineteenth century; in other words, what course did the process of transmission take so that the sixteenth-century offshoots of this tradition were linked with this early nineteenth-century psychiatry?

It is quite obvious that the classical German anthropological tradition paved the way for, and facilitated the reception of, the anthopological viewpoint in this form of psychiatry, but this does not alter the fact that the manner in which this (medical) anthropological tradition and motives from Aristotelian-scholastic psychology and anthropology[68] together contributed to the rise and spread of the institutional psychiatry (and its somaticist variant) which we have studied remains unclear. The problem is not so much one of how this so-called 'modern' anthropology managed to persist in medical thinking as the tradition of an *empirical* study of man in his entirety into the nineteenth century, but rather one of how we must envisage the route by which the above-mentioned specifically *philosophical* anthropological tradition was transmitted. This is a tradition of which we find only faint traces in sixteenth-century German 'psychology', one which was eclipsed by the spread of Cartesianism, and which possibly played a very modest role in the eighteenth- and nineteenth-century German tradition of anthropology (in the sense of the empirical study of man), but which, in any event, only revealed its presence again in the German philosophy of the second and third decades of the nineteenth century (Hegel) and in the more or less contemporary psychiatric-medical literature, particularly from the eighteen-thirties onwards.[69]

It is hard to see how the answer to the questions asked here can be anything but a confession of our ignorance. Nevertheless, however paradoxical it may sound, this ignorance has a positive side: we now at least see more clearly *which questions* we might usefully apply to the philosophical and history of science tradition. And that result, though far from spectacular, at least carries a promise of taking us further along the road of historical research.

NOTES

CHAPTER 1

[1] E. W. Ackerknecht, 1968, p. 71.

[2] Cf. K. Dörner, 1969, p. 359. "In spite of the fact that Griesinger moulded the future of psychiatry, not only as social institution and as therapeutic practice, but also by contributing essential elements to all subsequent schools of thought (modern pathological, clinical and psycho-analytical psychiatry), a comprehensive interpretation of his work is to date unavailable." Dörner's own discussion in the work quoted also does not (and probably does not intend to) provide such a "comprehensive interpretation".

[3] The distinction between institutional psychiatry and university psychiatry comes from K. Jaspers, 1959, p. 705.

[4] H. Damerow, 1844, p. III.

[5] Cf. D. Rössler, 1971, p. 374. I call this variant of anthropology 'more philosophical', because the reaction to the rise and spread of this type of anthropology, opposed as it was to the splitting into physical and psychic anthropology which had come about under the influence of Cartesianism, had its positive starting-point in the anthropological-philosophical idea of the cohesion of physical and spiritual man.

[6] The 1772 edition was followed in 1790 by his *Neue Anthropologie für Aerzte und Weltweise* (Vol. I), in which he characterised the earlier work as a "youthful misdeed" and disavowed any profound agreement of content between the old and the new work. According to M. Linden (1971, p. 53), Platner's *Antrhopologie* was above all important because of the response to the use of the word 'anthropology'. Thus, scholars began to incorporate the concept inaugurated by Platner in the systematics of the anthropological sciences, and themselves to deal with the complex of problems which fell within this compass under various titles: 'anthropologia philosophica' (J. A. H. Ulrich, 1785), 'true anthropology' (J. C. Wezel, 1784), 'philosophical anthropology in the narrow sense' (C. C. E. Schmid, 1791), 'True philosophical anthropology' (L. H. Jacobs, 1795[2]), 'anthropology of mankind in general' (G. S. A. Mellin, 1797, 1806) or, simply, 'anthropology' (G. A. Flemming, 1796, 1814[2]). (For bibliographical references, see M. Linden, op. cit.).

[7] In its Latin form (*anthropologia medica*) this had already been used in the *Philosophisches Lexicon* (J. Walch, 1733): general anthropology, i.e. 'anthropologia physica' (which, as was customary among physician-anthropologists in the seventeenth century, was limited to anatomy and physiology) is contrasted with '*anthropologia medica*', i.e. the special branch of anthropology concerned with the health and illness of the human body. The expression is also found in C. C. E. Schmid, L. H. Jacobs, J. H. Abicht, G. S. A. Mellin, J. K. Wezel and W. D. Snell, as well as in J. C. Loder and P. Usteri. Cf. M. Linden, op. cit., p. 152.

[8] Quoted in M. Linden, 1971, p. 156.

[9] M. Linden, op. cit., pp. 51, 155 n. 286.

[10] Kant, 1922a, p. 145.

[11] F. Hartmann and K. Haedke, 1963, pp. 82, 84.

[12] Cf. E. Platner, 1790, pp. 68–71. ("Some metaphysical reflections on the nervous spirit as an organ of the soul, and on the inherent possibility of its real communion with the soul").

[13] Cf. M. Linden, 1971, p. 55.

[14] Namely in as far as this last took the unity of body and soul as a philosophical starting-point.

[15] G. W. F. Hegel, 1970 [1830], p. 38.

[16] As M. J. Petry (1978, p. LII) remarks, anthropology, to Hegel, is the science which is concerned with "psychic states, closely dependent upon but more complex than purely physical ones, and not yet involving the full self-awareness of consciousness", that is to say, everything that can be regarded as belonging to the realm of the subconscious.

[17] Cf. H. Drüe, 1976, pp. 33–34.

[18] J. E. Erdmann, 1837, p. 128.

[19] Op. cit., p. 127.

[20] J. E. Erdmann, 1873, p. 11.

[21] Because the so-called somaticist variant of clinical psychiatry, which is dealt with later in this chapter, points in this direction, the subsequent influence of the Aristotelian example will be relatively heavily stressed in the historical outline in the Appendix (p. 201 ff.).

[22] A (partial) exception to this rule is J. E. Erdmann (1866), in which a few pages are devoted to anthropological authors like Heinroth and Steffens.

[23] In the words of O. Marquard (1971, p. 363): "by calling itself 'anthropology', the [German] School philosophy frees itself from the theologically-oriented metaphysical tradition and starts asking: how do we have to define man, if not (any longer) meta-physically and not (yet) by means of a mathematically-experimental natural science?" M. Wundt, 1964, p. 271, stresses that the "School . . . considerably overreaches itself" in its efforts to create "a philosophy for the world".

[24] 'Holistic' in the sense used here means directed and oriented towards the idea of man as a whole. This whole can be the individual totality of body and soul as conceived in the Aristotelian – Thomistic tradition (man as σύνολον), or the composite body – mind entity of the Cartesian tradition. In the former case 'holistic' is the same as 'synolistic', in the latter case it amounts to 'integral'.

[25] For this, cf. in particular M. Wundt, 1964, pp. 265 ff.

[26] Cf. M. Wundt, op. cit., pp. 265–66. "Nowadays we are no longer concerned with the science of being and its timeless, universally valid insights, but with the science of man in his constantly changing aspects. Instead of logic and metaphysics, anthropology is the leading science. While thinking so far [this refers to that of Wolff and his followers] has developed in the tension between world and God, which also encompasses man, the new thinking is centred in man and judges God and the world from this standpoint."

[27] For the history of philosophy literature, refer in the first place to the work by M. Wundt (1964) quoted earlier. For the history of science (psychology), M. Dessoir (1964) remains indispensable.

[28] The only time that Wundt (M. Wundt, 1964) uses the term anthropology it is in the less specific sense of the 'study of man', i.e. what he generally describes as the psychology of Enlightenment thinking; and there is also no reference to any specifically

anthropological literature belonging to this period in M. Dessoir (1964), as W. Sombart established as early as 1938.

[29] Cf. M. Linden, 1976. Linden (op. cit., pp. 36 ff.) states that from the middle of the eighteenth century onwards perceptible efforts were made by German physicians and philosophers to return to the doctrine of the *whole* man. In concrete terms this meant "that the physicians again concentrated more on the spiritual nature of man, whereas the philosophers, following medicine, focussed their interest on the problem of interpreting physiologically the correlation of both systems, united in man". In other words, 'anthropology' here is the *doctrine of psychophysical connection*, i.e. the connection of mind and body, a usage which, as we have already observed, is derived from Platner's pioneering work *Anthropologie für Aerzte und Weltweise* (1772). For further details on the historic roots of this form of anthropology see the Appendix (pp. 201 ff.).

[30] By this is meant what is known as 'romantic medicine' or the 'speculative medicine of the romantics' (cf. W. Leibbrand, 1956, completely rewritten edition of his *Romantische Medizin* [1937]), that is, a movement in German medical thinking, with Landshut, Bamberg and Munich as its centres, which exerted its greatest influence between about 1810 and 1815. From then on its significance was gradually reduced and virtually disappeared between 1830 and 1840 (cf. H. A. M. Snelders, 1973, pp. 146−47). Its star rose with Schelling's triumphant speculative natural philosophy which was brought to the notice of his contemporaries in a series of publications, the most important of which appeared from 1797 to 1800 (see F. W. J. Schelling, 1975 and 1975a), and in the founding of a number of journals (*Zeitschrift für speculative Physik* (1800/1801), *Neue Zeitschrift für speculative Physik* (1802)). Schelling's intensive involvement with medicine dates from his so-called Würzburg period (1803−36). (cf. G. B. Risse, 1972, pp. 154−55). His contributions to the *Jahrbücher der Medizin als Wissenschaft* were also written during this period (F. W. J. Schelling, 1958a and 1958b).

[31] I am thinking here particularly of J. C. A. Heinroth, whose 'personalism' (cf. Heinroth's letter to Damerow in H. Damerow (1844a) and H. G. Schomerus (1966)) prevents him, in my view, from being classed without further qualification as an adherent of the 'natural philosophical school' (as in A. Werner, 1909), or as a protagonist of (romantic) idealistic medicine (as in, for example, F. Hartmann and K. Haedke, 1963, p. 85).

[32] Cf. on this point, among others, O. Marquard, 1973a, pp. 98−99. The turning of philosophy towards medicine was mirrored by a no less striking turn by medicine towards philosophy: "to a quite astonishing extent ... the 'romantic philosophy of nature' has been written by physicians: Kielmeyer, Eschenmayer, Windischmann, Ritter, Treviranus, Oken, Troxler, Schubert, Baader, Carus (to mention only a few names) – these were after all physicians, pharmacists, professors of medicine".

[33] Cf. F. W. J. Schelling, 1958a [1806], p. 65. "Although natural scientists are all, each in his own way, priests and interpreters of certain forces of nature, the physician alone guards the holy fire burning in the centre: he observes the direct presence of God in the workings and the life of an organic body. Medical science is the crown and blossom of all natural sciences, in the same way as the organism in general, and the human organism in particular is the crown and blossom of the world."

[34] Cf. J. Bodamer, 1953, p. 516. Systematic analysis of these periodicals would, I suspect, yield valuable information concerning the relationship between the medical-anthropological tradition and the up and coming (somaticist) institutional psychiatry. In

this connection, the relationship between Nasse and Jacobi is particularly interesting (cf. p. 28 n. 139).

35 See the title-page of J. C. A. Heinroth, 1818.
36 J. C. A. Heinroth, 1822, p. 7.
37 J. C. A. Heinroth, 1822, p. 7.
38 J. C. A. Heinroth, 1822, p. 8.
39 This claim is implicit in what he says (op. cit., p. 373) about the 'reconciliatory method' he uses: "Only, the author of this anthropology has nowhere found that the reconciliatory method was followed".
40 Op. cit., pp. 369–89 (Appendix I: "On the standpoints of anthropological research"); and op. cit., pp. 389–401 (Appendix II: "On the advantages of objective thinking in anthropology").
41 J. C. A. Heinroth, 1822, p. 375.
42 Op. cit., p. 377.
43 Op. cit., pp. 378–79.
44 Here one must think in the first place of a work like J. F. Fries's *Handbuch der psychischen Anthropologie* (1820).
45 "Life eludes the dissecting scalpel, the psychological as well as the anatomical. A lifeless product of research will result when the wound, wherever it occurs, is not healed at once, or when the object of analysis is not immediately considered as an entirety instead of being dissected into increasingly smaller parts without grasping the vital unity. It will be an artificial and pitifully-produced artefact, put together from the torn elements of life, without any natural or spiritual truth, an unreal image of life, just as artificial flowers are mere imitations of real ones." (J. C. A. Heinroth, 1822, p. 380.)
46 J. C. A. Heinroth, 1822, p. 383.
47 For a possible identification of the target of Heinroth's criticism the following passage (which follows directly on from the one quoted in note 45) is important. "A striking example of this kind of an attempt, which failed totally, to construct man, as it were from parts of nature in its entirety, presented itself only a few days ago in all its nakedness, as a general warning. The best we could learn from it is that we have to start at home, with ourselves, if we want to understand that 'which is outside us' to any extent at all, as far as it is granted to us." (op. cit., p. 383). Regarded both in terms of content and chronologically, it is reasonable to assume that Heinroth had H. Steffens' *Anthropologie* (1822) in mind. It is clear that Heinroth's criticism concerning the standpoint of synthesis attacks speculative natural philosophy in general (cf. J. C. A. Heinroth, 1822, p. 392: the mind's task is to turn the material (of knowledge) into real, clear, evident knowledge. So-called natural philosophy is "the first flight of the mind towards this matter", but the mind ultimately cannot be content with this. It reaches its goal "not by a mere creation out of its own potential, by a mere *a priori* construction, in the way natural philosophy had imagined it in the first intoxication of its flight, but rather by a formation of the available matter, which is unable to shape itself.")
48 J. C. A. Heinroth, 1822, p. 387.
49 Op. cit., p. 388.
50 Op. cit., p. 389.
51 Cf. op. cit., pp. 370–72.
52 J. C. A. Heinroth, 1822, pp. 371–72.
53 Cf. J. W. Goethe, 1975a [1823].

[54] Cf. Goethe's review of J. C. A. Heinroth (1822) in *Kunst und Altertum V*, 2, 1825:
"The many excellences this work admittedly possesses are marred by the author himself
because he surpasses the limits set for him by God and nature. We too are convinced
anyway that anthropology could, might and should guide man into the forecourts of
religion, but no further. There the poet will meet him and one may hear, if one listens
carefully:

> In the clarity of our bosom an urge is found
> To submit gratefully, out of our own free will
> To a Higher, Purer Unknown.
> Solving [the enigma of] the eternal Unnamed;
> We call this piety . . . "

(Quoted in J. W. Goethe, 1975, pp. 569–70).

[55] p. 9.

[56] J. C. A. Heinroth, 1822, p. 430.

[57] Cf. op. cit., p. 429: "Without religion man and his existence are meaningless; religion
is the soul of his soul and the key which opens his entire 'mysterious being' ".

[58] Op. cit., p. 430.

[59] Op. cit., p. 430.

[60] J. C. A. Heinroth, 1822, p. 432.

[61] Op. cit., p. 460.

[62] There can be little objection to dealing with Heinroth's views in this order, because
there is a real continuity of anthropologically essential points from the earlier to the
later work; a point in favour of this procedure is that the later work is more explicit
regarding the philosophical and religious presuppositions which also formed the basis
of Heinroth's psychiatry.

[63] J. C. A. Heinroth, 1818, p. 4.

[64] Op. cit., p. 5.

[65] Op. cit., p. 7.

[66] "A voice of dissent (*Widerspruch*) raises itself against the I and its endeavours in the
heart of the being, conscious of itself. This dissenting voice, although within the I, does
not originate with the I but comes from a higher activity which we are accustomed to
call the conscience." (Op. cit., p. 7.)

[67] Op. cit., p. 11.

[68] Op. cit., p. 13.

[69] Op. cit., p. 14.

[70] Op. cit., p. 16.

[71] Op. cit., p. 19.

[72] Op. cit., p. 23.

[73] Op. cit., p. 23.

[74] Op. cit., pp. 24–25.

[75] Cf. J. C. A. Heinroth, 1818, p. 25. "The sinful man lives for the world, or for the I;
and basically for both. Whenever having or being takes hold of him, his will to have is
always aimed at his own existence and being, the one is inconceivable without the
other. The striving after having and being, as an aim in itself instead of as a means to
a higher end, is therefore sinful and a state of diseased humanity because it is a descent
from the realm of freedom for which man was born into the constrained state of the
animals and plants. Conscience . . . is meant to guide him into the realm of freedom and

sanctity for which he was created, when he clings, against his better judgement, out of a self-maintained indolence and inclination, to what is shown to him in his consciousness by a holy voice as not good and unjust, he disturbs his development, i.e. the revelation of life he is destined to realise, in short, the lawful order of being and life itself. His sinning against the highest life is, as far as he himself is concerned, a disturbance of life, a hampering and a restriction, i.e. a state of diseased humanity."

76 J. C. A. Heinroth, 1818, p. 34.

77 Op. cit., p. 35.

78 *Ibid.*

79 Op. cit., p. 38.

80 J. C. A. Heinroth, 1818, pp. 39–40.

81 To these last belong, in his view: "all febrile illnesses in general and brain and nerve diseases *as such* in particular. These are *phrenitis, paraphrenitis,* hydrophobia, febrile *delirium, catalepsis,* apoplexy and all somnolent conditions, illnesses of the senses, i.e. so-called *hallucinations,* epilepsy, St. Vitus' dance, ergotism, nightmares, hypochondria and hysteria. Yes, even so-called mental weaknesses and poor psychic habits have to be distinguished and separated from psychic disturbances in the proper sense". (Op. cit., p. 41.)

82 "*Continuous lack of freedom or lack of reason existing, as an illness or as a diseased state, independently and for itself even in apparent physical health, including the field of emotional, mental and volitionary diseases, perfectly summarises psychic disturbance*". (Op. cit., p. 42.)

83 It must be borne in mind here that within the field of 'romantic medicine' Heinroth's psychiatry was by no means unique. We find something resembling a sin-theory of illness defended in C. J. H. Windischmann, 1975 [1824]. J. N. von Ringseis, who is discussed in Chapter 3, is a kindred spirit. Cf. further W. von Siebenthal, 1950, pp. 62–83.

84 J. C. A. Heinroth, 1818, pp. 171 ff.

85 Op. cit., p. 174.

86 Heinroth speaks not only of '*Bedingungen*' (conditions) or elements, but also of '*ursachliche Momenten*' (causal factors) (op. cit., p. 194).

87 J. C. A. Heinroth, 1818, p. 174.

88 Op. cit., p. 175.

89 Op. cit., p. 176. A ("platonistic") viewpoint which Heinroth expressly claims to be original, since he speaks of it as "A way of thinking that has never yet been pursued, but which is the only one that leads us to the correct standpoint for a theory of mental disturbances."

90 Op. cit., p. 177.

91 Op. cit., p. 178.

92 Op. cit., p. 179.

93 Op. cit., p. 181.

94 *Ibid.*

95 Op. cit., p. 193.

96 *Ibid.*

97 Op. cit., p. 194: "The inception of psychic disturbances was not only compared with, but also considered identical to procreation".

98 Op. cit., pp. 194–95.

99 Op. cit., pp. 195–237.

[100] The stimulus is by definition 'external', that is, external in respect of the (psychic) *reaction* it provokes. In this sense, not only stimuli from outside, but also those from inside man (fantasies, emotions, thoughts, impulses) are 'external'. Cf. op. cit., pp. 211–12.

[101] What Heinroth and Jaspers have in common can only be seen if the two conceptions are considered from the viewpoint of the opposition to a dogmatic positivistic or materialistic denial of the religious and/or existential dimension. Apart from this, the two are diametrically opposed, because Heinroth refuses to make precisely that sharp distinction between the religious and existential, and science, which is the cornerstone of Jaspers' philosophy. Examples of the integration of the existential viewpoint into scientific enquiry in modern medicine and psychotherapy can be found, for example, in A. Jores (1950), V. E. von Gebsattel (1968), V. E. Frankl (1972).

[102] Cf. for example A. Kronfeld, 1927, p. 13.

[103] Cf. J. Bodamer, 1953, p. 520.

[104] Cf. W. Leibbrand & A. Wettley, 1961, pp. 492 ff. To my knowledge, these authors are the only ones in the existing historical literature on psychiatry who have noted this point: the chapter on Heinroth and Jacobi is aptly entitled *Kontrastierende Leib-Seele-Auffassungen* (Contrasting Body–Soul Conceptions).

[105] Cf. Plato, *Laws* 959. It is interesting to compare what Heinroth says about Plato's "anthropology" with what has been brought out about Heinroth's own anthropology in the preceding pages. The agreement between them is immediately obvious: " . . . Plato . . . in his writings, i.e. in Phaedo, Alcibiades, in the books of the Republic and on the Laws, firstly distinguishes physical life – as organic – from the mental life, the real life of man. The body is for him a tool, the soul that which uses this tool. Soul, as the more noble, precedes the body, according to him; and he is right in as far as he considers the idea of man (in this connection identified with the concept of soul) to be [ontologically] prior to the phenomenon of man". (J. C. A. Heinroth, 1822, p. 11). It is striking, too, that in Heinroth's discussion of the history of anthropology a page of text is devoted to Plato, whereas Aristotle is only *mentioned* in the bibliographical notes!

[106] Summary of an unpublished lecture given in Nijmegen in 1974.

[107] Cf. p. 19.

[108] C. J. de Vogel, op. cit.: "Socrates states in Krito as a self-evident principle that 'To live in a body which is in a hopelessly bad condition is a worthless life for a human being' (47e). We read the same thing in Gorgias (505a). The education of watchmen and philosophers in the State is based on the principle that body and mind must be equally developed. Anyone who is not prepared to undergo physical exertion is also, according to Plato, not suited to the intense intellectual efforts which philosophy demands. (Republic VII 535 c–d)."

[109] Considering the contrast with one-sided, spiritualistic Platonism, one could characterise Heinroth's position as Platonic. This term suggests, however, too far-reaching an agreement with the philosophical position of the historical Plato to be applicable without demur to *any* Christian Platonist. I therefore prefer to think of Heinroth as someone who represents a more platonic variant of Platonism than some traditional Christian Platonists. It should be noted in passing that Plato's anthropology was at this time considered relevant to anthropology-oriented psychiatry by others besides Heinroth. This can be deduced from the fact that an article like Sprengel's *Ueber Plato's Lehre*

von den Geisteszerrüttungen (On Plato's theory of madness) was included in a specifically anthropological, psychiatric journal, namely the *Zeitschrift für psychische Aerzte*, edited by Nasse (K. Sprengel, 1818). As in De Vogel's retouching of the too dualistic image of Plato's anthropology in the tradition of Christian Platonism and essentially in agreement with Heinroth's variant of Platonism, Sprengel lays particular emphasis on the significance of Plato's *Timaeus*.

110 Cf. W. Leibbrand & A. Wettley, 1961, p. 494.

111 Whether one referred to psychic *disorders* or psychic *diseases* depended on the variant of the somatic theory one defended. Jacobi talks about disorders because he denied that so-called 'psychic diseases' had an independent character. Friedreich, on the other hand, did not agree with him on this point and therefore speaks of psychic *diseases*. This controversy was later (in 1844) to take shape in the polemic between Jacobi and Nasse in the first volume of the *Allgemeine Zeitschrift für Psychiatrie*.

112 J. B. Friedreich, 1964 [1836], p. 86.

113 Examples are to be found in the historical review in Friedreich, op. cit., pp. 87 ff.

114 Op. cit., p. 90. The first work, chronologically, that Freidreich mentions in this context is Münch's *Praktische Seelenlehre* (1801).

115 L. Snell, 1872a, p. 416.

116 F. Nasse, 1818a.

117 It should be remembered that Heinroth's major work, the *Lehrbuch der Störungen des Seelenlebens*, appeared in the same year – 1818.

118 It was recognised not only by Friedreich (op. cit., p. 300) but also by, for example, L. Snell (1872a, p. 423) and W. Griesinger (1872c, p. 80) that Jacobi 1844a must rank as the trendsetter in the nineteenth-century somatic school of psychiatry.

119 Op. cit., p. 415.

120 Op. cit., p. 423.

121 J. Bodamer, 1953, p. 517. For that matter, this is hard to reconcile with the formulation he gives immediately before this, where he speaks of "the sharp contrast between the fundamental psychicist and somaticist concepts in theoretical psychiatry".

122 Cf. pp. 11–12.

123 J. B. Friedreich, 1964, p. 300.

124 J. Wyrsch, 1956.

125 An anthology of quotations which illustrate this can be compiled from the extensive quotes given in J. B. Friedreich, 1964, pp. 90–111. "If souls could be diseased, they could also die" (Münch, 1801); "Diseases are only possible, conceivable, in relation to living organisms" ... "Therefore there are no psychic diseases; one can only say that certain diseases disturb or annul the operations of the soul" (Ruland, 1801); "I have never been able to conceive a suffering of the soul itself ... " (Haslam, 1798); " ... I have never been able to conceive a disease of an incorporeal being like the soul" (Spurzheim, 1818); "In my opinion no presupposition is more indispensable for a correct judgement of insanity than the one which states that there is no immaterial insanity in the strict sense of the word" etc. "The only thing that counts is that we consider the soul as an independently existing, mysterious 'something', which ... can itself never become diseased" (Klose, 1824); "With [such] a distinction between an ideal and a real side of the soul, the answer to the question ... 'if it is at all possible for the soul to become diseased' appears to be superfluous. That part of the soul which is supernatural and eternal of course can never be susceptible to disease." (Stark, 1824).

126 W. Griesinger, 1872c, p. 105.

127 Cf. Appendix, p. 207.

128 If there is nothing about psychiatry in Aristotle, St. Thomas is a very different matter. It is true that there is no systematic treatment of psychopathology, but that does not alter the fact that in many places in his work (particularly in *Summa Theologiae secunda pars*) observations are made in passing which allow us to build up a picture of St. Thomas's psychopathology (cf. Appendix IV (emotional psychopathology) of St. Thomas Aquinas, 1965, pp. 156–63).

129 J. Wyrsch, 1956, pp. 29–30.

130 I am quoting from Aristotle, 1959, p. 55. St. Thomas's text reads as follows: "Cujus ratio est, quia nos videmus quod omnes debilitationes quae fiunt circa intellectum et sensum, non attingunt ad ipsam animam secundum se, sed proveniunt ex debilitate organi. Unde videtur quod intellectus et omnis anima sit incorruptibilis, et quod debilitatio in eius operationibus non sit ex eo quod ipsa corrumpatur, sed ex eo quod debilitantur organa. Si enim corrumperetur anima, maxime corrumperetur a debilitate quae est in senectute, sicut accidit in organis sensitivis quae debilitantur ex senecute, tamen anima non debilitatur ex hoc, quia si senex accipiat oculum juvenis, videbit ut juvenis. Quare senectus debilitat non quidem quod ipsa anima patiatur, seu virtus sensitiva, sed id in quo est. Sicut in aegritudinibus et in ebrietatibus non debilitatur, seu alteratur anima, sed corpus."

131 J. Wyrsch, 1956, p. 30.

132 Cf. Appendix, pp. 218–19.

133 In this connection one might think of J. B. Friedreich's theory.

134 Cf. Friedreich, 1964, pp. 86. 284–86, which is based on A. Combe, 1831.

135 Cf. W. Griesinger, 1872c, pp. 90, 100.

136 Quoted in W. Griesinger, 1872c, p. 95.

137 F. Nasse had already summed up the somaticist standpoint as a conception according to which one cannot refer to psychic disease, but only to psychic symptoms (or symptomatic insanity): psychic disorder is a symptom of the somatic dysfunction which causes it (F. Nasse, 1818a, pp. 132, 135).

138 W. Griesinger, 1872c, p. 90.

139 One is well-advised not to underestimate the real complexity of the somaticist tradition. Anyone who approaches the controversy between the psychicists and the somaticists by way of the work of the then influential "somaticist" F. Nasse will get an entirely different impression from that given by our comparison of Heinroth and Jacobi. This is bound up with the fact that Nasse viewed the conflict between psychicists and somaticists primarily as a conflict of (one-sided) *aetiological* points of view, and not primarily as a conflict on the level of philosophical *anthropological* presuppositions (concerning the soul-body relationship) which have secondary implications for psychiatric aetiology. The relativisation of the two points of view (conceived thus), as it occurs within the framework of his own psychosomatics, remains in this respect indirectly linked to the Cartesian postulate, something which cannot be said of either Heinroth or Jacobi.

Nasse, who did more than anyone else at that time to combat the Cartesian division of human nature into separate compartments of mind and nature, and the parting of the ways between psychiatry and medicine which this had provoked, and who defended the standpoint of a 'somaticist' variant of 'psychosomatic' (i.e. anthropological) medicine (cf., for example, F. Nasse, 1818 and 1818a, and F. Nasse, 1820) *approached* in his psychosomatic views the Aristotelian somaticists. One can acknowledge all this without thereby denying the fundamental difference between the two positions. (Of relevance to this difference is the polemic carried on by Nasse and Jacobi in Vol. 1 of the *All-gemeine Zeitschrift für Psychiatrie* (1844). Cf., for example, from this, M. Jacobi, 1844, pp. 371–72). The credit for directing the attention of modern scientific medical historians towards Nasse belongs to H. Schipperges, (cf. H. Schipperges, 1959).

140 In my view, Ackerknecht, 1968, p. 61, is certainly correct when he observes that 'romantic' psychiatrists like Heinroth and his colleagues were rightly known as 'psychicists' and not as psychologists: "What they [i.e. the "psychicists"] called psychological was largely moralistic, while strangely enough, what the French in this period designate as 'moral' was largely psychological".

141 Quoted in W. Leibbrand and A. Wettley, 1961, pp. 496–97.

142 Op. cit., p. 498.

143 Cf. J. B. Friedreich, 1964, p. 287, which draws on M. Jacobi, 1830.

144 E. Ackerknecht, 1968, p. 60.

145 This concept of formal psychic disorder has a polemic point which is directed against the psychicism of Heinroth and others. The physician (i.e. psychiatrist), says Jacobi, only has to do with the formal aspect of the soul and not with the "higher, truly mental activity in itself, the inner side of human life, the substance of the personality", since the *basis* of psychic disorder (i.e. that which makes the disorder a psychic disorder) is contained in the formal, cf. W. Leibbrand & A. Wettley, 1961, p. 498. The fact that Jacobi observes elsewhere of this same physician (psychiatrist) that he is "as such, somatologist, physiologist, physicist", who is concerned with "a physiology of psychical phenomena" must present a contradiction if viewed from any interpretation of the soul–body relationship other than the Aristotelian-scholastic interpretation. In fact, there is no contradiction between these two descriptions of the physician. The physician (psychiatrist) who occupies himself with the formal aspect of the soul (the psychic faculties) is the same one who deals with the physiology of psychic phenomena. Assuming the unity of soul and body, the difference between medicine directed towards the soul and that directed towards the body means no more than a difference between two poles of interest, two lines of approach.

146 This expression derives from the sarcastic description of Heinroth's sin-theory in J. B. Friedreich, 1964, p. 24: "Now we know all of a sudden how psychic diseases are developed, a problem which has already baffled so many physicians and philosophers. We do not need philosophy or physics, we need nothing at all; the great mystery is solved in a very simple way by this *Heinrothian* copulation theory. Evil approaches the soul . . . which is attracted to it . . . bends over to it and allows itself to be pulled down into the realm of darkness where the peculiar union is then consummated, which brings forth, as its fruit, psychic disease".

147 L. Snell, 1872, p. 424.

148 Jacobi's main early work appeared in M. Jacobi, 1822/1825 and 1830. The latest to which I had access was M. Jacobi, 1854.

149 L. Snell, 1817–92; director of the Eichenberg institution, and later of the institution in Hildesheim.

150 The syndrome described by Snell quickly achieved recognition in Germany (although not in France), and can be found under various names in the contemporary psychiatric

literature. 'Primary insanity' was officially acknowledged as a nosological entity in 1877, when Meynert put it to the vote at the Assembly of German Psychiatrists in Nürnberg. The meeting voted unanimously to recognise 'primary insanity' as an independent form of illness; cf. J. Bodamer, 1953, p. 531.

[151] L. Snell, 1852.

[152] The affinity with elements of psychoanalytic theory which sometimes emerges from some of his pronouncements is indeed striking. Snell appears to have been looking for the expression of 'deeper-lying' psychological laws in the psychopathological phenomena he describes. These laws, in the final analysis, he considered to be based in the structure of the brain. These remained, however, no more than suggestions (cf. L. Snell, 1852, p. 18, and L. Snell, 1860, pp. 545 and 553). He was, however, concerned less with causal explanation than with the descriptive definition of clinical pictures (as, for example, in L. Snell, 1872) and with extending practical diagnostic aids by describing characteristic symptoms. The heuristic significance of the comparison of hallucinations and word formation in disturbed patients with the dream life of the 'normal' individual did not escape him: "Apparently the faculty affected by illness [he is referring here to hearing] hereby acts with a certain independence and caprice in the case of the mentally ill [suffering from acoustic hallucinations]. Something similar occurs in normal mental life once in a while in dreams, as the dream fantasies are very similar to hallucinations and can often serve in the explanation and investigations of the latter. In the same way as absolutely unreasonable and impossible configurations sometimes appear in dreams, these may also contain peculiar word-formations" (L. Snell, 1852, p. 18).

The line of research pioneered by Snell and further developed at the beginning of this century by E. Bleuler and C. G. Jung at Burghölzli is being pursued today by the American psychiatrist, S. Arieti. The research carried out by Arieti and others is concerned with demonstrating – following on from the theories of Cassirer, Von Domarus and Goldstein – that the thought process of the schizophrenic can be understood not only in terms of content, from the viewpoint of conflict dynamics, but also – formally speaking – is characterised by retrievable regularities that are systematically related to psychodynamic development (cf. G. Benedetti, 1973, p. 450), while G. Bosch, for example, who is more interested in content, follows, in his study of infantile autism, the line of "a clinical and phenomenological-anthropological investigation taking language as the guide" (cf. G. Bosch, 1970).

[153] J. Bodamer, 1953, p. 530.

[154] This, strangely enough, is a point Bodamer does not pursue further.

[155] H. Ey, 1962, p. 1.

CHAPTER 2

[156] E. J. Dijksterhuis, 1975, pp. 456, 457.

[157] As for example in the article on mechanism in Winkler Prins, 1975, p. 778.

[158] E. Cassirer, 1973, p. 94.

[159] Cf. E. Cassirer, op. cit., p. 98.

[160] E. Cassirer, ibid.

[161] Cf. E. Cassirer, 1973, pp. 103 ff.

162 Amongst these were professors and statesmen, as well as people who were later to become famous, such as Søren Kierkegaard, Jacob Burckhardt, Friedrich Engels and Bakunin; cf. F. Copleston, 1965, p. 124.

163 We still find this misconception, for example, in J. Hirschberger, 1969.

164 In this shift of emphasis I am following M. Mandelbaum (1974, p. 5) (who on this point – with some reservations – is harking back to the work of Höffding [undated]), but unlike Mandelbaum I conceive the term positivism in a sense which includes both natural scientific positivism and the positivism of the humanities. In particular, I assume that what can be described as natural scientific positivism, with emphasis on the *experience ethos*, can be called (natural scientific) *naturalism*, when the stress lies on the implied *methodic* commitment. (For a more precise definition of the terms positivism and naturalism, cf. respectively pp. 60–61, 77 n. 275, 190 n. 689. Given these terminological restrictions, one can in any event agree with Mandelbaum's *formulation*, for example where he observes that individual thinkers, like Kierkegaard, who could be identified neither with positivism nor metaphysical idealism, but also representatives of materialism, like Feuerbach and Marx, who both constituted themselves – for all their mutual differences – as opponents of both idealism and positivism, fall outside this contrast between metaphysical idealism and positivism. (Kierkegaard only became influential in the twentieth century because of his 'rediscovery' by K. Jaspers, and the influence of Feuerbach and Marx in the nineteenth century was not based on their systematic philosophy, but on "other aspects of their thought", through which "they were intimately connected to the major intellectual developments of their time" (op. cit., p. 5)).

165 Cf. M. Mandelbaum, 1974, pp. 7–8.

166 In histories of nineteenth-century German philosophy, the famous cry of 'back to Kant' is generally attributed to Otto Liebmann, quoted from his work *Kant und die Epigonen* (1865), and is regarded as the starting signal for the Neo-Kantian movement. This turning towards Kant, however, does not only occur in the work of Liebmann's teacher Kuno Fischer, but is also found, as Cassirer (1973, p. 11) observed, in the work of Helmholtz – one of the earliest authors to take this line (the work in question is Helmholtz, 1855, pp. 87–88, 99, 116). This does not make Helmholtz a Neo-Kantian, but it does draw our attention to the fact that Liebmann's call came at a moment when the trend towards Kant's philosophy could scarcely be called new any longer.

167 The problem of knowledge was a major philosophical point for the positivists like Helmholtz and his followers, but not for the metaphysical idealists, and since in the opposition between metaphysical idealism and positivism the latter was to gain the upper hand, the problem of knowledge was to become a central one.

168 E. Cassirer, 1973, p. 24.

169 Op. cit., pp. 18–19.

170 This is true of the commencement of each of the phases. It is impossible to give strict chronological limits because the origins of the scientific initiative which distinguishes the second phase from the first overlap chronologically the end of the first phase. It is nevertheless still possible to indicate with a certain degree of accuracy when the shift in emphasis took place: the year 1847 marked a turning-point in this development, because it was in this year that the programme of *physiological reductionism*, prepared in the eighteen-forties by the combined efforts of Du Bois-Reymond, Helmholtz, Brücke and Ludwig, came to the fore. Cf. on this point D. H. Galaty, 1974, p. 295.

[171] At first glance one is inclined to look for the historical roots of Herbart's mechanistic psychology among the founders of British associationist psychology, Locke and Hume: it was, after all, Hume who conceived (empirical) psychology as a sort of representational mechanics analogous to Newton's mechanics. A closer examination of Herbart's psychology makes it seem less likely that the British empiricists exercised a primary and direct influence, for it strikes one that in the partly appreciative, partly critical discussion of earlier psychological theories in Herbart's major work, Locke figures alongside Leibniz, and Hume is not even mentioned (cf. J. F. Herbart, 1890b, Section 17 ff.).

[172] I. Kant, 1922b [1787], p. 374.

[173] I. Kant, 1922c [1788], p. 106.

[174] I. Kant, 1922c [1788], p. 103.

[175] It is in agreement with this that the (Kantian) conception of psychological freedom thus only has a place in the context of Kant's criticism of the *naturalisation* of freedom, as this takes place in the thinking of those who believe that freedom (I repeat) is a (psychological) property of the mind, "the explanation of which would only depend on a more accurate examination of the nature of the soul and the incentive of the will", that is to say, in Kant's view the (philosophical) psychological traditions of both empiricism and rationalism; criticism which aims at expressing the idea that there is a meaning of 'freedom' in which freedom (ontologically speaking) is not of the order of *nature* and (viewed in terms of the critique of knowledge) does not belong to the world of (in this case psychic) *phenomena*: freedom in the transcendental sense.

[176] *Kritik der reinen Vernunft* (Critique of Pure Reason) A341–A405: "On the paralogisms of pure reason" (I. Kant, 1922b [1787], pp. 273–90).

[177] Cf. "Psychology is, as far as human understanding is concerned, nothing but anthropology, and it can never become any more than this, i.e. a knowledge of man under the restriction that he understands himself as the subject of his inner sense" (I. Kant, 1923b [1804], p. 294).

[178] I. Kant, 1923a, pp. 332 ff.

[179] Cf. on this point, for example, E. Husserl, 1971 [1910/11] and the definition given there of what, to Husserl, is the essence of naturalism: " . . . What is characteristic of all forms of extreme and consequent naturalism . . . is on the one hand the *naturalisation of consciousness*, including all the intentional-immanent data of consciousness; on the other, the *naturalisation of ideas* and therewith of all absolute ideals and norms." (op. cit., p. 14).

[180] Including J. F. Herbart, R. H. Lotze, F. E. Beneke, F. A. Lange, J. Müller, H. Helmholtz, E. H. Weber and G. T. Fechner.

[181] In *Brett's History of Psychology* (in R. S. Peters, 1962, pp. 533–609) Kant, together with Herbart, Beneke and others, is rightly dealt with under the heading "Psychology becomes self-conscious".

[182] 'Subjective' refers, of course, to the subject of the consciousness.

[183] Thus Wundt's psychology culminates in the consideration of (the laws of) the two main forms of synthesis [*Verbindung*] distinguished by Wundt, i.e. association and apperception; cf. W. Wundt, 1911, pp. 492–500 (*Allgemeine Übersicht der Formen psychischer Verbindung*).

[184] The first German translation of Mill's *A System of logic, ratiocinative and inductive*

(London, 1843) was made in 1870 at the instigation of the celebrated German chemist Justus von Liebig.

[185] Cf. K. E. Rothschuh, 1953, p. 123.

[186] According to K. E. Rothschuh, op. cit., p. 123.

[187] In the words of K. E. Rothschuh, to whom the historiography of physiology owes this important correction, by which Carl Ludwig has acquired his rightful place in the history of physiology: "Müller was an extraordinary *forerunner* and pathfinder in experimental physiology, but he was not a great inaugurator and *discoverer* like C. Ludwig, to whom we are indebted for countless completely new insights. J. Müller may have been far greater than Ludwig in terms of spiritual eminence, education, versatility and universality, but none of this is relevant to the question of who has been of greater significance, i.e. who has had *the greatest influence on the development and course of subsequent trends in physiology.*" (op. cit., p. 122).

[188] Cf. K. E. Rothschuh, 1953, pp. 92 ff.

[189] The division of the course of Müller's development into these periods is taken from the monumental *Gedächtnissrede auf Johannes Müller*, with which E. Du Bois-Reymond honoured his teacher in 1858. (E. Du Bois-Reymond, 1887b.) This memorial address is invaluable for a reconstruction of Müller's scientific development.

[190] Referred to here are J. Müller, 1826 and 1826a.

[191] Quoted in E. Du Bois-Reymond, 1887b, p. 159.

[192] According to E. Du Bois-Reymond, 1887b, p. 167.

[193] Quoted in E. Du Bois-Reymond, 1887b, pp. 171–72, from Müller's foreword to his *Bildungsgeschichte der Genitalien* of 1830.

[194] E. Du Bois-Reymond, 1887b, p. 214.

[195] Anyone who undertakes a more thorough analysis of the philosophical background to Müller's thinking will wish to give serious attention to A. Werner, 1909. Werner's description of Müller's ties to the (Schellingian) philosophy of nature implies that Müller should not (like Du Bois-Reymond) be characterised as a *dogmatic* vitalist, but rather as the exponent of 'transcendental realism'. Cf. A. Werner, 1909, pp. 147–48. The attractive thing about this thesis is that it does not force us to accept that Müller's tolerant attitude towards the physical-chemical school in physiology, supported by his pupils and others, was *irreconcilable* with his basic philosophy. Cf. also A. Liebert, 1915, for the 'critical' element in Müller.

[196] E. Du Bois-Reymond, 1877b, p. 217.

[197] W. Griesinger, 1872, p. 13.

[198] One thinks here in the first place of the so-called Müller-Hall law of reflex action – one of the cornerstones of Griesinger's psychopathology as 'mechanical natural science'. Cf. J. Müller, 1844, pp. 608–23, particularly pp. 609–10 (for the difference between Müller's conception and that of M. Hall).

[199] E. Du Bois-Reymond, 1887b, pp. 218–19.

[200] E. Du Bois-Reymond, 1887b, p. 219.

[201] In the third part (Abschnitt) of T. Schwann, 1839.

[202] Quoted in E. Du Bois-Reymond, 1887b, p. 305; the full text of the letter is printed in op. cit, pp. 302–6.

[203] E. Du Bois-Reymond, 1887b, pp. 305–6.

[204] H. von Helmholtz, 1856/66; 1863.

[205] The theory of the 'specific energies of the senses' first appears in J. Müller, 1826a,

pp. 45, 52 ff., and J. Müller, 1826, pp. 5 ff., and is further found in J. Müller, 1844, pp. 667–68, in which, strikingly, the term 'energy' is no longer used. (I have been unable to consult Volume II of Müller's *Handbuch*.)

[206] P. Diepgen, 1959, p. 136. A list of the most important secondary literature is in E. G. Boring, 1957, p. 91. To the best of my knowledge, the only elucidation of the natural philosophical background to Müller's scientific *oeuvre* in general and the theory of the specific energies of the senses in particular is to be found in A. Werner, 1909, p. 140. According to Werner, Müller's empirical scientific work in all its diversity is nothing more than the carrying out of the research programme conceived by Müller in J. Müller, 1824, and the definitive formualtion of the law of the specific energies of the senses attributed to him is no more than "the result of subsequent work on the system of sensibility, and as such ... is the common property of the entire philosophy of nature school".

[207] J. Müller, 1826, p. 6.

[208] Cf. for this, for example, H. von Helmholtz, 1903c, pp. 392–93, and H. von Helmholtz, 1921a, pp. 115 ff.

[209] According to M. Dessoir, 1964, p. 401, Hebenstreit (1786) was the first German researcher in this field.

[210] Against the current overestimation of Müller's dependence on Kant (based on the parallel which has been drawn between Kantian subjectivism and Müller's fundamental ideas about the physiology of the senses), it must be stated emphatically that one can speak of no more than an *analogy* between Kant's *a priori* forms of *Anschauung* and Müller's notion of the 'innate energy' of our senses – the foundation of what is known as nativistic physiology. Müller was too much of a sensualist to regard Kant as an ally. As Du Bois-Reymond, 1887b, p. 210, observed, Müller rejected Kant's innate categories of the understanding and even so far in his sensualism as to doubt the apriority of the concept of causality, and to consider the capacity to form general concepts as the only original faculty of the human mind. A. Liebert, 1915, amongst others, stresses the Kantian element in Müller's thinking.

[211] H. von Helmholtz, 1903d, pp. 181–82.

[212] H. von Helmholtz, 1903b, p. 296.

[213] In Aristotle ἐνέργεια (ἐν ἐργῳ εἶναι) is the actuality as distinct from potentiality (i.e. the capability of existing or acting, δύναμις) (cf. Metaph. 1048a26 etc.) Seeing, knowing, living, etc. are 'energies'.

[214] E. G. Boring, 1942, p. 71. Although Müller's formulations on this point were explicit enough (cf. J. Müller, 1822 and 1826a, pp. 45–46) and A. Werner (1909, pp. 125, 141) also contains points of reference for a correct analysis in history of philosophy terms of Müller's use of the Aristotelian distinction between κάτα δύναμιν and κατ' ἐνέργειαν, in interpreting and criticising the theory of the specific energies of the senses subsequent generations of philosophers and scientists have systematically disregarded Müller's intention. Among present-day historians of science, to the best of my knowledge, only Riese shows any signs of being on the right track: "One cannot escape the conclusion that in the final analysis the law of the specific energies of the nerves rested on a metaphysical rather than an anatomical and physiological argument". (W. Riese, 1959, p. 123.) A more detailed examination of the philosophical background and *Wirkungsgeschichte* of Müller's work would also be able to throw light on the question of the extent to which the ambiguity in Müller's formulation of the theory of the specific

energies of the senses criticised by R. Weinmann (1895, pp. 50 ff.) is the result of bias in favour of psychophysical parallelism, which has prevented the recognition of the meaning of the concept of 'energy' which Müller intended. In this respect, Weinmann's criticism repeats, in my view, the same misunderstanding which had previously emerged in the criticism of Lotze and Wundt. (Cf. R. H. Lotze, 1848 [1842], p. 159; 1852, p. 188; W. Wundt, 1891.)

215 H. von Helmholtz, 1882a [1847].

216 Thus, for example, also in L. Kolakowski, 1972, pp. 10–11: "Positivism stands for a certain philosophical attitude to human knowledge"; " . . . Positivism is a normative attitude, regulating how we are to use such terms as 'knowledge', 'science', 'cognition', and 'information' ".

217 Antonio Aliotta, 1914, p. 53, quoted by M. Mandelbaum (1974, p. 376, note 23), who, like J. Passmore, 1966, for example, makes a distinction between positivism before 1870 and positivism after 1870. This breakdown into phases corresponds with the distinction between systematic or dogmatic positivism (Comte, Spencer) and critical positivism (Mach, Avenarius). (However, see also note 219.)

218 Described by A. Aliotta (see note 217) as: "searching criticism (of science itself) in order to eliminate any traces of metaphysics which might be sheltering themselves beneath the cloak of scientific theories".

219 Needless to say, the expression 'critical positivism' used hereafter thus stands for a historical manifestation which is distinct from dogmatic (or systematic) positivism on the one hand and from the empirio-criticism of Mach and Avenarius on the other. (The latter is also often called critical positivism – a usage that I, therefore, am not employing.) The view put forward by R. S. Cohen that Helmholtz should be described not as a positivist but as an empiricist (R. S. Cohen/Y. Elkana, 1977, pp. XII–XIII) is only tenable if one accepts as a starting-point the positivism of Mach and Kirchhoff (science does not explain *why* things happen as they do, but concentrates on the *how* of phenomena without penetrating to what lies hidden behind these phenomena). For the reasons I have already given I do not accept this.

220 H. von Helmholtz, 1903d, p. 189.

221 Op. cit., p. 189.

222 The most important of Helmholtz's epistemological writings (the chief of which are included in the selection made by Hertz and Schlick (H. von Helmholtz 1921)) is his treatise *Die Thatsachen in der Wahrnehmung* (1878).

223 Cf. M. Mandelbaum, 1974, p. 14.

224 As far as this is concerned, it is certainly no coincidence that two leading figures in neopositivism, Paul Hertz and Moritz Schlick, edited a collection of Helmholtz's epistemological writings, with a commentary, on the occasion of the hundredth anniversary of Helmholtz's birth (H. von Helmholtz, 1921).

225 H. von Helmholtz, 1882a [1847], *Einleitung* (Introduction).

226 E. Cassirer, 1973, p. 93.

227 "My original inclination drew me towards physics; external circumstances forced me to take up the study of medicine . . . " (H. von Helmholtz, 1903d, p. 169.)

228 Also referred to as J. R. (Julius Robert) von Mayer. The prefix 'von' is found in the names of many prominent scientists of this period, who had been raised to the nobility in recognition of their services to science. Among them, for example, were Lotze, Helmholtz and Liebig. Mayer was ennobled in 1867.

[229] Mayer was a close friend of Griesinger. In the edition of their correspondence (1842–45) Preyer writes (W. Preyer, 1889, pp. 127–28): "In 1864, when I asked him [i.e. Mayer] at the meeting of natural scientists in Giessen whether he felt more at home in the physical section or the medical one, he replied cheerfully that he did not really know; he was floating from one to the other, but he was pleased that both the physicists and the physicians were happy to consider him as one of themselves. He had registered in the physics section. That was twenty-two years after the first publication of his discovery."

[230] In the same vein, W. Preyer, 1889, IX, and M. Mandelbaum, 1974, p. 291.

[231] In H. von Helmholtz, 1882a.

[232] For the reductionism implied in this physicalisation, the (Kantian) background and the genesis of the reductionist programme, cf. D. H. Galaty, 1974.

[233] E. Cassirer, 1973, p. 102, makes the important observation that the formulation which Robert Mayer gave to this principle leaves open the possibility of a *non-mechanistic* interpretation: "He saw as the core of the principle the *equivalence* of heat and mechanical energy, without intending to interpret this as the identity of the two. Mayer wanted explicitly to eliminate the question of the 'nature' of heat."

[234] Cf. E. Du Bois-Reymond, 1848/1860 (his principal work) and E. Du Bois-Reymond, 1875/77.

[235] Strictly speaking, this refers to a short essay which was originally incorporated in the foreword to his major work *Untersuchungen über thierische Electricität* (Investigations into animal electricity), and was later included as a separate study in E. Du Bois-Reymond, 1887a.

[236] E. Du Bois-Reymond, op. cit., p. 24.

[237] Literature references in E. Du Bois-Reymond, op. cit., p. 9 (27), note 3. In this context it should be remembered that the first blow to undermine vitalism dated from as early as 1828, when F. Wöhler succeeded in creating an organic substance (urea) from inorganic components – an achievement which dashed the vitalists' hopes of finding a specific 'substance of life'.

[238] To mention just one of the arguments: according to the theory of life force, "it [i.e. life force] . . . should be transferred in reproduction without any loss and in that way be increased indefinitely", while this same life force "[would] come to a complete end in death . . . without any appropriate effects as a replacement, in order to make way for common physical and chemical forces." Neither of these theses can be reconciled with the idea of the *conservation of force*. (E. Du Bois-Reymond, 1887a, p. 19.)

[239] This refers to E. Du Bois-Reymond, 1886a and 1886b, respectively.

[240] E. Du Bois-Reymond, 1887a, p. 10.

[241] Op. cit., p. 25.

[242] What is meant is the superhuman mind described in Laplace, 1829 [1814²], p. 2 ff. "An intelligence which, at a given instant, would know all the forces which animate nature, and the respective situation of the beings which make it up; supposing moreover it were sufficiently immense to submit these given data to analysis, embracing in the same formula the movements of the largest bodies in the universe and those of the lightest atom: there would be nothing of which it was not certain, and the future, like the past, would be present to it. In the perfection which it has been able to give to astronomy, the human mind presents a feeble sketch of this intelligence."

[243] E. Du Bois-Reymond, 1886a, p. 123.

244 Op. cit., p. 125.

245 E. Du Bois-Reymond, 1886, pp. 391–97.

246 E. Cassirer, 1973, p. 94. In a similar vein A. Werner, 1909, p. 148.

247 The distinction made here between Helmholtz's conception of mechanism and that of Du Bois-Reymond is meant to be descriptive; that is to say, the distinction holds good independently of whether one is or is not of the opinion that the separation of the methodological and the metaphysical viewpoint can be rigidly sustained.

248 Cf. E. Lesky, 1965, p. 261.

249 Quoted in E. Lesky, 1965, p. 261.

250 Quoted in E. Lesky, 1965, p. 260.

251 Op. cit., p. 260.

252 S. Freud, 1972a, p. 35.

253 Op. cit., p. 290.

254 Cf., for example, P. Amacher, 1965.

255 H. F. Ellenberger, 1970, p. 541.

256 Quoted in K. E. Rothschuh, 1953, p. 119.

257 This demand for precision in the history of philosophy has often been ignored in the past. In the criticism of what later came to be called 'scientism', positivism and materialism were treated as the same thing, because they both found themselves in opposition to all traditional theology and both (with the exception of Feuerbach) defended the natural scientific ideal of knowledge. Positivism and materialism differ, however, in that materialism, unlike positivism, is a metaphysical position. (Cf. M. Mandelbaum, 1974, pp. 22–23.)

258 In general, the metaphysical position of materialism as it was defended in the nineteenth century can be described (with M. Mandelbaum, 1974, pp. 20 ff., cf. F. Gregory, 1977, X ff.) as a position which subscribes to to the following theses: (1) there is an independently existing world; (2) human beings are also material entities; (3) the human mind does not exist as an entity distinct from the body; (4) there is no God (or other non-human being) whose manner of existence is different from that of material entities. Precisely because the term 'materialism' is ambiguous in its application to certain philosophical trends around the middle of the nienteenth century in Germany, it is a good idea to make a distinction, as Mandelbaum (1974, p. 28) does, between materialism in the broader sense and materialism in the narrower or stricter sense. Such widely differing thinkers as Feuerbach, Marx and Engels can be regarded as materialists according to the broader definition, because and in so far as they opted for the primacy of being in the definition of the relationship between thinking and being, in opposition to Hegel's philosophical idealism. However, in a stricter definition of materialism, which implies, besides the material character of ultimate reality, that there exists one system of (scientific) categories, in terms of which all properties of all manifestations of the fundamental material substratum can be 'translated' and understood, it becomes possible to group together the Marxism of Engels and the 'naive' natural science materialism of Moleschott, Vogt and Büchner. The distinction between these two variants of materislism must therefore be defined in the sense that the basic science which produces the above-mentioned categories is physics, in the case of so-called natural scientific materialism, and dialectics, in the case of dialectic materialism (in the sense of the dialectic interpretation of man and nature).

259 This echoes Schopenhauer; cf. his criticism of the "fatuous, shallow materialists"

in A. Schopenhauer, 1962c, p. 302. Griesinger was familiar with this work, cf. W. Griesinger, 1871, p. 41.

[260] The publications most characteristic of this materialism appeared within the space of three years, from 1852 to 1855: they were J. Moleschott, 1852, C. Vogt, 1854, and L. Büchner, 1855.

[261] This is particularly true of H. Czolbe. Cf. R. H. Lotze, 1852, p. 41, where he regretfully concludes: "One cannot deny that the legitimate opposition to the vital force has been the spiritual movement, which has guided a large number of our contemporaries, as it were according to the law of inertia, far beyond their proper objective towards a denial of the existence of the soul." One could quote these words as a summing-up of Lotze's criticism of Czolbe although, remarkably enough, they were written several years before Czolbe's book (Czolbe, 1855) sparked off the controversy with Lotze. On this controversy cf. H. Breilmann, 1925.

[262] This applies particularly to Büchner, who, when discussing his own *Weltanschauung*, almost invariably describes it as 'materialistic or monistic', or just as 'monistic'.

[263] Moleschott's controversy with the chemist Liebig dates from as early as 1852, when his widely-read *Der Kreislauf des Lebens, physiologische Antworten auf Liebigs chemische Briefe* was published. (J. Moleschott, 1852, reprinted in D. Wittich, 1971, I, pp. 25–341.)

[264] As late as 1857, Wagner published *Der Kampf um die Seele vom Standpunkt der Wissenschaft* (R. Wagner, 1857).

[265] Quoted in C. Vogt, 1855, p. 49 (reprinted in D. Wittich, 1971, II p. 566).

[266] Quoted in Vogt, 1855, p. 86 (reprinted in D. Wittich, 1971, II p. 602).

[267] F. Ueberweg, 1923, p. 287.

[268] F. Gregory (1977, p. 213) comes to essentially the same conclusion when he remarks that "the overwhelming trademark of the scientific materialists, as far as the historian is concerned, is not their materialism, but their atheism, more properly, their humanistic religion".

[269] Cf. J. Moleschott, 1894, p. 209.

[270] Op. cit., p. 251.

[271] L. Büchner, 1904, p. 207.

[272] Cf. op. cit., p. 210.

[273] Cf., for example, J. Moleschott, 1894, pp. 172–76, 178–81, etc., L. Büchner, 1904, pp. 211, 212.

[274] Cf. J. Moleschott, op. cit., p. 208. This refers to Feuerbach's review of J. Moleschott, 1850 (in A. Schmidt (ed.), 1967, II, pp. 212–30). Feuerbach's dictum was meant as a summary of Moleschott's book. I should point out in passing that the comparison attributed to Vogt in fact occurs in the work of the French physician and philosopher P. J. G. Cabanis (1757–94). According to Cabanis (1844, p. 138), the brain is responsible for "the secretion of thought" – something with which Büchner, amongst others, was familiar (cf. L. Büchner, 1904, p. 252).

[275] In accordance with a usage of the term which is current today and was current around 1850, I conceive 'natural scientific naturalism' or, in short, 'naturalism', as the name for a not primarily ontological, but rather *methodological* monistic position, which, as such, must be distinguished both from metaphysical naturalism (for example, that of Aristotle, Spinoza, etc.) and from metaphysical materialism (which, like various other metaphysical positions, is certainly compatible with naturalism, but does not

coincide with it). (Cf. A. C. Danto, 1967.) In the course of Chapters 3 and 4 we shall see that different variants can be distinguished within this conception of naturalism (cf. summary in note 689).

276 The naive materialists were unable to awaken any positive response in contemporary German philosophical circles. The only exception to this was L. Feuerbach, and then only in a very limited sense (cf. S. Rawidowicz, 1964, pp. 331–34). There was, however, no lack of critical voices among the metaphysical idealists, positivists, and later, too, the Neo-Kantians.

277 Chapters 3 and 4 complete this explanation.

278 Hereafter referred to, respectively, as A. Schopenhauer, 1961a and 1961c.

279 O. Temkin, 1946, p. 324, mentions R. J. H. Dutrochet as a representative of a form of mechanistic physiology.

280 Cf., for example, E. Rádl, 1970b, pp. 65–73. Both these researchers have already been mentioned as forerunners of E. Du Bois-Reymond's criticism of the vitalistic idea of a life force; see p. 66.

281 Cf. E. Rádl, 1970b, p. 72, note 2: "... for fear that people might read an anti-clerical tendency into his work [i.e. *Mikroskopische Untersuchungen über die Übereinstimmung in der Struktur der Tiere und Pflanzen*, Berlin, 1839] he requested (and was granted) imprimatur from his archbishop."

282 In M. J. Schleiden, 1863.

283 For Schopenhauer's relationship to French physiology, see P. Janet, 1880. Cf. also F. Picavet, 1971, p. 437. Schmidt's identification of what Schopenhauer criticised as "our present day vogue-materialism, which has become real barber's apprentice and apothecary's pupil philosophy" (A. Schopenhauer, 1961c, p. 229) with the materialism of the "'materialism conflict' noisily preached in this country during the fifties by Büchner, Vogt and Moleschott ... " (A. Schmidt, 1977a, pp. 41–42 (note 57)) is untenable if only for chronological reasons. (The relevant passage in the third edition of *Die Welt als Wille und Vorstellung* (*II*) published in 1858 is identical to that in A. Schopenhauer, 1961a [1844].) As D. Wittich (1971, I: XV) rightly observes: "The major works of these materialistis were published during the years of reaction after 1848–49, or achieved their actual social importance at that time." The popular materialism referred to by Schopenhauer has its roots, I suspect, in the materialism of the so-called *Physiker* and *Chemiatriker* which flourished before the rise and spread of Schelling's natural philosophy. (This former materialism reached its peak in the system of J. C. Reil before he was 'converted' to Schelling's philosophy of nature.) Cf. A Werner, 1909, pp. 66, 68–69.

284 A. Schopenhauer, 1961c, p. 23.

285 A. Schopenhauer, op. cit., pp. 23–24.

286 A. Schopenhauer, 1961a, p. 62.

287 A. Schopenhauer, 1961c, p. 405. The (two) data of the consciousness of self, as data of direct knowledge, which form the basis of Schopenhauer's metaphysics of the will, are *Wille* and *Vorstellung* (Will and Idea).

288 A. Schopenhauer, 1961a, p. 63: "Materialism is thus the attempt to explain what is given to us directly from what is indirectly given".

289 A. Schopenhauer, 1961c, p. 406.

290 I shall refer hereafter consistently to 'Will' and 'Idea', rather than 'will' and 'idea', when discussing Schopenhauer's metaphysical concepts.

291 A. Schopenhauer, 1961, p. 397.

292 " ... matter [is] that by which *Will*, forming the inner essence of things, becomes observable, perceptible, *visible*." (*ibid*.)

293 *Ibid*.

294 *Ibid*., cf. op. cit., p. 398: "Matter consequently is the *Will* itself, but no longer in itself, rather in so far as it is *perceived*, i.e. assumes the form of objective Idea. Therefore what is objective matter is also subjective Will."

295 A. Schopenhauer, 1961c, p. 410.

296 Schopenhauer did not use the word 'improper' in connection with materialism, but the fact that he speaks of materialism *in the proper sense* and the whole context in which the above-mentioned distinction is made, support the use of this term.

297 Op. cit., p. 409.

298 Op. cit., p. 227.

299 Cf. op. cit., pp. 226–27, 410.

300 A. Schopenhauer, 1961c, p. 226.

301 Op. cit., pp. 227–28.

302 Op. cit., p. 228.

303 Op. cit., p. 224.

304 Op. cit., p. 228.

305 Op. cit., p. 229.

306 Op. cit., pp. 229–30.

307 R. H. Lotze, 1852, pp. 30–45. (*Die Einwürfe des Materialismus*) (The objections to materialism).

308 Op. cit., p. 30.

309 The controversy with the self-styled 'sensualist', H. Czolbe, the only 'real' materialist whom Lotze felt it was worth taking the trouble to criticise as an opponent, is of a later date (1855–56). H. Czolbe, 1855, was in a certain sense meant as a materialistic correction of R. H. Lotze, 1852. Lotze reacted in a review of Czolbe, 1855, (R. H. Lotze, 1891, pp. 238–50) and Czolbe replied in 1856 (H. Czolbe, 1856). Cf. H. Breilmann, 1925; F. Gregory, 1977, pp. 127, 131–32.

310 R. H. Lotze, 1852, p. 30.

311 Op. cit., p. 34.

312 Op. cit., p. 43.

313 *Ibid*.

314 A difference which is naturally explained by the different questions posed by the two scholars in the works we have quoted.

CHAPTER 3

315 Cf. J. Bodamer, 1953, p. 526.

316 Cf. H. Damerow, 1860.

317 J. Bodamer, 1953, p. 528.

318 Although it was forced to concede its leading position as a journal to Griesinger's *Archiv für Psychiatrie und Nervenkrankheiten* in 1867, the *Allgemeine Zeitschrift für Psychiatrie* was to continue to appear until the outbreak of the Second World War.

319 The distinction between 'institutional psychiatry' and 'university psychiatry' comes, as we have already remarked, from K. Jaspers, 1959, p. 705.

320 Quoted in R. Thiele, 1970, pp. 115–16.

321 Cf. K. E. Rothschuh, 1953, p. 101.

322 W. Griesinger, 1872d.

323 Magendie's criticism of vitalism was primarily directed against the work of X. Bichat. Bichat's vitalism differed from the vitalism of the Montpellier school (Barthez) in that it postulated not one general principle of life but a plurality of 'vital properties' (life-forces): cf. J. M. D. Olmsted, 1944, pp. 23–24.

324 Op. cit., pp. 202–3.

325 Cf. K. E. Rothschuh, 1953, p. 103; and further J. M. D. Olmsted, 1944, pp. 238–39, who refers to C. Bernard on Magendie: " . . . Magendie . . . was fond of saying that he had only eyes and ears but no brain when he was experimenting", with the variation that "he had eyes (for facts) but no ears (for theory)".

326 W. Griesinger, 1872d, p. 3.

327 Op. cit., pp. 3–4.

328 Op. cit., p. 4.

329 In Griesinger's work negation, one gets the impression, stands on the one hand for something like 'scientific thinking' (as distinct from the empirically-lacking, speculative fantasy of the philosophers and the 'thoughtless' observation of some researchers) and on the other – more specifically – for the way in which this thinking manifests itself as an activity of (critical) discrimination, analysis, on the level of the formation of scientific concepts. In this latter case, negation serves the 'reconstruction of concepts' which follows this analysis.

330 W. Griesinger, 1872d, pp. 4–5.

331 Or in his own words, the "literary roués, who may once, years ago, have enjoyed the embraces of science, and now hide their self-complacency in thoughts and convictions from us by parading the old adventures, 'O vanitatum vanitas', as experience" (op. cit., pp. 5–6).

332 Op. cit., pp. 5–7.

333 Op. cit., p. 6.

334 To prevent misunderstanding: this physiological materialism (Cabanis, Richerand, Magendie, Bichat, and others) was not a mechanistic materialism but a 'vitalistic' materialism; cf. O. Temkin, 1946.

335 W. Griesinger, 1872e.

336 C. A. Wunderlich, 1869, p. 119: "this remarkable article also determined the attitude of the Archiv [i.e. the Archiv für physiologische Heilkunde] towards the prevailing schools of thought of that time".

337 W. Griesinger, 1872e, pp. 30–31.

338 Op. cit., p. 65.

339 W. Griesinger, 1872e, p. 58.

340 Op. cit., p. 59.

341 Op. cit., p. 58.

342 F. J. V. Broussais (1772–1838); cf. note 524.

343 W. Griesinger, 1872e, p. 65.

344 Op. cit., p. 65: "Systems come and go; the facts of physiology will remain forever. If the physiological way of thinking keeps trying to recognise the inner course and

242   NOTES

connection of the phenomena in the healthy and diseased organism in all its aspects,
it avoids all one-sidedness without neglecting any possible relationship."
345 Op. cit., p. 65.
346 Op. cit., pp. 65–66.
347 W. Griesinger, 1872e, p. 66.
348 Just under half of Griesinger's total *oeuvre* is devoted to studies in the field of
general pathology, and internal medicine in particular. This is reflected in his career –
he was several times professor of internal medicine: 1843 Tübingen, Kiel; 1854 Tübingen,
1860 Zürich. His experiences during a stay in Egypt in 1850 as personal physician to
the Viceroy and director of the *Conseil de santé* in Cairo took concrete shape in his
Klinische und anatomische Beobachtungen über die Krankheiten von Egypten which,
according to Wunderlich (1869, p. 147), was scarcely less important than his major
psychiatric work.
349 In 1847 he even took over the editorship; cf. W. Griesinger, 1872g.
350 Wunderlich, his friend and former colleague at Tübingen, made it clear that this
was Griesinger's own conviction in the introduction to Griesinger's *Gesammelte Abhandlungen*, which Wunderlich published after Griesinger's death. "Griesinger himself later
found that his exposition of mental diseases did not exhaust the specifics of the occurrence and relationships of mental disorders; rather it represents only *one* approach and
more or less contains only the general pathology of mental diseases." (W. Griesinger,
1872, II, p. V.)
351 As a reminder, in or around 1842 we find (1) Helmholtz's graduation (1842);
(2) J. von Liebig, 1841, J. R. Mayer, 1842/5 and 1845, K. von Rokitansky, 1842, the
first edition of M. Schleiden, 1849, the first part of R. Wagner, 1842/53, the first edition
of R. H. Lotze, 1848; (3) the foundation of the *Archiv für physiologische Heilkunde*
(1842).
352 R. Wagner, 1842/53.
353 W. Griesinger, 1872f.
354 Cf. A. Hirsch, 1962, p. 847.
355 I am referring to Lotze's so-called theory of local signs, a theory about spatial
perception, which had a great deal of influence.
356 Born in 1817, Lotze graduated after studying medicine and philosophy (Leipzig)
in the medical and philosophical faculty at Leipzig in 1838 with a *Dissertatio de futurae
biologiae principiis philosophicis*. He established himself as a *Privatdozent* in medicine
in 1839 and in philosophy in 1840. His *Metaphysik* (1841) was followed in 1842 by his
Allgemeine Pathologie und Therapie als mechanische Naturwissenschaften. In the same
year he was appointed professor extraordinary in philosophy while maintaining his
function as *Privatdozent* in the medical faculty, and produced the first of his three
contributions to Rudolf Wagner's *Handwörterbuch der Physiologie*, namely the article
Leben, Lebenskraft (followed by the articles *Instinct* (1844) and *Seele und Seelenleben*
(1846)). In 1844, at the instigation of Rudolf Wagner, he was appointed as Herbart's
successor at Göttingen, where he remained until 1881, that is, until his appointment
at Berlin. His *Allgemeine Physiologie des körperlichen Lebens* appeared in 1851, *Medicinische Psychologie oder Physiologie der Seele* in 1852, *Logik* in 1853, the three-volume
*Mikrokosmus. Ideen zur Naturgeschichte und Geschichte der Menschheit. Versuch
einer Anthropologie* between 1856 and 1884, and *System der Philosophie* between
1874 and 1879. He died shortly after his appointment to Berlin in 1881.

357 Griesinger – like Lotze – was born in 1817.
358 R. H. Lotze, 1848 [1842].
359 R. H. Lotze, 1842.
360 R. H. Lotze, 1842, p. IX.
361 R. H. Lotze, 1842, p. X.
362 Cf. R. H. Lotze, 1842, p. XIV: "[Such a] specific arrangement of causes, which allow the rule to be applied, can only be learned empirically because it is not a necessary one but only one of the many possibilities." For the concept of cause used by Lotze cf. op. cit., p. XIII, in which 'cause' is used in the sense of 'real premise'; the passage which follows makes it clear that such a premise not only *describes* reality, but *is* reality. "In the most accurate sense of the word, however, a cause is never anything but a reality. Its characteristics, when connected in a certain way with the characteristics of another, equally really present thing, form together with the latter the complete set of antecedent conditions from which a consequence originates. This consequence, because of the reality of the premises, is a real event (*ein wirkliches Ereignis*) too, a *Wirkung* (effect). The causes are therefore nothing but a vehicle for the realisation of the abstract parts of the ground, and the entire structure of the process is only a repetition, in the field of reality, of these relations between ground and consequence".
363 Cf. op. cit., p. XV: unlike a cause (which is always a thing), an end is never something which exists, but is (as long as it is unfulfilled) a "should-be" . . . "The end therefore is a *legislative power*, which the masses of nature never obey unless they are forced from the start and pressed into a determined course by the medium of the causes which form the *executive power*" (my italics). The interchange of the concepts of cause and end, according to Lotze, has led physiology until very recently "to consider the investigation of causes and that of ends as two different approaches, one of which is properly focused on one part of the phenomena, the other on another, one of which always gives results where the other fails; however, both principles are of a more general nature; they want to be employed equally on each object of investigation, but each in quite a different sense . . . Anyone who explains one phenomenon from its causes, another from its ends, is answering totally different questions." (Op. cit., p. XV.) The teleological approach does not lead to an explanation (i.e. a 'real', natural scientific explanation) but only to the "means, whose mutual relationships [by which is meant the system of causal factors] give this explanation." (Op. cit., p. XVI.) In Lotze's view teleology has to be conceived as a 'heuristic maxim'.
364 R. H. Lotze, 1842, p. XVII.
365 R. H. Lotze, 1848, p. 3.
366 The natural *historical* school in medicine must not be confused with the so-called natural *philosophical* school. (Cf. P. Diepgen, 1959, pp. 27–28, and B. Wehnelt, 1943, pp. 86–88.) The former was characterised by its defence of (1) the ontological conception of disease and (2) the so-called 'parasitic' theory of disease which, going back to Parcelsus, was resurrected in the nineteenth century by Novalis ("As man is a parasite in the living organism of the world, so disease is a parasite in man"); the natural history school was concerned with studying this parasitic relationship in terms of natural history. In the case of some of the representatives of this school (for example, Novalis, K. J. H. Windischmann, J. N. von Ringseis) this theory was coupled with the sin-theory of disease. (Cf. W. von Siebenthal, 1950, pp. 62–83.) One also comes across the occasional name from among the members of this natural history school (such as Windischmann)

among the adherents of the natural philosophical school, that is to say, among those physicists, chemists, biologists (physiologists), physicians, etc. who were inspired by Schelling's philosophy of nature. (On this last point, with particular reference to the inorganic sciences, cf. H. A. M.Snelders, 1973; with the emphasis on biology and medicine, cf. A. Werner, 1909.)

367 R. H. Lotze, 1843, p. 3.

368 R. H. Lotze, 1848, p. 9.

369 The fact that the 'science' of pathology, as it is conceived in a general sense in Lotze's work and realised in a more specific application in that of Griesinger, still contained a strongly *aprioristic* element did not escape subsequent generations, and has led to criticism which insists that Lotze's work was still not scientific enough and Griesinger's was still too philosophical to be able to meet the (methodological) requirements of experimental natural science. Cf. E. G. Boring, 1957, p. 266, on Lotze's contribution to psychology, and E. Westphal's obituary on Griesinger (1868/9).

370 R. H. Lotze, 1848, p. 3.

371 With the exception of the last one, these terms are not used by Lotze.

372 R. H. Lotze, 1848, p. 4. This description goes further than that in the article *Leben, Lebenskraft*, where the teleological aspect is explained only in terms of the means–goal relationship.

373 Op. cit., p. 12.

374 Lotze apparently assumes a *hierarchy* of perspectives or views of natural phenomena. The teleogical view is the highest, because "it investigates the ground which has brought forth the given interconnection to have the form it has" (1842, p. 12). The "nomological" view, which provides us with the "simple, general conditions" (op. cit., p. 4) or "mathematical principles" (op. cit., p. 12) on which the realisation of these phenomena is based, occupies (we may assume) the second place in this hierarchy, while the third place goes to the empirical view.

375 Op. cit., p. 13.

376 Op. cit., p. 13.

377 Op. cit., p. 21; cf. R. H. Lotze, 1842, p. xv.

378 W. Griesinger, 1872f.

379 W. Griesinger, 1872f, p. 110.

380 Not entirely new, in so far as one can see in this – with Griesinger, 1872f, p. 117 – a revival of seventeenth-century iatromechanical medicine.

381 Op. cit., p. 109.

382 For Lotze's position on this, cf. R. H. Lotze, 1848, pp. 55–56.

383 This probably explains why, with a few insignificant exceptions, Griesinger did not refer again later to Lotze's work (i.e. after his review of Lotze's work of 1842 (R. H. Lotze, 1848 [1842]) in the 1843 article we have quoted), and why Lotze is totally silent on the subject of Griesinger.

384 For this dualism cf., for example, R. H. Lotze, 1846, p. 149: "We may . . . state that psychology, as a separate science, can effectively be based on the assumption of a soul which has a character of its own, which makes it inevitable that we adhere to the incomparability of mental phenomena and corporeal processes and to the unity of the consciousness." For the interactionist aspect of this dualism cf., for example, R. H. Lotze, 1848, pp. 64 ff, where the problem of the body–soul interaction is "solved" by means of the "assumption . . . of a physical force in the substance of the central

organs, whose intensity is variable in such a way that its variations correspond with the ideal processes in the mind, according to certain general arrangements".

385 Although it is not an exhaustive list, the following summary of the publications I consulted represents the most important literature on Griesinger: E. H. Ackerknecht, 1968, pp. 65–73; H. J. Dietze & G. E. Voegele, 1965; K. Dörner, 1969, pp. 356–79; E. von Feuchtersleben, 1846; K. F. Flemming, 1846; K. Jaspers, 1959, pp. 382, 403; R. Kuhn, 1957; M. Lazarus, 1869; O. M. Marx, 1972; A. Mette, 1976; R. Thiele, 1970; E. Westphal, 1868/69; C. A. Wunderlich, 1869; G. Zeller, 1968.

386 W. Griesinger, 1872a.

387 W. Griesinger, 1872b.

388 Griesinger wrote his work during the time he was working in his friend Wunderlich's clinic for internal medicine in Tübingen.

389 In 1847 he was appointed professor extraordinary at Tübingen in the fields of general pathology, materia medica, and the history of medicine. See C. A. Wunderlich, 1869, p. 119.

390 We have already discussed this more extensively and in more general terms in Chapter 2.

391 Cf. W. Griesinger, 1872a, p. 3: "When we invite the reader, in what follows, to accompany us away from the demands of the medical topics of the day towards the quiet banks between which the current of psychic phenomena flows, it will scarcely be necessary to assure him in advance that we shall not use the hand of philosophy, with its dreaded school idiom, to guide us, but only the simple and generally understandable light of those perceptions and concepts which belong to empirical physiology".

392 Cf. W. Griesinger, 1872a, p. 3: "The development and explanation of those phenomena of organisms, called the psychic ones, are, in our opinion, the prerogative of the natural scientist alone, precisely because of their organic character; and the fact that the same concepts and laws, which modern physiology has created and developed for a number of other phenomena of organic matter have been found useful for even the most specific of those phenomena, will be shown in more detail hereafter."

393 W. Griesinger, 1872a, p. 4.

394 Cf. E. H. Ackerknecht, 1968, p. 65.

395 W. Griesinger, 1872a, pp. 10–11.

396 Op. cit., p. 11.

397 W. Griesinger, 1872a, p. 12.

398 Op. cit., p. 25.

399 Op. cit., p. 13.

400 Cf. op. cit., p. 25: "In the same way as, from the scattered, centripetal impressions in the spinal cord a condition of the organ in its entirety emerges that regulates the average motor activity or tone, a state of apparent rest is formed in the (large?) brain from the entire mass of scattered and mutually combined ideas, which regulates the force and the normal course of psychic movements, i.e. striving."

401 Op. cit., p. 15.

402 W. Griesinger, 1872a, p. 25. Freedom as the goal of the self-actualisation of man (the human 'organism') is understood by him as "the possibility of allowing other ideas and strivings to act on those which are current at that moment and of calling, within certain limits, those other ideas to the fore spontaneously, not arbitrarily, however, but according to precisely determined laws" (op. cit., p. 26). A similar freedom also exists

in animals, but human freedom is 'infinitely greater' than the freedom which man finds
even in the most highly-developed animals.

[403] Op. cit., p. 11.

[404] W. Griesinger, 1872a, p. 11.

[405] Op. cit., p. 11.

[406] I am using the third edition (W. Griesinger, 1871), which is an unchanged reprint
of the second edition of 1861.

[407] E. H. Ackerknecht, 1968, p. 71.

[408] I am leaving out of account the significance of his idea of the unitarian psychosis
which, taking up where Zeller left off, acquired a new shape in his work, and which has
survived up to the present day in the work of Neumann, H. Ey and K. Menninger; (cf.
H. Ey, P. Bernard, C. Brisset, 1974, p. 565). I shall discuss this more extensively later in
connection with his relationship to Herbart. Equally, I am ignoring for the moment the
fact that among the twentieth-century anthropologically oriented psychiatrists (Binswanger, etc.) Griesinger is remembered as the man who brought the depersonalising
attitude in psychiatry into vogue, and thus has the dubious distinction of having put
psychiatry on the wrong track. I shall go into this further in the discussion of the relationship between Binswanger and Griesinger (p. 154).

[409] K. Jaspers, 1959, pp. 708–9.

[410] Op. cit., p. 382; cf. p. 403. Jaspers' historical view of Griesinger's position in this
context, and his criticism of it, are certainly not unconnected with his own philosophical
'prejudice', namely his psychophysical parallelism; admittedly, according to Jaspers,
in many cases there are connections between physical and mental changes, in the sense
that mental changes can certainly be regarded as the result of physical changes; moreover,
there exists no mental occurrence without some physical basis or other. "However, we
know of no physical process in the brain which is identical, as its 'counterpart', as it
were, to the morbid mental process" (p. 382).

[411] Op. cit., p. 382.

[412] The relative right which Jaspers here attributes to the standpoint "diseases of the
mind are diseases of the brain" is based on history of science considerations: "From a
historical point of view, the sovereignty of the dogma 'diseases of the mind are diseases
of the brain' has worked in a *stimulating*, as well as in a *harmful* way. Research into
the brain was stimulated. Every institute currently has its anatomical laboratory. What
was harmed was specifically psycho-pathological research; unintentionally many a
psychiatrist got the feeling that once we understand the brain really well, we will understand the life of the soul and its disturbances too." (K. Jaspers, 1959, p. 382.)

[413] More on this hereafter (pp. 109 ff.)

[414] W. Griesinger, 1871, p. 1.

[415] p. 27.

[416] Cf. W. Griesinger, 1872c, pp. 96 ff. (i.e. Griesinger's review (1844) of M. Jacobi,
1844b).

[417] As, for example, J. B. Friedreich; cf. J. B. Friedreich 1964 [1836], pp. 258–84
(§ IX) and, summarising, against Jacobi (op. cit., p. 298): "The seat of every psychic
disease is after all that organ which also determines psychically normal life, i.e. the
brain. Its suffering, be it an idiopathic or a consensual one, is the primary condition for
psychic alienation, although this can have its remote cause in any of the other organs."

[418] Cf. R. H. Lotze, 1848, p. 8: "The prospect of such an exact physiology [i.e. as a

"mechanical natural science"] is extremely remote." What is true of physiology is true *a fortiori* of pathology.

[419] "*Pathological* facts prove as clearly as do physiological ones that the brain alone can be the seat of normal and abnormal mental action" (W. Griesinger, 1871, p. 3).

[420] "While we are forced by facts to refer having of ideas and willing to the brain, still, however, nothing can be assumed as to the relation existing between these mental acts and the brain, the relation of soul to matter. From an empirical point of view the unity of soul and body is indeed a fact [*sic!*] primarily to be maintained, and the *a priori* investigation of the possibility of soul apart from body, of a bodiless soul, confining itself to abstract considerations of its immateriality and oneness as distinguished from the manifold of matter, must be entirely dismissed." (Op. cit., pp. 5–6.)

[421] W. Griesinger, 1871, p. 8: "Although, however, every mental disease proceeds from an affection of the brain, every disease of the brain does not, on that account, belong to the class of mental diseases."

[422] By 'psychic or mental diseases' must be understood "all those affections of the brain in which anomalies, derangements of the having of ideas and of the will, *constitute the most striking symptoms*." (My italics.) In the case of 'mental diseases', in general the sensory and motor cerebral functions are also impaired, says Griesinger, but this impairment is subordinate, "the psychic anomaly appears to be the main thing". Conversely, 'ordinary diseases of the brain' (the more circumscribed inflammations, abcesses, brain tumours, tubercular meningitis, etc.) are not called mental diseases even though, as a rule, they go together with more or less serious disorders in psychic activity "because other cerebral symptoms, those of disturbed sensation and movement in general, greatly predominate: *a potiori fit denominatio*" (op. cit., p. 9).

[423] W. Griesinger, 1871, p. 9.

[424] *Ibid.*

[425] Cf. L. Binswanger, 1955a, p. 89: "What almost immediately gave Griesinger's design the character of a lasting psychiatric constitution was in the first place the circumstance that he guarded against one-sidedness and exaggeration without compromising in the bad sense of the word. He displayed an excellent eye for *what was possible at that time* and in particular for the idea and the well-being of the entire enterprise of psychiatry."

[426] W. Griesinger, 1871, p. 6.

[427] W. Griesinger, 1871, p. 6.

[428] Op. cit., p. 6.

[429] Cf. the continuation of the passage quoted above, which reads as follows: "What shall now be said of the flat and shallow materialism which would overturn the most general and most valuable facts of human consciousness because it finds no palpable trace of them in the brain? Empirical perception, in ascribing the phenomena of feeling, intellect and will to the brain as its function, leaves not only the actual contents of the life of the human soul untouched in all its riches, and, in particular, maintains energetically the fact of free self-determination; it leaves also, naturally, the metaphysical question, what it may be that enters as soul-*substance* in this relation of the feeling, having of ideas, and willing, which takes the form of psychical existence, etc." (W. Griesinger, 1871, pp. 6–7.)

[430] Griesinger's criticism of (metaphysical) materialism is, if I am right, in fact directed against the *metaphysical apriorism* inherent in it, and thus, strictly speaking, is applicable to philosophy in general, in the meaning which this had for Griesinger, that is, as

metaphysics. On this point, as far as he was concerned, materialism and spiritualism (idealism) were on a par with each other (cf. W. Griesinger, 1871, p. 6), something which is easily overlooked because Griesinger elaborates his criticism of philosophy one-sidedly, i.e. only with respect to materialism. This one-sidedness is undoubtedly connected with the fact that 'spiritualistic' philosophy was of no importance whatsoever to Griesinger's own mechanistic cause.

431 Griesinger makes no systematic distinction between soul and *mind*, but usually refers to the psychic, the soul, the consciousness, in a sense which embraces both soul (in the narrow sense) and mind. Cf., for example, the Table of Contents in Griesinger, 1871: "Elementary disorders of *psychic* diseases" (my italics) are subdivided into 'mental', 'sensory' and 'motor' elementary disorders. See also note 422.

432 What Griesinger understands by the *form* of the life of the soul is not explained. One would not be far wide of the mark if one stated that the concept of the form of the life of the soul embraced everything which was susceptible to natural scientific treatment (conceptualisation, 'explanation') in the sense of Griesinger's mechanistic psychology.

433 W. Griesinger, 1871, p. 7: "The elementary phenomena which occur in the nerve masses must always be identical in all men, especially if they be considered (as is now believed by many) as essentially electrical, necessarily in the highest degree simple, consisting of plus and minus. How could the endless variety of thoughts, feelings, and desires, not only of individual men, but of whole ages, proceed from these *alone and immediately*?"

434 W. Griesinger, 1871, p. 7. A fairly pointless piece of advice, if it means one must work on the assumption that the dawn of this era can only break when man ceases to be man and has reached the level of that "angel who explains everything to us"! One might like to compare this, out of interest, with Lotze's diametrically opposed view of the possibility of an ideal ('mechanical') psychology, as conceived by Griesinger, in R. H. Lotze, 1896, p. 165. The passage ends with the words: "The gulf between the last state of the material elements which we can achieve and the dawning of sensation remains as great, and *there is hardly anyone who would cherish the idle hope that a more developed science will find a mysterious bridge where the impossibility of any continuous transition whatsoever is forced upon us with the simplest clarity*." (My italics.)

435 The idea that Lotze left room for teleological explanation in *biology*, as D. H. Galaty believes (D. H. Galaty, 1974, p. 300), is based on a misunderstanding. Lotze actually stressed that the teleological view in biology could have no more than heuristic significance. Cf., for example, R. H. Lotze, 1842, p. xvi: "Teleological viewpoints therefore never give the explanation itself, but only lead back to the means whose mutual relationships give this explanation."

436 It accords with our interpretation of the difference between Lotze's position and that of Griesinger, that in Griesinger's work there is no hint of, or reference to, an explicit foundation of the teleological approach in the factum of the unity of the consciousness, such as we find in Lotze (cf. R. H. Lotze, 1912, p. 481).

437 Also according to L. Binswanger, 1955a, p. 86.

438 K. Jaspers, 1959, p. 708. Jaspers' description of Griesinger's work and his judgement of its significance in the history of psychiatry is certainly one-sided, but this does not detract from the fact that his judgement is extremely significant as evidence of the recognition of Griesinger's *descriptive* talents.

[439] p. 103.

[440] See the foreword to the first edition (1845) of his *Pathologie und Therapie der psychischen Krankheiten*, pp. IV–V, where Griesinger otherwise says no more than that, in his description of the main points in psychology, in order to avoid excessive detail in the presentation of his own psychological views, he has had to make use of "the forms and terms of a recognised psychology (Herbartian)". In the later editions of this work the reference to Herbart is omitted.

[441] M. Lazarus, 1869, pp. 777–78.

[442] I should like to remark, in passing, that in 1809, i.e. five years after Kant's death, Herbart succeeded Kant in the chair of philosophy at Königsberg.

[443] Cf. J. F. Herbart, 1891a, pp. 146–270 (*Einleitung in die Metaphysik*). For a brief description of the approach and purpose of metaphysics cf. op. cit., pp. 148–49: "It will be shown that these concepts (i.e. those of the common view of life), which are really forced on us by experience, are nevertheless *unthinkable*; that we cannot *store* the datum as such, in the way we find it, that we consequently have to remould it in our thinking, to submit it to a necessary change, because we cannot get rid of it. This is precisely the aim of metaphysics."

[444] R. H. Lotze, 1846, pp. 248 ff. It should be observed in passing that the far-reaching agreement between (the psychology of) Herbart and (that of) Lotze which is usually suggested in history of psychology summaries needs to be revised. Lotze himself was fully aware of the difference, witness his words: "I share none of his [Herbart's] convictions on any of their essential points" (op. cit., p. 248).

[445] Op. cit., p. 249.

[446] In the following passage I am following Lotze's outstanding summary, op. cit., p. 249.

[447] An 'idea' distinguishes itself from another 'idea' by its quality and force or intensity. Two qualitatively different ideas can be in harmony and thus strengthen each other, or they can be in disharmony and come into dynamic opposition whereby the stronger idea 'suppresses' the weaker; the suppression can go so far that the suppressed idea changes its state to that of a possible, i.e. latent, idea (it becomes 'unconscious').

[448] Cf. W. Griesinger, 1872, II, p. 1.

[449] The title reads: *Psychologie als Wissenschaft, neu gegründet auf Erfahrung, Metaphysik und Mathematik* (J. F. Herbart, 1890b [1824] and 1892b [1825]).

[450] The fact that Herbart nonetheless understands his psychology as "(newly) based on experience" does not mean that he wanted to isolate his work from "Kant's *a priori* psychology", as E. G. Boring (1957, p. 252) mistakenly believes (Kant did not defend any '*a priori* psychology' – he only criticised it!), it is intended to prevent his psychology – the psychology of a thinker who, *as opposed* to Kant, defended the right of ('realistic') metaphysics to exist – from becoming identified with the *aprioristic* psychology of pre-Kantian philosophy.

[451] Cf. the introduction to his *Psychologie als Wissenschaft* where he writes: "The purpose of this work consists in establishing a mental research, resembling the investigation of nature; as far as this everywhere presupposes the completely regular interconnection of phenomena, and searches for it by a sorting of the data, by careful inferences, by daring, tested and amended hypotheses, and finally, if at all possible, by considerations of a quantitative nature and by calculation. *The observation that mental theories, in more than one respect, are open to calculation has put me on the track of the research*

presently under consideration; and the further I pursue it, the more I am convinced that only in this way can the disproportion between our knowledge of the outer world and the uncertainty about our innermost self be reduced. Only in this way can we properly process the material offered to us by self-observation, by our social intercourse with other people, and by history." (My italics.) (J. F. Herbart, 1890b, p. 185.)

[452] The most important work in this area is still M. Dorer, 1932. A more recent work is J. Buelens, 1971, which does justice to Griesinger's significance for nineteenth-century psychiatry, and psychoanalysis in particular. For the relation of Herbartianism and psychoanalysis, see also O. Andersson, 1962.

[453] According to L. J. Pongratz, 1967, p. 214, and also to G. A. Lindner, 1858.

[454] E. H. Ackerknecht, 1968, p. 71.

[455] Thus Freud quotes Griesinger in his *Traumdeutung* (*Interpretation of Dreams*) where he is concerned with supporting the thesis which is so crucial to this major work, i.e. that wish-fulfilment is the common structure of dream and psychosis. He refers there to Griesinger's "subtle discourse, showing most clearly that wish-fulfilment in a dream, and in psychosis, are of the same character" (S. Freud, 1973a, p. 95.) Of scarcely less importance is S. Freud, 1973b, p. 230, where Freud (referring to Griesinger) speaks about some cases of hallucinatory psychosis in which the occurrence which brings about insanity must be denied.

[456] Cf. L. Binswanger's criticism of the book by M. Dorer which we have quoted. L. Binswanger, 1955a, p. 85, note 2.

[457] I am referring here to L. Jülicher, 1961.

[458] Cf. D. Wyss, 1972, p. 142.

[459] S. Freud, 1972b, p. 86.

[460] J. F. Herbart, 1890b, p. 307.

[461] Cf. M. D. Altschule, 1965.

[462] Cf. U. Schönpflug, 1976, p. 6.

[463] J. F. Herbart, 1890b, p. 241.

[464] J. F. Herbart, 1891c, p. 406.

[465] J. F. Herbart, 1891c, p. 405. The edition published by G. Hartenstein [J. F. Herbart, 1900, p. 142] gives "to move" (*versetzen*) instead of "to decompose" (*zersetzen*).

[466] J. F. Herbart, 1890b, p. 239.

[467] J. F. Herbart, 1890b, pp. 241–42.

[468] J. F. Herbart, 1890b, p. 247.

[469] J. F. Herbart, 1892b, p. 320.

[470] For his criticism of faculty psychology, and the mechanistic-psychological reinterpretation of the forms of space and time, cf., for example, J. F. Herbart, 1891c, pp. 324–25 (Section 33–35) beginning with the words: "*Space* and *time* became the objects of a very wrong theory, in so far as they have been considered as the characteristic, only and independently given *forms of sensibility*". For his criticism of the categories ('logical forms'), cf. op. cit., pp. 325–26) (Section 36). For Herbart's general attitude towards faculty psychology classification (which in his view explained nothing): "When the assumption of a *faculty we possess* is added to the unscientifically-developed concepts about *what takes place within us*, then psychology becomes mythology. While nobody will admit he seriously believes it, the most important investigations are nevertheless made to depend on it in such a way that nothing meaningful remains when that basis is taken away."

(J. F. Herbart, 1900, p. 8, which is the same as J. F. Herbart, 1891c, p. 303). Cf., however, also J. F. Herbart, 1900, p. 12, where the relative right of the hypothesis of the faculties of the soul is defended when it concerns, as it does in Herbart's *Lehrbuch*, an "easy, almost popular representation".

[471] Cf. J. F. Herbart, 1891c, p. 302: "Self-observation mutilates the facts of consciousness as soon as they are apprehended. It tears them from their necessary connections and submits them to a tumultuous abstraction that does not come to rest until it arrives at the highest generic concepts, *having ideas, feeling*, and *desiring*. The manifold observed is now subordinated to the latter as far as possible by determination (i.e. in the wrong way for an empirical science)."

[472] This concerns primarily (1) the infinite regress in the self-representing ego, and (2) the contradiction which is contained in the idea of the subject-object identity, or more precisely, the identity of the representing ego-subject and the represented ego-object. Cf. J. F. Herbart, 1890b, p. 42.

[473] J. F. Herbart, 1892b, pp. 140–42.

[474] Leibniz distinguishes perception as "the interior state of the monad, representing external things", and apperception as "*conscience*, or the reflexive knowledge of this interior state". (G. W. Leibniz, 1961a, p. 600.)

[475] J. F. Herbart, 1892b, pp. 140–42.

[476] R. H. Lotze, 1846, p. 249.

[477] W. Wundt, *Grundzüge der physiologischen Psychologie* (1880[2]), II, p. 236, quoted in W. Janke, 1971, p. 455.

[478] J. F. Herbart, 1891c, p. 405.

[479] J. F. Herbart, 1891c, p. 363.

[480] Op. cit., p. 364.

[481] *Ibid.*

[482] Apart, that is, from all those negative determinations, not specific to the soul, which pertain to the soul as a 'real'.

[483] *Ibid.*: "The acts of self-preservation of the soul are *ideas* (at least partly and as far as we know them)", etc.

[484] W. Griesinger, 1871, pp. 22–60.

[485] Herbart's denunciation of physiology did not refer to physiology *per se*, but to early nineteenth-century physiology, which was still too 'philosophical' (by which we can understand 'unscientific') to be really interesting (cf., for example, his polemic with Rudolphi in J. F. Herbart, 1892b, pp. 46–50). Griesinger would have agreed with Herbart's criticism. Herbart, who died in 1841, did not live to see the immense leap forward made by 'scientific' physiology in Germany during the eighteen-forties with work of Johannes Müller and his followers.

[486] Of the 563 pages in his *Psychologie als Wissenschaft* (J. F. Herbart, 1890b and 1892b), fifty-three pages — the last paragraphs of the whole work — are devoted to the discussion "of the connection between body and soul" (Section 153–59) and "of those mental conditions on which the body has a perceptible influence" (Section 160–68). The celebrated *Lehrbuch der Psychologie* (J. F. Herbart, 1891c) takes fewer than five pages for the paragraphs on the connection of soul and body (Sections 120–33 and 158–66).

[487] Cf. L. Jülicher, 1961, p. 33. The passage quoted there is quite clear: "We may find that we would not have any particular reason to collect information from unusual

emotional states. The wealth of opinion we can daily form regarding ourselves is just as large as its handling is difficult and protracted. It should also be possible for us, when we know the general laws for the phenomena within us, to understand and explain the mental states of others, even if they are very far from normal, by means of such laws, rather than by merely transferring our own feelings." (J. F. Herbart, 1890b, p. 195.)

488 See p. 119.

489 "Between several simple, mutually different beings [by which is meant "reals"] there is a relationship which one may, with the assistance of an example from the physical world, describe as *pressure* and *counter-pressure*. In the same way as pressure is an impeded movement, that relation exists because something *would* be changed in the simple quality of that particular being, *if* it did not resist and maintain itself in its own quality against the disturbance. Such acts of self-preservation are the only things that truly *occur* in nature; and this is the relationship between occurring and being." (J. F. Herbart, 1891c, p. 364.)

490 Cf., for example, "And the soul, it is true, lives in a body, and there are also corresponding states between one and the other. But nothing corporeal takes place in the soul, nothing purely spiritual, which we could account for by our ego, occurs in the body; the affections of the body are not ideas of the ego and our pleasant and unpleasant feelings do not directly lie in a promoted and impeded organic life." (J. F. Herbart, 1892b, p. 56.)

491 J. F. Herbart, 1892b, p. 287.

492 This vision presupposes a reductionistic translation of the phenomenon of the will which — typical of Herbart as it is — is very difficult to follow because it takes us a long way away from what we usually understand by 'will': willing is an internal state of our being (in this case our soul–real), more precisely it is a state of *ideas*; these ideas are, however, acts of self-preservation of the soul, namely those acts of self-preservation of the soul "which, in willing, return to a less inhibited state". J. F. Herbart, 1892b, p. 228.

493 I quote his explanation in full here: "We . . . know at least that the soul is joined at one end to the nerves, which forms the general condition for all causality; furthermore, that the nerve, which presents itself as a *coherent* thread, must be a chain of simple entities in a state of imperfect interconnection; finally that the slightest change in the inner state of one of the entities has an influence on the disturbances and consequently on the self-preservation of all entities in the chain. This is always to be expected in such a chain. *This influence can therefore propagate through space, moving along a nerve fibre* (only not through *empty* space) *without itself being of a spatial nature at all.* It need not reveal itself therefore as a movement at all, either of the nerves themselves or of something within the nerves; the nerves can be greatly affected without even the slightest movement. If there appears to be something magical in this, it is because it has not been made clear how a simple, non-spatial thing can get involved in spatial relations, indeed can even fill space. This is a matter for discussion in general metaphysics. At the end, where the nerve passes into the muscle, a movement of the muscle will now have to occur with considerable mechanical force. This entails much that is not known, but nothing that is strange or incomprehensible. In the nerves there are disturbances and acts of self-preservation of each of the elements; something similar must be foremost in all the simple entities which together form the muscle; and as the muscle and the nerve are interdependent, the states of self-preservation must be oriented one towards the other. Experience teaches us, however, that from the changed inner state of the

muscle external changes also result, i.e. a coming together of its parts. This does not imply anything unheard-of, anything that does not already occur in the rudiments of chemistry. The attraction of the elements in a chemical solution takes place with tremendous violence compared with mechanical forces; nevertheless it happens without any real spatial force. It is, in an entirely understandable way, only the necessary result of the inner states of the solvent and the soluble body. Is it therefore a miracle when a muscle contracts because the inner states of its parts are changed by the inner states of the nerves, and the latter by an inner state of the soul?" (J. F. Herbart, 1892b, pp. 288–89.)

494 Herbart tells us nothing here about precisely what this 'conforms to' implies, but the inference to be drawn from this figure of speech is clear. The chain reaction of the total series of mutually affecting systems of self-preservation, which connects the will and the muscle contraction, is *functional* in character, or perhaps one should say: if a muscle contraction M appears to us as a functional response to an act of will W, this can be explained in more 'scientific' terms such that the process of the causing of M by W can be analytically solved in the sum of the functional reactions with which the affected systems of self-preservation react to the causal influence of the system immediately preceding them in the series.

495 Cf. J. F. Herbart, 1892b, pp. 300–1. "As far as the organic body is concerned, it may never come as a surprise that a teleological view interferes, something which is highly suspicious in other fields of physics. Throughout the living body reigns a higher art. The very combining of elements which, detached from the living body, tend to decompose rapidly, causes legitimate surprise. When, however, so many other wonders reveal themselves everywhere in this organism; when, to mention but one example, the structure of the crescent-shaped valves in the main arteries so obviously shows the mark of purposeful arrangement, it can be no surprise when we have to designate *the sub-ordination of the nerve-system to the soul as something which cannot be understood from general relationships in nature but only by presupposing a special arrangement whose origin is in that particular art by which the higher animals were brought into existence at all.*"

496 Cf., for example, J. F. Herbart, op. cit., p. 287. "Surmounting all real vital force in the elements is the purely ideal, artistic unity of the living being, i.e. its beauty and functionality. This exists only for the observer; it points him, however, upwards to the greatest of artists who by the most sublime wisdom used the structural possibilities of the elements to give it first and foremost a *value*. Without religious considerations the investigation of nature can, it is true, be commenced, but not completed; and the latter will always be and remain the pillar of religion, while everything that is based on fanatical inner contemplation will lend itself to becoming the toy of changeable opinions, together with this fanaticism."

497 I am referring to the passage quoted earlier, where Griesinger defends the right of a 'teleological view' of the factum of reflex self-regulation (W. Griesinger, 1872a, p. 13).

498 The similarity between Herbart's scheme of disturbance and self-preservation which influenced the idea of a functional mechanical reaction, and the reflex action model favoured by Griesinger, must not be allowed to tempt us to assume a deep-seated *philo-sophical* affinity between the two scholars. On closer analysis the similarity tells us nothing more than that both of them endorsed the mechanistic ideal of science. There is nothing absurd in this when one considers that the commitment to the mechanistic

view, as we have observed, left room for a number of variations in the definition of the mind-body relationship, so that mechanism as it appears in the work of Herbart, Lotze and Griesinger, is in each case mechanism in a different philosophical setting.

[499] Cf. Section 28 of his *Pathologie und Therapie der psychischen Krankheiten* (W. Griesinger, 1871, pp. 48–49), which is central to his (ego) psychology.

[500] Cf. for this point and the explanation which follows the important passage in W. Griesinger, 1871, pp. 48–49: "The ego (I) is an abstraction in which traces of all former separate sensations, thoughts and desires are contained, as it were, bundled together, and which in the progress of the mental processes, supplies itself with new material; but this assimilation of the new ideas into the pre-existing ego does not happen at once – it grows and strengthens very gradually, and that which is not yet assimilated appears in man at the outset as an opposition to the *I*, as *a you*. Gradually it confines itself no longer to a single complexity of ideas and desires which represents the ego, but there are found several such masses of ideas united, organised, and strengthened; two (and not only two) souls then dwell within man, and this changes or is divided according to the predominance of the one or of the other mass of ideas, both of which may now represent the ego. Out of this, internal contradiction and strife may result; and such actually occurs within every thinking mind. In happy harmonious natures this conflict is spontaneously and rapidly brought to an end, since in all these various complexes of perceptions a number of general fundamental intuitions which recur in all of them are developed in common, even though they may be obscure and are not clearly expressible. These give a harmonising fundamental direction to all the spheres of thought and will. Faith on the one hand and empiricism on the other may serve as examples of such various fundamental directions. It is the highest object of self-education not only to acquire such general and solid fundamental directions, but to elevate them gradually as much as possible by thought into consciousness, and so, in the firm possession of such, to attain the well thought-out first premises of all thought and will adequate to the particular individual nature."

[501] The difference I am trying to bring out here is not that the teleological viewpoint and the teleological 'discourse' with terms like *self-education, self-control*, etc. are absent from Herbart's work (which is by no means the case, not even in his psychology: cf., for example, J. F. Herbart, 1891c, pp. 417–24, *Von der Selbstbeherrschung, insbesondere von der Pflicht, als einem psychischen Phänomen* (On self-control, particularly on duty as a psychical phenomenon)), but that Griesinger, precisely because he did not take over Herbart's metaphysics along with his psychology, could use such terms without this necessarily implying that disqualification of everyday (self)experience which Herbart, as a metaphysicist, was bound to defend.

[502] W. Griesinger, 1871, pp. 211–415; cf. K. Dörner, 1969, p. 372.

[503] Cf. p. 114.

[504] In respect of this last point, K. Dörner (1969, p. 372) even refers to an "unreconciled duality of approach" kept open by Griesinger.

[505] In Feigl we see how the conviction that "states of direct experience which conscious human beings 'live through' . . . are identical with . . . aspects of neural processes in those organisms" is combined with what he himself describes as a " 'double knowledge' theory" (quoted in R. J. Bernstein, 1971, p. 205). There exists, in his view, "one reality, which is represented in two different conceptual systems – on the one hand, that of physics and on the other hand . . . that of phenomenological psychology". (H. Feigl, 1973, p. 41.)

506 G. Zeller, 1968, p. 332.
507 The article by G. Zeller (1968), which we have quoted, is a partial exception to this.
508 Griesinger's significance in this respect is certainly greater than that of the friends who were most closely associated with him in these efforts of renewal, Wunderlich and Roser, with whom he shared the publication and editorial responsibility of the *Archiv für physiologische Heilkunde* (from 1842 onwards). (See p. 93.) For the role played by the alumni of the Tübingen school in this medical 'reformation', cf. G. Sticker, 1939. (As far as I can discover, the projected sequel to Sticker's publication never appeared.)
509 G. Zeller, 1968, p. 322.
510 Cf. p. 36 and the quotation from H. Ey (1962) given there.
511 E. A. Zeller (usually referred to as A. (Albert) Zeller, 1804–77).
512 W. Griesinger, 1845, p. vi.
513 C. Westphal, 1868/69 p. 761.
514 L. Binswanger, 1955, p. 88.
515 In this context, special reference should be made to the important part which Griesinger played in the humanising of psychiatric practice (defence of the no(n)-restraint method, abolition of outdated coercive methods, reorganisation of institutional life, etc.). For Griesinger's standpoint in practical psychiatry the following publications are relevant to a greater or lesser degree: W. Griesinger, 1872, I, pp. 254–331, and 1871, pp. 457–537.
516 G. Zeller, 1968, pp. 330–1.
517 According to G. Zeller, 1968, p. 331. Cf. also P. Vliegen, 1973, p. 148, where he uses, amongst other things, an unpublished – prize-winning – essay by G. Zeller, entitled *Die Geschichte der Einheitspsychose vor Kraepelin* (1961).
518 Johann Christian Reil (1759–1812) – primarily known in the history of psychiatry for J. C. Reil, 1799, and J. C. Reil, 1803.
519 J. Guislain, 1838.
520 Cf. E. A. Zeller, 1840. (Cf. G. Zeller, 1968, p. 330.)
521 In the history of psychiatry literature (J. Bodamer, G. Zeller, etc.), when Zeller is distinguished from the other somaticists because of his emphasis on the psychosomatic thesis, the authors are overlooking the fact that and the extent to which *all* somaticist psychiatry which is based on the thesis of the bodily soul, etc. essentially thinks 'psycho-somatically'.
522 In the reconstruction attempted by G. Zeller (1968) the relationship between A. Zeller and Autenrieth in fact remains a largely unsubstantiated hypothesis which, as far as I can see – apart from the agreement as to the fundamental psychiatric principles described – is based solely on the fact that A. Zeller studied at Tübingen, that is to say, probably studied under Autenrieth, and that the model of the polarity of cerebral and nervous ganglia that he uses in his psychiatric nosology stems from the work of J. C. Reil. Reil was a friend of Autenrieth's and they published the (*Reilsche*) *Archiv für Physiologie* together. Zeller is supposed to have become familiar with Reil's ideas through Autenrieth's work. Any more detailed reconstruction of something like an 'old Tübingen school' would also need to look at the mediating role played by Herbart's work; cf., for example, J. F. Herbart, 1892b, pp. 293, 295, 299.
523 Cf. W. Griesinger, 1871, pp. 212–13.
524 Cf. for Broussais, E. H. Ackerknecht, 1953. F. J. V. Broussais (1772–1832), known

as one of the earliest defenders of the pathological-anatomical standpoint and responsible for the medical reorientation in the spirit of his (at the time extremely influential)
system of 'physiological medicine' (1816), developed a pathological-monistic theory
based on the hypothesis of a universal disease (gastro-enteritis) to which all diseases
were reduced.

525 For Herbart, cf. note 552; for Lotze, who sides with M. Jacobi on the issue of
psychopathology, cf. R. H. Lotze, 1852, p. 603 (Section 500).

526 E. A. Zeller, 1838.

527 According to Hermann Zeller (1921, p. 211) who draws on the correspondence
between Zeller and Jacobi.

528 Cf., for example, E. A. Zeller, 1838, p. 532. "It is, however, as inconceivable that
there can be a healthy mental life without a healthy body, at least as far as its psychic
functions are concerned, as that there can be a real psychic disorder without a disorder
of the body."

529 As J. Bodamer (1953, p. 523) does.

530 Cf. E. A. Zeller, 1838, p. 515, where Zeller, in opposition to the current dualistic
misconception, puts forward the idea that "mind, soul and body [are] only the life of
the inseparable force of man". Besides this, Zeller also uses the term 'soul' in a wider
sense, that is to say in a connotation which also embraces the mind – in this case understood as 'the higher part of the soul' (p. 554). This (broader) sense of 'soul' is presupposed wherever, in general, the relationship between 'body and soul' is discussed.

531 E. A. Zeller, 1838, p. 554.

532 E. A. Zeller, 1838, p. 554.

533 In the contrast between spiritual and natural life something is after all expressed
of that Christian-inspired dualism which in the final analysis is difficult to reconcile
with the monistic ('romantic' Spinozist) tendency described.

534 Op. cit., pp. 554–55.

535 Cf. W. Leibbrand & A. Wettley, 1961, p. 498.

536 E. A. Zeller, 1838, p. 522.

537 Op. cit., p. 535: "The soul becomes diseased in its own way, from one cause or
another, but it really does become diseased, as every single case of psychic disturbance
indeed shows us. On the other hand, its becoming diseased entails the body's becoming
diseased too. This cannot be otherwise, because both, body and soul, are always united
within the one higher common concept of life."

538 Op. cit., p. 537.

539 Zeller does not, however, go so far as to state that (therefore) the soul can become
diseased: in this respect he shows himself to be a good somaticist and makes the distinction between 'psychic disease' (Gemüthskrankheit) and 'disorder of the mind'.
Cf. E. A. Zeller 1844, p. 37, " . . . there cannot be a psychic disease without a disorder
of the mind."

540 E. A. Zeller, 1838, p. 539.

541 The psychicists are never mentioned as such or by name in these passages, but they
are the undoubted target of the criticism (I am restricting myself chiefly to some quotations from Zeller's article of 1838). Op. cit., p. 551 (spiritual temptations can also be
determined by somatic suffering): "Moreover, the observation and experience that
such struggles and temptations are very common phenomena in the lives of the greatest
religious heroes and of the most noble people is in itself enough to contradict and

destroy the view that insanity always has its direct root in sin." And even more emphatically, it is a sin against humanity "when one postulates the assertion that insanity originates in human guilt" (p. 557). "If there were a direct link between a higher degree of sin, stupidity and irrationality and the genesis of psychic disorder, one would have to build asylums for the rational and leave the world to be inhabited by the insane." etc. (p. 558). Cf. E. A. Zeller, 1844, p. 22: "The fact that insanity is a disease and not a mistake can best be observed from this fatal outcome of psychic disturbance. Nobody dies by mistake or by immorality alone, however great these may be."

542 Cf. Hermann Zeller, 1921, p. 215: "Zeller saw the practice of psychiatry as his real task in life."

543 Quoted in H. Zeller, op. cit., p. 214.

544 E. A. Zeller, 1844, pp. 12–13.

545 Griesinger went to Winnenthal in 1840 and stayed there for two years, according to C. Wunderlich, 1869, pp. 117–18. Cf. the title of E. A. Zeller, 1844: *Bericht über die Wirksamkeit der Heilanstalt Winnenthal vom 1 März 1840 bis 28 Febr. 1843.*

546 E. A. Zeller, 1844, p. 13.

547 E. A. Zeller, 1844, p. 13. If this explanation of Zeller's remarks is correct, this implies that his first point of criticism of the 'mechanical and atomic view of psychic life', namely its misjudgement of the existence of spiritual life, is somewhat modified here. What the mechanists are prepared to recognise as mind or reason is, at best, a *reduced* concept of reason – reason in so far as it is settled in our established habits and attitudes, i.e. precisely that reason which Zeller wanted to bring out in opposition to Jacobi. However, the confrontation with the (physiological) mechanists probably forced Zeller to adopt a conceptual differentiation which is not made explicit by a corresponding terminological differentiation in the passage quoted above. It is nevertheless in my view, contained in the text, and I have rendered it above in the distinction between 'natural' reason (reasonableness) and reason in that sense which we ascribe only to conscious beings.

548 E. A. Zeller, 1844, pp. 26–27.

549 Cf. E. A. Zeller, 1844, p. 26: "Recently, people have also wanted to eliminate the concept of feeling (*Gemüth*) and have represented it as obscure and undeveloped ideas. This, however, has caused considerable confusion, as people considered the continuous cycle of our spiritual activity between idea and feeling, the repetition and transformation of feelings into ideas and the arousing of the feelings by ideas, to be one and the same process, only divided into different steps. The feelings accompanying the ideas, and the pleasure and sadness they cause, however, are not the same as conceptions but are the directly experienced changes and movements in the state of the soul itself" . . . "Ideas and feelings are such entirely different processes in our inner selves, by their very nature, that they have been given different names right from the outset in the languages of every nation."

550 Op. cit., p. 27.

551 Cf. also op. cit., pp. 31–32.

552 In 1865 L. Snell delivered his lecture *Monomanie als primäre Form der Seelenstörung* (L. Snell, 1865) to the conference of Natural Scientists. In this lecture, on the basis of a description of what would nowadays be diagnosed as paranoid hallucinatory schizophrenia, he depicted monomania or insanity as a primary mental disorder. According to Snell, this monomania (as a primary mental disorder) could be distinguished from mania and melancholia as an independent, i.e. irreducible nosological form. Snell's thesis meant,

however, a breaking through of the psychopathological systematics contained in the Zeller–Griesinger theory of unitarian psychosis. Because Griesinger's version of the unitarian psychosis theory was inseparably linked to (the categories of)Herbart's psychology (as, conversely, "[Herbart's doctrine], by its nature, could only lead to a unitarian psychosis in psychiatry and theoretically did not permit of separate clinical pictures at all' (J. Bodamer, 1948, p. 308)), in Griesinger's case, the recognition of Snell's primary insanity or monomania meant a choice in favour of (the standpoint of) clinical empiricism and against the unitarian pretensions of the Herbartian system of psychological concepts. We must indeed acknowledge that, although Griesinger appears as the great defender of theory and systematics against the atheoretical or antitheoretical scepticism of the French physiologists (Magendie and his followers) – this is the aprioristic element in his work which was criticised by C. Westphal (1868/69, p. 766) – when it came to the point he was too much of a clinician not to abandon his own theory (or parts of it) if it proved to be incompatible with clinical experience.

[553] A particularly good illustration of this point is Griesinger's criticism of the dualism which formed the basis of Jacobi's psychiatric theory, that is to say, his criticism of "the assumption of an absolute difference, a total separation, between mental health and illness" or, as we read elsewhere, of the thesis "that in insanity the soul is only changed in its anthropological relationships, never, however, in its characteristic inner psychological and moral being". (W. Griesinger, 1827c, pp. 90, 92.)

[554] I should like to remark in passing that the problem concealed here is not dissimilar to that which left its mark on the theology of Friedrich Schleiermacher, whose ideas are thought to have had a decisive influence on Zeller's religious convictions and view of life. P. Vliegen (1973, p. 148) states as a fact that "[Schleiermacher's] theology and religious philosophy made a lasting impression on him [i.e. Zeller]". He does not, however, substantiate this assertion in any way whatsoever, nor does he say anything about where we can find evidence of such an influence. Hermann Zeller (1921, p. 211) does tell us that Zeller married Maria Reimer in 1828. Her father's house was a meeting-place for well-known personalities of the time, one of whom was Schleiermacher, who even lived in the house and was a friend of Reimer's. The young couple was joined in matrimony by Schleiermacher. This information makes it clear that Zeller knew Schleiermacher personally, and it is therefore reasonable to assume that he was familiar with Schleiermacher's theological and religious philosophical views. It tells us nothing about the nature of the influence. The publications by Zeller which I consulted gave too few points of reference to permit of more than the above-mentioned suggestion.

[555] L. Binswanger, 1959, published without indication of place or date, but probably written in 1932 (see the beginning of the study). According to H. Spiegelberg, 1972, p. 196, note 5, this was published privately in 1959. The word 'anthropological' is used here and subsequently in the looser sense of 'directed towards the human individual'. In this sense, institutional psychiatry, to a certain extent Griesinger's psychiatry, and Binswanger's so-called 'existential analytical' psychiatry are all 'anthropological', in spite of all their mutual differences.

[556] To distinguish him from his grandson, who bore the same name, I shall refer to him as 'the elder' Ludwig Binswanger.

[557] This line has been continued up to the present day, because when Ludwig Binswanger surrendered the directorship of Bellevue in 1956 it was his son Wolfgang who took over from him. (Cf. C. Wenin and J.-P. Deschepper, 1966, p. 694.)

558 L. Binswanger, 1959, p. 10.
559 L. Binswanger, 1959, p. 11.
560 *Ibid.*
561 Op. cit., p. 16.
562 In this context, his lecture *Lebensfunktion und innere Lebensgeschichte* (1928) (L. Binswanger, 1961a) is particularly important. We shall discuss this further hereafter.
563 L. Binswanger, 1961a, p. 60.
564 Op. cit., p. 52.
565 In L. Binswanger 1955b [1924] this was (still) implied.
566 L. Binswanger, 1961a, p. 53; cf. p. 57.
567 The suggestion (for it is not much more than this) which emerges from his learned reflections on the history of ideas is that whereas Aristotle's theory of soul must be regarded as the origin of the functional view, and St. Augustine's *Confessiones* can stand as an unsurpassed illustration of the principle of the 'inner life history', the *relationship* between life as function and life as inner history can still best be conceived according to Hegel's example (*Enzykl.* par. 405, 406), namely as a relationship between 'genius' (which equates to Binswanger's 'psychic–physical function complex') and the 'self-conception of genius' (Binswanger's "spiritual person and its history"). (L. Binswanger, 1961a, p. 71.)
568 It is certainly not superfluous to emphasise again the correctness of this historical insight: as late as 1957 Jacob Wyrsch could still write that the psychicists and somaticists "did not [recognise] a history of the soul in the sense of individual development, of an inner history of life". (J. Wyrsch, 1957, p. 35.)
569 Cf. L. Jülicher, 1961, p. 40.
570 Cf. R. Kuhn, 1957.
571 L. Binswanger, 1955a, p. 84, alludes to the second revised edition of Griesinger's *Pathologie und Therapie der psychischen Krankheiten* of 1861.
572 L. Binswanger, 1955a, p. 88.

CHAPTER 4

573 Or, in Wolff's very general formula, which Schopenhauer chose (*Ontologia*, par. 70): "Nihil est sine ratione, cur potius sit quam non sit", quoted in A. Schopenhauer, 1962a, p. 15.
574 A. Schopenhauer, 1962a, p. 14.
575 See note 573.
576 A. Schopenhauer, 1962a, p. 15.
577 Cf. A. Schopenhauer, op. cit., p. 185: "Every science has one of the forms of the 'principle of [sufficient] reason' which is first and foremost its guide."
578 The most important source for this is A. Schopenhauer, 1961c, p. 165 ('*Zur Wissenschaftslehre*').
579 According to Schopenhauer's scheme (1961c, p. 165), this subgroup is made up of (1) general: physiology of plants and animals together with their auxiliary science, anatomy; (2) specific: botany, zoology, zootomy, comparative physiology, pathology, and therapy.
580 These terms were not used by Schopenhauer. In the subgroup of natural sciences in

the narrower sense (i.e. sciences of *inorganic* nature) Schopenhauer includes (a) general: mechanics, hydrodynamics, physics, chemistry; (b) specific: astronomy, geology, technology, pharmacy. Under "human" (or, if you will, motive) sciences, he includes (a) general: ethics, psychology; (b) specific: law, history.

[581] Schopenhauer (born 1788, died 1860) published his dissertation *Über die vierfache Wurzel des Satzes vom zureichenden Grunde* in 1813. His major work, *Die Welt als Wille und Vorstellung*, dates from 1819 (the second, "generally improved and considerably enlarged" edition appeared in 1844).

[582] P. Janet, 1880.

[583] Cf. note 292.

[584] A. Schopenhauer, 1961c, p. 397.

[585] A. Schopenhauer, 1961c, p. 320.

[586] Op. cit., p. 319.

[587] Op. cit., p. 320.

[588] Op. cit., p. 320.

[589] Cf., for example, A Schopenhauer, 1961c, p. 352. The empirical view ("objective view") with regard to the world of animals is "first of all zoological, anatomical, physiological", and it only becomes philosophical through connection with the "subjective" view and from the higher standpoint thus achieved.

[590] Op. cit., p. 317.

[591] Op. cit., p. 353.

[592] Op. cit., p. 369.

[593] Op. cit., p. 353.

[594] To put it explicitly: if it is true that the Will objectivises itself in the organism (body) and therefore *a fortiori* in (the nervous system of) the brain, in which the organism achieves its "efflorescence", and if it is true that the intellect (thinking) is a function of the brain and thus in the final analysis is based on the body as a living organism, then it follows that the whole "objective" world, i.e. the world of Ideas, perishes with the organism. In other words, the world (as Idea) is dependent on the intellect, but the intellect is a function of the brain which, in its turn, exists as an objectivisation of the Will, that is to say, as Idea.

[595] Cf. E. Cassirer, 1973, p. 430.

[596] Op. cit., p. 429.

[597] Of importance in this context are: (1) the first chapter (*Vom Sehn*) and *Über das Sehn und die Farben* (1816) (reprinted as A. Schopenhauer, 1962b) and (2) — even more detailed on this point — Section 21 of *Über die vierfache Wurzel des Satzes vom zureichenden Grunde* (1813) (reprinted as A. Schopenhauer, 1962a, pp. 67 ff.).

[598] A. Schopenhauer, 1962, p. 67.

[599] Op. cit., p. 70.

[600] Or, as it is called in his 1816 work (op. cit., p. 204) the "*Verständigkeit*" of *Anschauung*.

[601] Op. cit., pp. 101–2.

[602] Op. cit., p. 102.

[603] Op. cit., p. 68.

[604] Op. cit., p. 103.

[605] Op. cit., p. 204: "*Anschauung*, i.e. knowledge of an *object*, is only reached when the intellect relates every impression received by the body, to its cause, and places it in

the *a priori* perceived space where the causal activity (*Wirkung*) originates. In this way it acknowledges the cause as an active one (*wirkend*), as *real* (*wirklich*), i.e. as an idea of the same kind and class as the body."

[606] More detailed on this point: Section 21 of *Über die vierfache Wurzel* (A. Schopenhauer, 1962a, pp. 67–106). Schopenhauer specifies the activity of the intellect in respect of sensations in four directions: (1) reversal of the impression which the object leaves on the retina, so that what is at the bottom on the retina becomes the top and vice versa; (2) creation of one perceptual idea from a double sensation (because it is mediated through two different eyes); (3) construction of bodies on the basis of planes, i.e. construction of the third dimension; (4) knowledge of the distance between objects and ourselves.

[607] Cf. E. Cassirer, 1973, p. 433.

[608] Cf. A. Schopenhauer, 1961c, pp. 352–53: "[Cabanis] ... whose excellent work *Des Rapports du physique au moral de l'homme* has become a physiological milestone for this [i.e. objective] physiological approach." Further op. cit., pp. 337 ff., where Schopenhauer praises Bichat's work *Sur la vie et la mort* (1802) and comments on it as a physiology which is entirely in agreement with his metaphysical insights concerning the way and the form in which the Will objectivses itself in the body: "Therefore anyone who wants to understand me should read him [i.e. Bichat], and anyone who wants to understand him better than he understood himself, should read me" (op. cit., p. 338).

[609] Cf. A. Schopenhauer, op. cit., pp. 521 ff. Albrecht von Haller lived from 1708 to 1777. His work *Elementa physiologiae corporis humani* (8 parts), which Schopenhauer was familiar with, dates from 1757–66.

[610] In respect of this point concerning the (historical) genesis of Schopenhauer's theory of knowledge I find Mandelbaum's explanation (1974, p. 315) convincing.

[611] Cf., for example, E. Cassirer, 1973, p. 413.

[612] I am referring here not so much to the fact that Goethe's theory in fact assumes only two fundamental colours (yellow and blue), that is to say, one polarity, from which all other colours can be 'deduced', and that Schopenhauer assumes three polarities, three pairs of complementary colours (namely red–green, orange–blue, yellow–violet), but to the fact (underlined by Schopenhauer) that whereas Goethe essentially restricts himself to assembling 'materials for a future theory of colour' and is thus content with a sort of 'phenomenological' description of the phenomena of colour which can be seen as a contribution to a future *physical* theory of colour, Schopenhauer is aiming to provide a 'real' *theory*, a *physiological* one, which can *explain* the phenomena described by Goethe and thus supplement and justify scientifically Goethe's insights. Cf. A. Schopenhauer, 1962b, pp. 198–200.

[613] It is therefore remarkable that Schopenhauer, some 38 years after the appearance of his theory, should have found occasion to publish a revised, expanded, but essentially unchanged edition of his early work. This occurred in 1854, that is to say, at a time when the 'romantic' polarity idea in sensory physiology and psychology was a thing of the past and – from the viewpoint of the development of science – a work such as his could be little more than stillborn. We may, however, assume that the increased interest in the work of a 26-year-old was not based on a belated recognition of its scientific importance, but was rather a result of the interest in Schopenhauer's philosophy in general which was aroused after about 1850 and which brought him the recognition which he had sought in vain for the last forty years. (Cf. for this last point A. Schopenhauer, 1962b, p. 195 (Preface to the second edition).)

[614] We find a similar (intermediate) position in the case of J. Frauenstädt who –
relying on Kant and Schopenhauer – showed himself to be a (Neo-Kantian) critic
almost ten years before Lange's *Geschichte des Materialismus*. Cf. J. Frauenstädt, 1856.
On Frauenstädt cf. F. Gregory, 1977, pp. 31–33.
[615] Cf. E. H. Ackerknecht, 1957, pp. 129, 131.
[616] E. Lesky, 1965, p. 136.
[617] Cf. pp. 56–57.
[618] The most important of these – all of which appeared after the death of Johannes
Müller, that is, after the victory of the mechanistic viewpoint in medicine was a fact
– are K. von Rokitansky, 1859, 1862 and 1867.
[619] Cf. E. H. Ackerknecht, 1958, p. 41; E. Lesky, 1965, p. 139; W. Leibbrand, 1953–
54, M. Neuburger, 1934.
[620] K. von Rokitansky, 1859, pp. 136–37.
[621] Op. cit., pp. 138–39.
[622] Op. cit., pp. 139–40.
[623] K. von Rokitansky, 1867, p. 112.
[624] Op. cit., p. 113.
[625] Op. cit., p. 114.
[626] Op. cit., pp. 126 ff.
[627] Op. cit., p. 130.
[628] Op. cit., p. 130.
[629] Op. cit., p. 133.
[630] Op. cit., pp. 107–8.
[631] Cf. op. cit., p. 140.
[632] Op. cit., pp. 134–35.
[633] Op. cit., p. 135.
[634] Op. cit., p. 131.
[635] Op. cit., p. 135.
[636] Op. cit., p. 136.
[637] Rejection, for that matter, of arguments which *de facto* go no further than to support the *transcendental*-idealistic position; cf. K. von Rokitansky, 1867, pp. 137–39.
[638] For M. J. Schleiden (1804–81), cf. his *Ueber den Materialismus der neueren deutschen Naturwissenschaft, sein Wesen und Geschichte* (1863), in which, calling himself
a "true disciple of Kant", he gives it as his opinion that "an antidote to materialism
[in the sense in which it denies mind, freedom and God] can only be found in a complete, empirical psychological foundation and in the detailed development towards a
logic that is based on it" (op. cit., p. 57). Schleiden is alluding here to the (anthro-
pological-psychological) Kantianism of Fries, which he had defended at length as early
as 1842 in the great methodological introduction to his major work *Grundzüge der
Wissenschaftlichen Botanik* (cf. M. J. Schleiden, 1849, pp. 4 ff., 17 ff., 24 ff., 28, 29 ff.,
etc.). For R. Virchow (1821–1902), cf. E. H. Ackerknecht, 1957, pp. 41–42, and the
references to the literature given there. Virchow used Kant's limitation of the human
possibilities of knowledge as a basis for his agnosticism, largely anticipated Du Bois-
Reymond's arguments in favour of agnosticism, and preferred to be called a 'realist',
rather than a 'materialist'.
[639] Quoted in L. J. Pongratz, 1967, p. 76.
[640] F. A. Lange, 1974, II, pp. 776–849.

641 Not least through the medium of the Neo-Kantian H. Vaihinger. Cf. H. Vaihinger, 1876, pp. 18–29, 54–68, 103–19, 178–97, and his *Philosophie des Als Ob*, which was inspired by Lange's interpretation of Kant, in H. Vaihinger, 1913 (on Lange, op. cit., pp. 753–71).

642 L. J. Pongratz, 1967, p. 91.

643 Cf. pp. 48–50.

644 Cf. G. Weiss, 1928.

645 In Lange's terminology, 'mathematical psychology' is equivalent to the psychology of Herbart (and the Herbartians). His criticism of this 'mathematical psychology' in no way relates to the application of mathematics in it, but to the 'bizarre' use it is put to. Cf. F. A. Lange, 1865, and his commentary, F. A. Lange, 1974, II, pp. 818–23.

646 F. A. Lange, 1974, II, p. 819.

647 Cf. p. 119.

648 W. Dilthey, 1968a, pp. 154 ff.

649 F. A. Lange, 1974, II, p. 822.

650 Op. cit., p. 823.

651 Op. cit., p. 823.

652 Op. cit., p. 828.

653 Cf., for example, op. cit., p. 828: "It belongs to the central truths of a new epoch of humanity now dawning – not that, with *Comte*, we should abolish speculation, but certainly that we should once and for all assign it its place, that we should know what it can and cannot do for knowledge."

654 F. A. Lange, 1974, II, p. 835.

655 Quoted in L. J. Pongratz, 1967, p. 76, without reference, probably taken from F. A. Lange, 1870.

656 F. A. Lange, 1974, II, pp. 829–30.

657 Op. cit., pp. 835–36.

658 F. A. Lange, 1974, II, p. 836.

659 Op. cit., pp. 835–36.

660 Op. cit., p. 827.

661 Op. cit., p. 887.

662 If this interpretation is correct, one can go a step further and ask oneself whether, with his methodological prescription, Lange is not in fact adopting a definition of (the object of) psychology – (the object of) psychology includes (only) those phenomena which fall within the area of application of the somatic method – which belies the alleged metaphysical neutrality of his method. It seems to me – partly on the basis of other consdierations which will be discussed later in this chapter – not too far-fetched to assume that Schopenhauer's metaphysics had an influence on this point, and to state that this influence, with his theory of the Will which objectivises itself and is expressed in phenomena, not only provided impulses to the later 'philosophy of life', but also – in a more concealed fashion – carried through into the ontological presuppositions of Lange's methodological ideal.

663 What is meant here is research in the sense of psychophysics, i.e. that experimental science which, prepared by J. Müller, was developed in the work of, in particular, E. H. Weber, G. T. Fechner and H. Helmholtz, and which was directed at the study of the relationship between physical stimuli and the phenomena of consciousness (primarily the processes of perception).

664 We should not lose sight here of the connection between the psychophysical approach and the metaphysical position of (psychophysical) materialism: where early (experimental) natural scientific psychology successfully tackled psychophysical questions, at the very least sympathy with psychophysical materialism seemed to be scientifically justified.

665 Needless to say, we have here a question of (1) a development in natural scientific psychology from psychophysics (Fechner, Weber, Helmholtz) to physiological psychology (Wundt), and *parallel* with this (2) a shift in the metaphysical presuppositions of this psychology, from psychophysical materialism to (dualistic) psychophysical parallelism.

666 To prevent misunderstanding I should like to point out that the fact that the (natural scientific) psychology which – with Wundt – emancipated itself from physiology constituted itself as *physiological* psychology is not a contradiction: this psychology is called 'physiological' because of its method, not because of its object of investigation. Cf. W. Wundt, 1908, p. 4.

667 Cf. K. Vorländer, 1903, p. 457.

668 F. A. Lange, 1974, II, p. 449.

669 H. Cohen expresses himself in a similar vein on this point in the *Biographisches Vorwort* to the 1914 edition: "Therefore the history of materialism became a justification of idealism, that is real idealism, which had proved its fertility right from the start in the history of critical thought. Transcendental idealism was taught and preached as the victory over and the end of materialism." (F. A. Lange, 1914, I, p. vii.)

670 Not without justice, Hermann Cohen, his colleague in Marburg, remembered him as an "apostle of Kantian *Weltanschauung*", (F. A. Lange, 1914, I, p. VI).

671 F. A. Lange, 1974, I, p. 72.

672 Cf. op. cit., II, p. 987: "One thing is certain, that man needs to supplement reality with an ideal world of his own creation, and that the highest and noblest functions of his mind cooperate in such creations."

673 Cf. op. cit., II, p. 981: "The unity which makes the facts into a science and the science into a system is a product of free synthesis, and springs therefore from the same source as the creation of the ideal."

674 Op. cit., II, p. 981.

675 Cf. op. cit., II, p. 983: "The same principle which rules absolutely in the sphere of the beautiful, in art and poetry, appears in the sphere of conduct as the true ethical norm which underlies all the other principles of morality, and in the sphere of knowledge as the shaping, form-giving factor in our picture of the world."

676 Namely as the one who "with divinatory spiritual power, has taken the innermost part of his [i.e. Kant's] doctrines and purified them of scholastic dross" (op. cit., II, p. 510).

677 *Ibid.*

678 Op. cit., II, pp. 510–11.

679 Cf. op. cit., II, p. 509–10: "Kant would not understand what Plato before him would not understand, that the 'intelligible world' is a *world of poetry*, and that precisely upon this fact rests its worth and nobility. For poetry, in the high and comprehensive sense in which it must be taken here, cannot be regarded as a capricious playing of talent and fancy with empty imaginations for amusement, but is a necessary offspring of the soul, arising from the deepest life-roots of the race, and a complete counterbalance to the pessimism which springs from an exclusive acquaintance with reality."

680 Op. cit., p. 511.
681 Op. cit., II, p. 988.
682 Cf. op. cit., II, pp. 989–90.
683 Op. cit., II, p. 992.
684 The 'pagan' religious thought of Schiller's poem is – paradoxically – even "nearer to the traditional life of Christian faith than the rationalising dogmatism which arbitrarily maintains the notion of God, and abandons the notion of redemption as irrational". F. A. Lange, 1974, II, p. 990.
685 Cf. op. cit., p. 992: "This predominance of form in belief also betrays itself in the remarkable fact that the believers, in varying and even mutually hostile confessions, show more agreement with one another, betray more sympathy with their most zealous opponents than with those who appear indifferent in matters of religious controversy."
686 Cf. H. Vaihinger, 1913, p. 755.
687 F. A. Lange, 1974, II, p. 455.
688 Cf. op. cit., p. 459.
689 As a reminder: *methodological* materialism (or in short, naturalism) in this context contrasts in the first place with *dogmatic* (metaphysical) materialism. Within this naturalism I have thus far distinguished between (1) idealistic naturalism (e.g. Rokitansky), (2) pragmatic agnostic naturalism (Helmholtz, Du Bois-Reymond), (3) identity theory oriented naturalism (Griesinger).
690 F. A. Lange, 1974, II, p. 459.
691 Op. cit., p. 456.
692 F. A. Lange, 1974, II, p. 850.
693 Op. cit., II, p. 571 (note 25).
694 Op. cit., p. 572.
695 Op. cit., p. 572. The expression "cause of the *a priori*" – meaningless from the logical viewpoint – is only understandable from what Schopenhauer would call the "objective (i.e. physiological) viewpoint", which Lange evidently presupposes here.
696 Op. cit., p. 573.
697 Op. cit., p. 572.
698 Cf. F. A. Lange, 1974, II, p. 478: "The fact that we experience at all is, in any case, conditioned by the organisation of our thinking, and this organisation is present *before any experience*".
699 Cf. op. cit., p. 479.
700 Op. cit., p. 480.
701 *Ibid.*
702 Op. cit., p. 481.
703 Op. cit., II, p. 481.
704 Op. cit., p. 494.
705 *Ibid.*
706 Cf. p. 186.
707 F. A. Lange, 1974, II, p. 852.
708 Op. cit., p. 852, cf. also p. 867.
709 Op. cit., p. 852.
710 *Ibid.*
711 Op. cit., p. 867.
712 Op. cit., II, p. 868. Cf. also the summary of his standpoint given by Lange in op.

266 NOTES

cit., II, p. 864: "1. The sense-world is a product of our organisation. 2. Our visible (bodily) organs are, like all other parts of the phenomenal world, only pictures of an unknown object. 3. The transcendental basis of our organisation therefore remains just as unknown to us as the things which act upon it. We always have before us only the product of the two."

713 See previous note.

714 H. Cohen, 1914, p. vii.

715 Op. cit., pp. vii–viii.

716 What critical philosophy (read: correctly understood critical philosophy) is about is not the principles of human reason, but "the foundations of the sciences, determining scientific validity"; our organisation, Cohen remarks, is "a question of psychology, as far as it can be a question at all" (H. Cohen, 1914, p. viii).

717 Cf. p. 195.

718 If, in the history of nineteenth-century philosophy in Germany, Lange (like Helmholtz who is similar in some respects) is not numbered among the Neo-Kantians *in the strict sense*, a judgement of this kind undoubtedly reflects the opinion of these Neo-Kantians themselves. The slightly older Lange was, it is true, an important supporter of the cause of critical philsophy. He was, however, not (yet) considered 'pure' enough to do the utmost justice to the logical meaning of the Kantian critique of knowledge. Cf., for example, K. Vorländer, 1903, pp. 457 ff.

719 Cf. for the relationship of this materialism to Lange's Neo-Kantianism, L. Büchner, *Kant und F. A. Lange*, in L. Büchner, 1884, pp. 331–42.

720 Indeed, as philosophical 'logos' of methodological materialism. Needless to say, Lange is not concerned with a conception according to which, for example, the method of the natural sciences (the 'materialistic' method) has to hold good as a method of philosophy (Brentano's standpoint). The 'logos' concerned here is a transcendental one, albeit in that meaning of 'transcendental' (inadmissible in Neo-Kantian logical 'purism'), in which the word is more or less synonymous with "relating to the structure of the cognitive apparatus (the 'psychic–physical organisation')". (Cf. our observations on Helmholtz's use of the term 'transcendental' on p. 62).

APPENDIX NOTES

1 Works in the fields of the history of psychology, psychiatry, physiology, medicine and anthropology on the one hand, and philosophy on the other do exist (in some cases there are large numbers of works, in others far fewer). In contrast, the history of anthropology in the sense of empirical studies focusing on man in his entirety or conceived as a totality has still not received comparable treatment. For a *general* orientation, the literature survey given in W. Sombart 1938, which includes a brief reference to earlier literature (op. cit., p. 98, n. 1), is still indispensable. At the present time there is also a small number of sometimes one-sided and/or chronologically limited studies which do not offer much more than information (which is often faulty) on the (so-called anthropological) literature and the conceptual history. Of these publications I should like to mention the following – which I used: F. Hartmann & K. Haedke, 1963; O. Marquard, 1965 and 1971; M. Linden, 1976. Of an entirely different order, but certainly just as inspiring, is the work of G. Gusdorf, particularly his monumental work G. Gusdorf,

1968/78, compared with which the earlier G. Gusdorf, 1960, gives the impression of being a preliminary study. Gusdorf's work, however, gives at one and the same time both more and less than is required. More, in the sense that it provides a sort of comprehensive survey of the origins of the (i.e. all) human sciences in the context of (the development of) Western thinking; less, in the sense that he too does not recognise the history of anthropology in the above-mentioned sense of the word as an independent theme (it is highly significant that he too still includes that which is relevant to the history of anthropology under the headings 'psychology', 'medicine', 'psychiatry' and 'anthropology').

2 Cf. Aristotle, Hist. an. I,6,490b31 ff.

3 ἡ νοητική ψυχή, Latin: anima intellectiva, therefore sometimes also translated as intellectual soul'. Cf., for example, Aristotle, De Anima, 429a28. The fact that Aristotle, in determining the existence of this 'reasoning soul', oversteps the bounds of the 'bio-logical' because and in so far as (the function of) the soul can no longer be interpreted as a specification of the general principle of life (ψυχή) is dealt with later, in the discussion of his theory of the soul.

4 In this summary of the characteristics of (corporeal) man I am following E. Zeller, 1963, pp. 564 ff.

5 Aristotle, Gen. An. II, 4, 737b26.

6 Real in the sense of the individually existing things, πρῶται οὐσίαι.

7 Aristotle, De Anima, 412a27–8 (cf. 412b5).

8 This 'synolistic' (= anthropological) implication of Aristotle's psychology, according to R. E. Brennan, 1946, p. 48, "impressed Aquinas so much that when he came to write his own matured views on psychology, he incorporated them into his Treatise on Man. Now the focal point of analysis becomes man rather than soul".

9 The expression νοῦσ ποιητικός does not, it is true, originate from Aristotle himself, but it is justified by the descriptions of it which he gives. (Latin: intellectus agens.) Of the new literature on the Aristotelian νοῦσ–doctrine, I refer particularly to H. Seidl, 1971, which is critical of Ross and those of like mind.

10 παθητικός, i.e. susceptible to impressions. According to D. Ross (1961, p. 46) the νοῦσ παθητικός is not called 'passive' because it is itself passive (all thinking is active), but because it is dependent on sensory perception. Its function (according to Ross, op. cit., p. 46) consists of understanding universals in so far as this presupposes the observation of individual things; in other words, provided in this case that the universals are understood as actually present in perceptible things.

11 The νοῦς ποιητικός must, according to Ross (op. cit., pp. 46–47), be understood as that faculty of (ideative) 'abstraction' that is directed towards the understanding of universals which are not present in perceptible things, such as the perfect square or the perfect circle, which can never be observed as such in perceptible, imperfect squares and circles. The νοῦς ποιητικός does not create ex nihilo, but causes the universal, the essence, to appear as a result of its 'abstractive' activities, comparable with the light which "changes potential colours into actual colours". (De Anima, 430 a 16–17.)

12 In the history of scholastic philosophy this 'naturalistic' school is often referred to as 'Alexandrinist', after Alexander of Aphrodisias (second century B. C.).

13 The negative valuation of the individual, characteristic of the ontological thinking of the ancients – only the universal, the essential 'is' in the full meaning of the word; the individual, on the other hand, belonging to the sensory, temporal world, does not

actually count ontologically – still played an important part in Aristotle's thinking. In Hellenism (Stoa: Panaetius) and later ancient thinking (Plotinus) the concept of individuality was gradually to acquire a positive meaning – a development which paved the way for the transition to the (early) Christian concept of the person (Boethius: persona est *individua substantia* naturae rationalis). Cf. H. Heimsoeth, 1958, pp. 172–74.

[14] A. M. S. Boethius, 1343: "Quocirca si persona in solis substantiis est, nec in universalibus, sed in individuis constat, reperta personae est definitio: *Persona est naturae rationalis individua substantia.*" (My italics.)

[15] The barely concealed "Platonistic" dualism of incarnate and non-incarnate soul which rendered St. Thomas's avowed hylomorphism, (Aristotelianism) problematic, was clearly to emerge in the somaticist psychiatry of the first half of the nineteenth century in Germany (cf. Chapter 1).

[16] Cf. G. Gusdorf, 1969, p. 179.

[17] J. Burckhardt, 1977, p. 284: "In addition to the discovery of the world, the culture of the Renaissance adds a still greater achievement by discovering and bringing to light for the first time the entire, complete content of man" (to which Burckhardt adds in a note "These striking words were taken from Volume 7 of Michelet's History of France (Introduction)").

[18] As does, for example, F. Coppleston, 1963, pp. 57–58.

[19] Consider, for example, the 'violent tenor of life' which is stressed by J. Huizinga, 1979, pp. 9 ff., as a characteristic of (late) mediaeval society, to get an idea of the relativity of such a description of the Renaissance.

[20] Cf. E. Cassirer, 1977, pp. 100 ff. on the changing of the Prometheus motif in the Renaissance.

[21] From Pico's *Oratio de hominis dignitate* (Oration on the dignity of man), 1487, in Pico della Mirandola, 1942, p. 106: (The Creator is speaking to Adam) – "Nec te caelestem neque terrenum, neque mortalem neque immortalem fecimus, *ut tui ipsius quasi arbitrarius honorariusque plastes et fictor, in quam malueris tute formam effingas.*" (My italics.)

[22] B. Groethuysen thus looks at the Renaissance, not entirely without justification, under the title: *Die Grundlagen der modernen Anthropologie* (The foundations of modern anthropology), B. Groethuysen, 1928, pp. 99ff.

[23] Cf. E. Cassirer, 1977, p. 4: "The philosophy of the Quattrocento is and remains essentially theology, especially in its most important and successful achievements. Its total content is compressed within the three great problems: God, freedom and immortality. It was these which brought about the conflict of opinions in the school of Padua, the argument between the 'Alexandrinists' and the 'Averroists'. They are also the core of all the speculations of the Florentine Platonist circle."

[24] Cf. W. Windelband and H. Heimsoeth, 1958, p. 307.

[25] Needless to say, we are concerned here with the so-called *naturalistic* variant of anthropology, that is to say, the empirical study of man which was the primary interest of physicians and which, as far as its object was concerned, was somatically oriented.

[26] For the historical background to so-called Latin Averroism, which gained a firm foothold in Padua from the end of the thirteenth century onwards, in particular, and the background to the subsequent development of naturalistic Italian Aristotelianism (Padua, Bologna, Pavia, etc.) which, supported by sixteenth-century thinkers like Pomponazzi and Zabarella, paved the way for the advent of modern natural science and anticipated

Spinoza, in general, cf. P. O. Kristeller and J. H. Randall Jr., 1941; J. H. Randall Jr., 1961; 1962, pp. 53–55 and 65–88. More detailed on some points, but largely overlapping op. cit. 1962, are parts of E. Cassirer, P. O. Kristeller, J. H. Randall Jr. (eds.), 1971, namely the *General Introduction* (by Kristeller and Randall) and the introduction to the English translation of Pomponazzi's *De immortalitate animae* (by Randall).

[27] G. Gusdorf, 1969, p. 182.

[28] From the introduction to Ficino's translation of Plotinus, quoted in J. Hirschberger, 1969, p. 17.

[29] On the question of the Arabic medical tradition in the (Latin) Middle Ages in general, the most important medico-historical work to date is that of H. Schipperges, 1964. The fact that Italian, chiefly medical, naturalism had a noteworthy counterpart in the medical naturalism of the sixteenth-century Spanish Renaissance (the most important representatives of mediaeval Arabic medicine up to the thirteenth century were Spanish Arabs, including Averroës!) was recently elucidated by C. G. Noreña, 1975, Chapter IV: *Juan Huarte's naturalistic philosophy of man.*

[30] We are referring here primarily to Cardanus (Cardano) d. 1576, Telesius (Telesio) d. 1588, Patritius (Patrizzi) d. 1597, Campanella d. 1639, and – in a class of his own – Giordano Bruno, 1548–1600.

[31] Cardano's *De rerum varietate*, 1553, is strictly speaking a sort of encyclopaedia in which a few chapters are devoted to man (*'Humana natura'*; *'Hominis mirabilia'*).

[32] In W. Dilthey, 1977.

[33] In my view, Dilthey's greatest contribution to the history of anthropology is his evaluation of the significance of pantheistic religious thought as an *anthropological* motif – cf., amongst others, W. Dilthey, 1977a, p. 340, where Bruno is described as "the foremost representative of modern pantheistic religious thinking".

[34] Thus we see how, on the one hand, alongside the literary philosophical output of the group of Neo-Platonist humanists who, led by Petrarch, and fundamentally linked to St. Augustine as their great model, explored the depths of individual-subjective inwardness, one finds the more systematic psychology, developed with an empirical-scientific intention in mind, of an author like the Spanish humanist Juan Vives (1492–1540); while, on the other hand, Montaigne's work, as the most important French contribution to a "theory of the conduct of life" (Dilthey), was to become the beginning and the basis of the humanistic tradition of the French moralists. This in turn was to make its influence felt in that practical psychological tradition of eighteenth-century German philosophy of enlightenment, which was to culminate in Kant's *Anthropologie in pragmatischer Hinsicht.*

[35] O. Marquard, 1971. cf. M. Wundt, 1964, particularly on the significance of Christian Thomas (op. cit., pp. 37 ff.) and on 'the theory of man' in the period 1750–80 (op. cit., pp. 26 ff.) For discussion of the (practical) psychology of this period cf. M. Dessoir, 1964.

[36] As we have pointed out, there is still no systematic source study of any significance relating to the empirical study of man in this (transitional) period. For details of the literature see Appendix note 1.

[37] The full title is: *Anthropologium de hominis dignitate, natura et proprietatibus, de elementis, partibus et membris humani corporis etc. de Spiritu humano etc. de anima humana et ipsius appendiciis.* Quoted in O. Marquard, 1965, p. 225.

[38] E. Rádl, 1970a, p. 110.

[39] C. Singer and E. Underwood, 1962, p. 92.

[40] Cf. T. Ballauf, 1954, p. 125.

[41] E. Rádl, 1970a, p. 125.

[42] The influence of Aristotle has always been strongest, however, in the field of (general) biology. The supersession of Aristotelianism by Galilean natural science thus meant that biology declined and the work of Caesalpino was consigned to oblivion. As Rádl, op. cit., p. 126, remarks: "It was not until biology prepared to rouse itself for a new life that Linnaeus fell back upon Caesalpino — Linnaeus who had founded the new science of botany".

[43] Cf. C. Singer and E. Underwood, 1962, p. 121.

[44] According to E. Rádl, 1970a, p. 135; also cf., in particular, W. Pagel, 1967, which can currently be considered as the leading monograph on Harvey.

[45] C. Singer and E. Underwood, 1962, p. 112; cf. further G. Gusdorf, 1969, pp. 139–57. With emphasis on the scientific methodological aspect: W. A. Wallace, 1972, pp. 159–210; pp. 184–93 refer specifically to Harvey. This is a work which deserves attention as a more recent supplementation of and support for J. H. Randall's thesis about the significance of Paduan naturalism as the cradle of modern natural science and scientific methodology (cf., in particular, J. H. Randall Jr., 1961).

[46] This last point was a factor in Harvey's approach to the problem of the circulation of the blood. The ancient theory about the manufacture and movement of the blood proved to be absolutely untenable, if only on the basis of quantitative considerations concerning the amount of blood which is pumped out of the heart per heartbeat, in a single systole. This led to the adoption of a new theory in which the circulation of the blood was compared to a complicated machine with one part (the heart as the motor of blood circulation) causing the movement, and the other parts following passively.

[47] W. Pagel, in particular, has (again) brought out the Aristotelian basis to Harvey's thinking and calls "Harvey's *critical allegiance to Aristotle*" (my italics) the "keynote" of his book (W. Pagel, 1967, p. 19).

[48] K. E. Rothschuh, 1968, pp. 72–74, quotes at length from the second letter to Jean Riolan (1649), which not only throws light on Harvey's methodology of physiology but also strikingly illustrates this (conscious) dissociation from Aristotelian views.

[49] Still indispensable for knowledge of this poorly-treated period in the history of German philosophy are E. Weber, 1907; P. Petersen, 1921; M. Wundt, 1939. Very valuable bibliographically, although limited to sixteenth-century psychological literature, is H. Schüling, 1967.

[50] *Corpus Reformatorum* (ed. Bretschneider), Vol. XIII, pp. 5–178. Quoted in P. Petersen, 1921, p. 80.

[51] P. Petersen, 1921, p. 80.

[52] Cf. St. Thomas Aquinas, 1953, pp. 438b–562b (*Summa Theologiae* I, pp. 75–90).

[53] This refers to a collective work containing the writings of various authors on the essence and origin of the soul (man), published by Goclenius in Marburg in 1579, under the title: ψυχολογια, *hoc est: De hominis perfectione, animo et in primis ortu hujus commentationes ac disputationes quorundam theologorum et philosophorum nostrae aetatis.*

[54] The term 'psychologia' occurs for the first time, according to H. Schüling, 1967, p. 125 (cf. op. cit. p. 7) in Joh. Thomas Freigius, 1579 (pp. 761–71: *De psychologia*, and pp. 1147–1283: *De anthropologia*). The "correction" given in L. J. Pongratz,

1967, p. 17, concerning the widespread misconception that one must look to Melanchthon for the origin of the word 'psychology' can thus be considered as out of date.
55 I am referring to M. Hundt, 1501, in which the anthropological synopsis is given
from a primarily medical point of view and the emphasis is placed on human anatomy
and physiology. In the words of F. Hartmann and K. Haedke, 1963, p. 49: "Structure
and function, anatomy and physiology, physiological and psychological significance,
philosophical and theological interpretation are viewed as coinciding with the aim of
learning better to preserve the health of the body through knowledge."
56 Cf. M. Linden, 1976, pp. 1–3. Casmann understands anthropology as part of natural
philosophy ("anthropologia est pars Zoographiae, hominis naturam explicans" it says
in a work dated 1605). Cf. M. Linden, op. cit., p. 6.
57 O. Casmann, 1594 and 1596.
58 F. Hartmann and K. Haedke, 1963, p. 67, refer in this context to a "brilliant conclusion to this epoch of natural scientific medical anthropology".
59 Thus F. Hartmann and K. Haedke, 1963, pp. 49–50, remark, with reference to the
anthropology of Girolamo Cardano: "One is reminded that the first period of modern
science was a time of collecting new observations and that, especially in medicine,
curiosities, abnormalities and deformities were of particular interest".
60 F. Hartmann and K. Haedke, op. cit., p. 67.
61 This expression comes from H. Plessner, 1975, p. 37.
62 In Germany the influence of Cartesianism was minimal. Aristotelianism dominated
Protestant scholasticism and the philosophy of Wolff and his followers. According to
Wundt, in Germany strict Cartesianism was "defended, *if at all*, only among members
of the Reformed Church" (my italics). (M. Wundt, 1964, pp. 109, 319). In the Netherlands, on the other hand, the influence of Cartesianism was extremely great (and the
Aristotelian influence correspondingly small). Cf. C. L. Thijssen-Schoute, 1954; for the
significance of Descartes in medicine cf. G. A. Lindeboom, 1979.
63 This reduction necessarily followed from the ontological dualism of *res cogitans*
and *res extensa* in Descartes' interpretation. The mind has no extension; only that which
is 'extended' is susceptible to scientific (i.e. mathematical) definition – that is, as far
as man is concerned, only his body. A science of man is therefore necessarily a science
of corporeal man. For Descartes' anthropology cf. R. Descartes, 1953 [1664].
64 Cf. M. Linden, 1976, pp. 16 ff., 19 ff. In order to do justice to Descartes' significance
for the history of anthropology (in the sense of the empirical study of man) it is as well
to remember that it was ultimately the 'halving' of traditional anthropology sanctified
by Descartes which, with its opposition to the traditional (synoptic/synolistic) view,
created the justification for the specialisation and limitation of anthropology as a scientific discipline, and was the cause of a development which continued until well into the
nineteenth century, through which physical anthropology and psychical anthropology,
each with one-sided totalitarian pretensions, stood opposed to each other. (Cf. F. Hartmann and K. Haedke, 1963, pp. 67–68.) Note: the paradoxical implications of the
Cartesian anthropological figure should not go unremarked. The orientation towards
the 'modern' concept of nature did, it is true, make it possible to approach man (like
nature) from the perspective of a quantitative-mechanistic ontology, but the result of
this – Descartes' mechanistic anthropology – was, strange as it may sound, an 'anthropology' from which man had disappeared. Animals (according to Descartes) have no
mind and can therefore be conceived without difficulty as *animal machine*; but man as

homme machine – the theme of (mechanistic) anthropology – is man as (= in as far as) animal, in other words, in as far as the *specifically human* element (the 'mind') is left out. Considered thus, it is even good Cartesianism to agree with Gusdorf when he says that Descartes' mechanistic anthropology is not anthropology. (Cf. G. Gusdorf, 1960, p. 93: "Mechanistic anthropology may perhaps be mechanics; it is not anthropology".)
[65] Cf. I. Kant, 1923a [1798], pp. 3–4.
[66] E. Platner, 1772 and 1790. A nice example of this medical trend in 'philosophical' anthropological thinking, although it is not given the title of 'anthropology', is the dissertation by the young physician Friedrich Schiller, *Versuch über den Zusammenhang der thierischen Natur des Menschen mit seiner geistigen* (1780) (Essay on the relation between the animal and the spiritual nature of man). In this work the relationship between mind and body is compared to that between two string instruments placed side by side. If one strikes a cheerful or a mournful note on one instrument, the corresponding string on the other instrument will resonate. "This is the wonderful and remarkable sympathy which virtually transforms the heterogeneous principles of man into one essence; man is not soul and body, man is the most profound mixture of both these substances." (F. Schiller, 1962a [1780], p. 64.)
[67] For a more detailed and reasoned account of this parallel, see Chapter 1.
[68] Leaving aside for the moment other influences of a less direct nature such as those of the Schellingian philosophy of nature.
[69] Thus, for instance, our present knowledge of the history of Thomistic thinking still has, generally speaking, too many gaps in it to enable us to reconstruct something like a tradition which goes up at least to the beginning of the nineteenth century. If, recently, any attempt at filling some of these gaps has been made, such as that concerned with throwing light on the (relative) significance and the *survival* of Thomism in the Italian Renaissance (cf. for this P. O. Kristeller, 1974), then this certainly helps us to extend a little further the main line of Thomistic tradition *in general*. However, it brings us no nearer to finding an answer to the more specific questions of how – that is to say, along which historical channels – and to what extent Thomistic anthropology may have affected nineteenth-century German somaticist psychiatry. It is indeed surprising and highly significant of the current state of affairs in research into the *history* of Thomism, that among the vast amount of literature listed in bibliographical works in the field of Thomism so little emerges which is of help here. Cf. in this connection P. O. Kristeller's remarks (1974, p. 31) and the recent bibliographical information (up till 1978) contained in T. L. Miethe and V. J. Bourke, 1980.

BIBLIOGRAPHY

Ackerknecht, E. H.: 1953, 'Broussais, or a forgotten medical revolution'. *Bulletin of the History of Medicine* 27, 320–343.

Ackerknecht, E. H.: 1957, *Rudolf Virchow, Arzt-Politiker-Anthropologe*, Stuttgart.

Ackerknecht, E. H.: 1968, *A Short History of Psychiatry*, 2nd edn, New York and London.

Aliotta, A.: 1914, *The Idealistic Reaction against Science*, London.

Altschule, M. D.: 1965, 'Broussais and Griesinger: the introduction of egopsychology in psychiatry', in M. D. Altschule (ed.), *Roots of Modern Psychiatry. Essays in the History of Psychiatry*. Second revised and enlarged edition, New York.

Amacher, P.: 1965, 'Freud's neurological education and its influence on psychoanalytic theory', *Psychological Issues* IV, no. 4, Monograph 16.

Andersson, O.: 1962, *Studies in the prehistory of psychoanalysis. The etiology of psychoneuroses and some related themes in Sigmund Freud's scientific writings and letters 1886–1896*, Stockholm.

Aristotle: 1959, *Aristotle's De Anima in the Version of William of Moerbeke and the Commentary of St. Thomas Aquinas*, translated by K. Foster and S. Humphries, with an introduction by I. Thomas, 3rd edn, London.

Aristotle: 1960, *Metaphysics*, W. Jaeger (ed.), Oxford.

Aristotle: 1965, *De Generatione Animalium*, H. J. Drossaart Lulofs (ed.), Oxford.

Aristotle: 1965, *Historia Animalium*, 1 (Books I–III), with an English translation by A. L. Peck, London and Cambridge, Mass.

Aristotle: 1974, *De Anima*, W. D. Ross (ed.), Oxford.

Ballauf, Th.: 1954, *Die Wissenschaft vom Leben. Eine Geschichte der Biologie. Vol. I: Vom Altertum bis zur Romantik*. Freiburg and München.

Benedetti, G.: 1973, 'Schizophrenie', in Chr. Müller (ed.) 1973, 440–458.

Bernstein, R. J.: 1971, 'The challenge of scientific materialism', in D. M. Rosenthal (ed.) 1971, 200–222.

Bichat, X.: 1829, *Recherches physiologiques sur la vie et la mort*, Cinquième édition, revue et augmentée de notes pour la deuxième fois par F. Magendie, Paris.

Binswanger, L.: 1955, *Ausgewählte Vorträge und Aufsätze, Vol. II: Zur Problematik der psychiatrischen Forschung und zum Problem der Psychiatrie*, Bern.

Binswanger, L.: 1955a, 'Freud und die Verfassung der klinischen Psychiatrie' (1936), in L. Binswanger 1955, 81–104.

Binswanger, L.: 1955b, 'Welche Aufgaben ergeben sich für die Psychiatrie aus den Fortschritten der neueren Psychologie?' (1924), in L. Binswanger 1955, 111–146.

Binswanger, L.: 1959, *Zur Geschichte der Heilanstalt Bellevue zu Kreuzlingen 1857–1932*, (n.d.).

Binswanger, L.: 1961, *Ausgewählte Vorträge und Aufsätze, Vol. I: Zur phänomenologischen Anthropologie*, 2nd edn, Bern.

273

Binswanger, L.: 1961a, 'Lebensfunktion und innere Lebensgeschichte' (1928), in L.
   Binswanger 1961, 50—73.
Bodamer, J.: 1948, 'Zur Phänomenologie des geschichtlichen Geistes in der Psychiatrie',
   Der Nervenarzt 19, 299—310.
Bodamer, J.: 1953, 'Zur Entstehung der Psychiatrie als Wissenschaft im 19. Jahrh-
   undert', Fortschritte der Neurologie und Psychiatrie und ihrer Grenzgebiete, Heft 11,
   511—535.
Boethius, A. M. S.: 'Liber de persona et duabus naturis contra Eutychen et Nestorium',
   in Migne, Patrol. Lat. 64, 1337—1354.
Boring, E. G.: 1942, Sensation and Perception in the History of Experimental Psy-
   chology, New York.
Boring, E. G.: 1957, A History of Experimental Psychology, 2nd edn, New York.
Borst, C. V. (ed.): 1973, The Mind—Brain Identity, 2nd edn, London and Basingstoke.
Bosch, G.: 1970, Infantile Autism. A Clinical and Phenomenological-Anthropological
   Investigation Taking Language as the Guide. (German edn 1970, Der frühkindliche
   Autismus), Berlin, Göttingen and Heidelberg.
Breilmann, H.: 1925, Lotzes Stellung zum Materialismus, unter besonderer Berücksichti-
   gung seiner Controverse mit Czolbe, Diss. Munich, Telgte.
Brennan, R. E.: 1946, Thomistic Psychology. A Philosophic Analysis of the Nature of
   Man, 7th edn, New York.
Büchner, L.: 1855, Kraft und Stoff. Empirisch-naturphilosophische Studien in allgemein
   verständlicher Darstellung, Leipzig.
Büchner, L.: 1884, Aus Natur und Wissenschaft. Studien, Kritiken und Abhandlungen
   in allgemeinen verständlicher Darstellung, 3rd edn, Vol. II, Leipzig.
Büchner, L.: 1904, Kraft und Stoff, oder Grunzüge der natürlichen Weltordnung. Nebst
   einer darauf gebauten Moral oder Sittenlehre, 21st edn (1st edn 1855), Leipzig.
Buelens, J.: 1971, Sigmund Freud, kind van zijn tijd. Evolutie en achtergronden van
   zijn werk tot 1900, Meppel.
Burckhardt, J.: 1977, Die Kultur der Renaissance in Italien. Ein Versuch. W. Goetz
   (ed.), 10th edn, Stuttgart.
Cabanis, P. J. G.: 1844, Rapports du physique au moral de l'homme, 8th edn, (1st edn
   1802), Paris.
Cardano, G.: 1557, De rerum varietate libri XVII, Basel.
Casmann, O.: 1594, Psychologia anthropologica, sive animae humanae doctrina, Vol.
   I, Hannover.
Casmann, O.: 1596, Psychologia anthropologica, sive animae humanae doctrina, Vol. II,
   Fabrica humani corporis, Hannover.
Cassirer, E., P. O. Kristeller and J. H. Randall Jr. (eds): 1971, The Renaissance Philoso-
   phy of Man. Selections in Translation, 12th edn, Chicago and London.
Cassirer, E.: 1973, Das Erkenntnisproblem in der Philosophie und Wissenschaft der
   neueren Zeit, Vol IV: Von Hegels Tod bis zur Gegenwart (1832—1932), 3rd edn
   (reprint of the 2nd edn, 1957) Darmstadt.
Cassirer, E.: 1974a, Das Erkenntnisproblem in der Philosophie und Wissenschaft der
   neueren Zeit, Vol. I, 4th edn (reprint of the 3rd edn, 1922) Darmstadt.
Cassirer, E.: 1974b, Das Erkenntnisproblem in der Philosophie und Wissenschaft der
   neueren Zeit, Vol. II, 4th edn (reprint of the 3rd edn, 1922) Darmstadt.

Cassirer, E.: 1974c, *Das Erkenntnisproblem in der Philosophie und Wissenschaft der neueren Zeit*, Vol. III: *Die Nachkantische Systeme*, 3rd edn (reprint of the 2nd edn, 1923) Darmstadt.

Cassirer, E.: 1977, *Individuum und Kosmos in der Philosophie der Renaissance*, 5th edn (reprint of the 4th edn, 1927) Darmstadt.

Cohen, H.: 1914, 'Biographisches Vorwort' in F. A. Lange 1914, pp. iii–xi.

Cohen, R. S. and J. Elkana (eds.): 1977, *Hermann von Helmholtz. Epistemological Writings*. The Paul Hertz/Moritz Schlick centenary edition of 1921, with notes and commentary by the editors, newly translated by Malcolm F. Lowe, with an introduction and bibliography, Dordrecht and Boston.

Combe, A.: 1831, *Observations on Mental Derangement*, Edinburgh.

Copleston, F.: 1963, *A History of Philosophy*, Vol. 3: *Late Mediaeval and Renaissance Philosophy*, Part II, *The Revival of Platonism to Suarez*, New York.

Copleston, F.: 1965, *A History of Philosophy*, Vol. 7: *Modern Philosophy*, Part I, *Fichte to Hegel*, New York.

Czolbe, H.: 1855, *Neue Darstellung des Sensualismus*, Leipzig.

Czolbe, H.: 1856, *Entstehung des Selbstbewusstseins, Eine Antwort an Herrn Professor Lotze*, Leipzig.

Damerow, H.: 1844, 'Einleitung', *Allgemeine Zeitschrift für Psychiatrie* 1, I–XLVIII.

Damerow, H.: 1844a, 'Heinroth' (including a letter from J. Chr. A Heinroth to Damerow of March 16th, 1842), *Allgemeine Zeitschrift für Psychiatrie* 1, 156–159.

Damerow, H.: 1860, 'Ueber die Grundlage der Mimik und der Physiognomik als freier Beitrag zur Anthropologie und Psychiatric', *Allgemeine Zeitschrift für Psychiatrie* 17, 385–452.

Danto, A. C.: 1967, 'Naturalism', in P. Edwards 1967, V, 448–450.

Descartes, R.: 1953, 'Traité de l'homme (1649), in A. Bridoux (ed.), *Oeuvres et lettres de Descartes*, 807–873, Paris.

Dessoir, M.: 1964, *Geschichte der neueren deutschen Psychologie*, 2nd edn (reprint of the 1st edn, 1902, Berlin) Vo. II, Amsterdam.

Diepgen, P.: 1959, *Geschichte der Medizin. Die historische Entwicklung der Heilkunde und des ärztlichen Lebens*. Vol. II, Part I: *Von der Medizin der Aufklärung bis zur Begründung der Zellularpathologie (etwa 1740–etwa 1858)*, 2nd ed, Berlin.

Dietze, H. J. and G. E. Voegele: 1965, 'Wilhelm Griesinger's Contributions to Dynamic Psychiatry', *Diseases of the Nervous System* 26, 579–582.

Dilthey, W.: 1968, *Gesammelte Schriften*, Vol. V, Part I: *Die Geistige Welt. Einleitung in die Philosophie des Lebens. Abhandlungen zur Grundlegung der Geisteswissenschaften*, 5th edn, Stuttgart and Göttingen.

Dilthey, W.: 1968a, 'Ideen über eine beschreibende und zergliedernde Psychologie' (1894), in W. Dilthey 1968, 139–240.

Dilthey, W.: 1977, *Gesammelte Schriften*, Vol II: *Weltanschauung und Analyse des Menschen seit Renaissance und Reformation*, 10th edn, Stuttgart.

Dilthey, W.: 1977a, 'Der Entwicklungsgeschichtliche Pantheismus nach seinem geschichtlichen Zusammenhang mit den älteren pantheistischen Systemen' (1900), in W. Dilthey 1977, 312–390.

Dörner, K.: 1969, *Bürger und Irre. Zur Sozialgeschichte und Wissenschaftssoziologie der Psychiatrie*, Frankfurt.

276 BIBLIOGRAPHY

Dorer, M.: 1932, *Historische Grundlagen der Psychoanalyse*, Leipzig.

Drüe, H.: 1976, *Psychologie aus dem Begriff. Hegels Persönlichkeitstheorie*, Berlin and New York.

Du Bois-Reymond, E.: 1848/1860, *Untersuchungen über thierische Elektricität*, Vol. I: 1848; Vol. II/1: 1849, Vol. II/2: 1860.

Du Bois-Reymond, E.: 1875/1877, *Gesammelte Abhandlungen zur allgemeinen Muskel- und Nervenphysik*, Vol. I: 1875; Vol. II: 1877, Leipzig.

Du Bois-Reymond, E.: 1886, *Reden. Erste Folge: Litteratur–Philosophie–Zeitgeschichte*, Leipzig.

Du Bois-Reymond, E.: 1886a, 'Ueber die Grenzen des Naturerkennens' (1872), in E. Du Bois-Reymond 1886, 105–140.

Du Bois-Reymond, E.: 1886b, 'Die sieben Welträthsel' (1880), in E. Du Bois-Reymond 1886, 381–417.

Du Bois-Reymond, E.: 1887, *Reden. Zweite Folge: Biographie–Wissenschaft–Ansprachen*, Leipzig.

Du Bois-Reymond, E.: 1887a, 'Ueber die Lebenskraft' (1848), in E. Du Bois-Reymond 1887, 1–28.

Du Bois-Reymond, E.: 1887b, 'Gedächtnissrede auf Johannes Müller' (1858), in E. Du Bois-Reymond 1887, 143–334.

Dijksterhuis, E. J.: 1964, *The Mechanization of the World Picture*, translated by C. Dikshoorn, 2nd edn, Oxford.

Edwards, P. (ed.): 1967, *Encyclopedia of Philosophy*, 7 vols, New York and London.

Ellenberger, H.: 1970, *The Discovery of the Unconscious. The History and Evolution of Dynamic Psychiatry*, 2nd edn, New York.

Erdmann, J. E.: 1837, *Leib und Seele, nach ihrem Begriff und ihrem Verhältniss zu einander. Ein Beitrag zur Begründung der philosophischen Anthropologie*. Halle.

Erdmann, J. E.: 1866, *Grundriss der Geschichte der Philosophie. Zweiter und Letzter Band. Philosophie der Neuzeit*, Berlin.

Erdmann, J. E.: 1873, *Grundriss der Psychologie*, 5th edn, Leipzig.

Ey, H.: 1962, 'Kurt Schneider ou le primat de la clinique', in H. Kranz 1962, 1–5.

Ey, H., P. Bernard and Ch. Brisset: 1974, *Manuel de psychiatrie*, Paris.

Feigl. H.: 1973, 'Mind–body, *not* a pseudo-problem', in C. V. Borst 1973, 33–41.

Feuchtersleben, E. von: 1846, 'Die Pathologie und Therapie der psychischen Krankheiten, für Ärzte und Studirende, dargestellt von Dr. Wilh. Griesinger' (review of Griesinger 1845), *Zeitschr. der k. k. Gesellsch. der Ärtzte zu Wien* 3, 144–160.

Flemming, K. F.: 1846, 'Die Pathologie und Therapie der psychischen Krankheiten, für Aerzte und Studirende, dargestellt von Dr. Wilh. Geiesinger' (review of W. Griesinger 1845), *Allgemeine Zeitschrift für Psychiatrie* 3, 296–311.

Frankl, V. E.: 1972, *Der Wille zum Sinn. Ausgewählte Vorträge über Logotherapie. Mit einem Beitrag von Elisabeth S. Lukas*, Bern, Stuttgart, and Wien.

Frauenstädt, J.: 1856, *Der Materialismus. Eine Erwiderung auf Dr. L. Büchner's 'Kraft und Stoff'*, Leipzig.

Freigius, J. Th.: 1579, *Quaestiones physicae, in quibus methodus doctrinam physicam legitime docendi describendi rudi Minerva descripta est libris XXXVI*, Basel.

Freud, S.: 1972a, *Gesammelte Werke*, Vol. XIV, 5th edn, Frankfurt a.M.

Freud, S.: 1972b, *Gesammelte Werke*, Vol. XVI, 4th edn, Frankfurt a.M.

Freud, S.: 1973a, *Gesammelte Werke*, Vol. II/III, 5th edn, Frankfurt a.M.

Freud, S.: 1973b, *Gesammelte Werke*, Vol. VI, 6th edn, Frankfurt a.M.

Friedreich, J. B.: 1964, *Historisch-kritische Darstellung der Theorien über das Wesen und den Sitz der psychischen Krankheiten* (reprint of the Leipzig 1836 edn), Amsterdam.

Fries, J. F.: 1820, *Handbuch der psychischen Anthropologie oder der Lehre von der Natur des menschlichen Geistes*, Vol. I, Jena.

Galaty, D. H.: 1974, 'The philosophical basis of mid-nineteenth century German reductionism', *Journal of the History of Medicine* 29, 295, 316.

Gebsattel, V. E. von: 1968, *Imago hominis. Beiträge zu einer personalen Anthropologie*, 2nd edn, Salzburg.

Goclenius, R.: 1590, ψυγολογια, *hoc est: De hominis perfectione, animo et in primis ortu huius commentationes ac disputationes quorundam theologorum et philosophorum nostrae aetatis*, Marburg.

Goethe, J. W.: 1949, 'Schriften zur Farbenlehre', *Johan Wolfgang Goethe, Gedenkausgabe der Werke, Briefe und Gespräche*, Vol. 16, 9–837, A. Speiser (ed.), Zürich.

Goethe, J. W.: 1975, Werke, Vol. XIII: *Naturwissenschaftliche Schriften*. D. Kuhn and R. Wankmüller (eds), 7th edn, München.

Goethe, J. W.: 1975a, 'Bedeutende Fördernis, durch ein einziges geistreiches Wort' (1823), in J. W. Goethe 1975, 37–41.

Gregory, F.: 1977, *Scientific Materialism in Nineteenth Century Germany*, Dordrecht and Boston.

Griesinger, W.: 1845, *Die Pathologie und Therapie der psychischen Krankheiten für Aerzte und Studirende*, Stuttgart.

Griesinger, W.: 1871, *Die Pathologie und Therapie der psychischen Krankheiten für Aerzte und Studirende*, 3rd edn, Braunschweig.

Griesinger, W.: 1872, *Wilhelm Griesinger's gesammelte Abhandlungen. Vol. I: Psychiatrische und nervenpathologische Abhandlungen. Vol. II: Verschiedene Abhandlungen*, Berlin.

Griesinger, W.: 1872a, 'Ueber psychische Reflexactionen; Mit einem Blick auf das Wesen der psychischen Krankheiten' (1843), in W. Griesinger 1872, I, 3–45 (originally publ. in *Archiv für physiologischen Heilkunde* 2, 1843, 76 ff.).

Griesinger, W.: 1872b, 'Neue Beiträge zur Physiologie und Pathologie des Gehirns' (1844), in W. Griesinger 1872, I, 46–79 (originally publ. in *Archiv für physiologische Heilkunde* 3, 1844, 69 ff.).

Griesinger, W.: 1872c, 'Recension über: M. Jacobi, die Hauptformen der Seelenstörungen, in ihren Beziehungen zur Heilkunde, nach der Beobachtung geschildert' (1844), in W. Griesinger 1872, I, 80–106 (originally publ. in *Archiv für physiologische Heilkunde* 3, 1844, 278 ff.).

Griesinger, W.: 1872d, 'Theorien und Thatsachen' (1842), in W. Griesinger 1872, II, 1–8 (originally publ. in *Archiv für physiologische Heilkunde* 1, 1842, 652 ff.).

Griesinger, W.: 1872e, 'Herr Ringseis und die naturhistorische Schule' (1842), in W. Griesinger 1872, II, 14–67 (originally publ. in *Archiv für physiologische Heilkunde* 1, 1842, 43 ff.).

Griesinger, W.: 1872f, 'Bemerkungen zur neuesten Entwicklung der allgemeinen Pathologie' (1843), in W. Griesinger 1872, II, 96–112 (originally publ. in *Archiv für physiologische Heilkunde* 2, 1843, 270 ff.).

Griesinger, W.: 1872g, 'Vorwort bei der Uebernahme der Redaction des Archivs für physiologische Heilkunde' (1847), in W. Griesinger 1872, II, 113–121 (originally publ. in *Archiv für physiologische Heilkunde* 6, 1847, 1 ff.).

Groethuysen, B.: 1928, *Philosophische Anthropologie*, München and Berlin.

Guislain, L.: 1838, *Abhandlung über die Phrenopathien oder neues System der Seelen-störungen auf praktische und statistische Boebachtungen und Untersuchung der Ursachen, der Natur, der Symptome, der Prognose, der Diagnose und der Behandlung dieser Krankheiten*. Transl. from the French by Dr. Wunderlich. Preface and additions by Dr. Zeller, Stuttgart and Leipzig.

Gusdorf, G.: 1960, *Introduction aux sciences humaines. Essai critique sur leurs origines et leur développement*, Paris.

Gusdorf, G.: 1966–1978, *Les sciences humaines et la pensée occidentale*, Paris.

Gusdorf, G.: 1969, *Les sciences humaines et la pensée occidentale III: La révolution galilé-enne*, Vol. II, Paris.

Gusdorf, G.: 1978, *Les Sciences Humaines et la Pensée Occidentale VIII: La Conscience Révolutionnaire. Les Idéologues*, Paris.

Haller, A. von: 1757–1766, *Elementa physiologiae corporis humani*, 8 Vols, Lausanne.

Hartmann, F. and K. Haedke: 1963, 'Der Bedeutungswandel des Begriffs Anthropologie im ärztlichen Schrifttum der Neuzeit', *Marburger Sitzungsberichte* 85, Heft 1–2, 39–99.

Hegel, G. W. F.: 1970, *Enzyklopädie der philosophischen Wissenschaften im Grundriss* (1830), E. Moldenhauer and K. M. Michel (eds), Frankfurt a.M..

Heinroth, J. Chr. A.: 1818, *Lehrbuch der Störungen des Seelenlebens oder der Seelenstö-rungen und ihrer Behandlung. Vom rationalen Standpunkt aus entworfen, Zwey Theile. Erster oder theoretischer Theil*, Leipzig.

Heinroth, J. Chr. A.: 1822, *Lehrbuch der Anthropologie. Zum Behuf academischer Vorträge, und zum Privatstudium. Nebst einem Anhange erläuternder und beweisfüh-render Aufsätze*, Leipzig.

Heimsoeth, H.: 1958, *Die sechs grossen Themen der abendländischen Metaphysik und der Ausgang des Mittelalters*, 4th edn, Darmstadt.

Helmholtz, H. von: 1856–1866, *Handbuch der physiologischen Optik*, 3 Vols, Leipzig.

Helmholtz, H. von: 1863, *Die Lehre von den Tonempfindungen als physiologische Grundlage für die Theorie der Musik*, Leipzig.

Helmholtz, H. von: 1882, *Wissenschaftliche Abhandlungen*, Vol. I, Leipzig.

Helmholtz, H. von: 1882a, 'Ueber die Erhaltung der Kraft' (1847), in H. von Helmholtz 1882, 12–76.

Helmholtz, H. von: 1903, *Vorträge und Reden*, Vols I and II, 5th edn, Braunschweig.

Helmholtz, H. von: 1903a, 'Ueber das Sehen des Menschen' (1855), in H. von Helmholtz 1903, Vol. I, 87–117.

Helmholtz, H. von: 1903b, 'Die neuere Fortschritte in der Theorie des Sehens' (1858), in H. von Helmholtz 1903, Vol. I, 265–367.

Helmholtz, H. von: 1903c, 'Ueber das Ziel und die Fortschritte der Naturwissenschaften' (1869), in H. von Helmholtz 1903, I, 367–401.

Helmholtz, H. von: 1903d, 'Das Denken in der Medicin' (1877), in H. von Helmholtz 1903, Vol. II, 165–191.

Helmholtz, H. von: 1921, *Schriften zur Erkenntnistheorie*, P. Hertz and M. Schlick (eds), Berlin.

Helmholtz, H. von: 1921a, 'Die Tatsachen in der Wahrnehmung' (1878), in H. von Helmholtz 1921, 109–152.

Herbart, J. F.: 1890a, *Sämtliche Werke*, Vol. V, K. Kehrbach (ed.), Langensalza.

Herbart, J. F.: 1890b, *Psychologie als Wissenschaft neu gegründet auf Erfahrung, Meta-*

physik und Mathematik. Erster synthetischer Theil (1824). See J. F. Herbart 1890a, 177–402.

Herbart, J. F.: 1891a, *Sämtliche Werke*, Vol. IV, K. Kehrbach (ed.), Langensalze.

Herbart, J. F.: 1891b, *Lehrbuch zur Einleitung in die Philosophie* (1813, 4th edn 1837). See J. F. Herbart 1891a, 1–294.

Herbart, J. F. 1891c, *Lehrbuch zur Psychologie* (1816, 2nd edn 1834). See F. J. Herbart 1891a, 295–436.

Herbart, J. F.: 1892a, *Sämtliche Werke*, Vol. VI, K. Kehrbach (ed.), Langensalza.

Herbart, J. F.: 1892b, *Psychologie als Wissenschaft neu gegründer auf Erfahrung, Metaphysik und Mathematik. Zweiter analytischer Theil* (1825). See F. J. Herbart 1892a, 1–338.

Herbart, J. F.: 1900, *Lehrbuch zur Psychologie*, G. Hartenstein (ed.), 6th edn (3rd edn 1850, Leipzig), Hamburg and Leipzig.

Hertz, P. and M. Schlick (eds): 1921, *Hermann von Helmholtz Schriften zur Erkenntnistheorie*, Berlin.

Hirsch, A. (ed.): 1962, *Biographisches Lexicon der hervorragenden Ärzte aller Zeiten und Völker*, Vol. III, München and Berlin.

Hirschberger, J.: 1969, *Geschichte der Philosophie*, Vol. II: *Neuzeit und Gegenwart*. (8th edn, Basel, Freiburg, and Wien).

Höffding, H.: *History of Philosophy*, 2 vols, (n.d.).

Huizinga, J.: 1979, *The Waning of the Middle Ages. A Study of the Forms of Life, Thought, and Art in France and the Netherlands in the Fourteenth and Fifteenth Centuries*, translated by F. Hopman, Harmondsworth (originally published in Dutch, 1919).

Hundt, M.: 1501, *Anthropologium de hominis dignitate, natura et proprietatibus, de elementis, partibus et membris humani corporis etc. de Spiritu humano etc. de anima humana et ipsius appendiciis*, Leipzig.

Husserl, E.: 1971, *Philosophie als strenge Wissenschaft*, W. Szilasi (ed.), 2nd edn, Frankfurt a.M. (originally published in *Logos* 1, 1910–1911).

Jacobi, M.: 1822–1825, *Sammlungen für die Heilkunde der Gemüthskrankheiten*, Vol. I: 1822; Vol. II: 1825, Elberfeld.

Jacobi, M.: 1830, *Beobachtungen über die Pathologie und Therapie der mit Irreseyn verbundenen Krankheiten*, Elberfeld.

Jacobi, M.: 1844, 'Bemerkungen über die Bedeutung des Ausdrucks "Seelenstörung" in der Psychiatrie, und über die Mitwirkung der Geistlichen bei Behandlung von Irren, durch *Nasse's* Schrift: "Über die Behandlung von Gemüthskranken und Irren durch Nichtärzte" veranlasst', *Allgemeine Zeitschrift für Psychiatrie* 1, Heft 3, 353–422.

Jacobi, M.: 1844a, *Die Hauptformen der Seelenstörungen in ihren Beziehungen zur Heilkunde nach der Beobachtung geschildert*, Vol. I: *Die Tobsucht*, Leipzig.

Jacobi, M.: 1851, *Naturleben und Geistesleben. Der Sinnenorganismus in seinen Beziehungen zur Weltstellung des Menschen: La divina commedia*, Leipzig.

Janet, P.: 1880, 'Schopenhauer et la physiologie française. Cabanis et Bichat', *Revue des deux mondes* 39, 35–39.

Janke, W.: 1971, 'Apperzeption', in J. Ritter 1971, 448–455.

Jaspers, K.: 1959, *Allgemeine Psychopathologie*, 7th edn, Berlin, Göttingen, and Heidelberg.

Jores, A.: 1950, *Vom Sinn der Krankheit*, Hamburg.

Jülicher, L.: 1961, *Die Psychologie Johann Friedrich Herbarts und ihre Bedeutung für die Psychiatrie des 19. Jahrhunderts*, D. Thesis, Bonn.

Kant, I.: 1922a, *Kant's gesammelte Schriften. Bd. X. Zweite Abteilung: Briefwechsel. Erster Band. 1747–1788*, 2nd edn, Königl. Preuss. Ak. d. Wiss., Berlin and Leipzig.

Kant, I.: 1922b, 'Kritik der reinen vernunft', 2nd edn (1787), in *Immanuel Kants Werke*, E. Cassirer (ed.), Vol. III, Berlin.

Kant, I.: 1922c, 'Kritik der praktischen Vernunft', (1788), in *Immanuel Kants Werke*, E. Cassirer (ed.), Vol. V, 1–176, Berlin.

Kant, I.: 1923a, 'Anthropologie in pragmatischer Hinsicht' (1798, 2nd edn 1800), in *Immanuel Kants Werke*. E. Cassirer (ed.), Vol. VIII, 3–228, Berlin.

Kant, I.: 1923b, 'Logik. Ein Handbuch zu Vorlesungen' (1800), in *Immanuel Kants Werke*, E. Cassirer (ed.), Vol. VIII, 325–452.

Kant, I.: 1923c, 'Ueber die Fortschritte der Metaphysik seit Leibniz und Wolff' (1804), in *Immanuel Kants Werke*, E. Cassirer (ed.), Vol. VIII, 233–321, Berlin.

Kirchhoff, Th. (ed.): 1921, *Deutsche Irrenaerzte. Einzelbilder ihres Lebens und Wirkens*, Vol. I, Berlin.

Knittermeyer, H.: 1929, *Schelling und die romantische Schule*, München.

Kolakowski, L.: 1972, *Positivist Philosophy – from Hume to the Vienna Circle*, Harmondsworth.

Kolle, K. (ed.): 1970, *Grosse Nervenärzte*, Vol. I: *21 Lebensbilder* (1956), 2nd edn, Stuttgart.

Kranz, H. (ed.): 1962, *Psychopathologie heute. Kurt Schneider zum 75. Geburtstag gewidmet*, Stuttgart.

Kristeller, P. O. and J. H. Randall, Jr. 1941, 'The study of the philosophies of the Renaissance', *Journal of the History of Ideas* 2, 449–496.

Kristeller, P. O.: 1974, 'Thomism and the Italian Thought of the Renaissance', in P. Mahoney (ed.) 1974, 29–91.

Kronfeld, A.: 1927, *Die Psychologie in der Psychiatrie. Eine Einführung in die psychologischen Erkenntnisweisen innerhalb der Psychiatrie und ihre Stellung zur klinisch-pathologischen Forschung*, Berlin.

Kuhn, R.: 1957, 'Griesinger's Auffassung der psychischen Krankheiten und seine Bedeutung für die weitere Entwicklung der Psychiatrie', *Bibl. Psychiat. Neural.* 100, 41–67.

Kyper, A.: 1660, *Anthropologia, corporis humani, contentorum et animae naturam et virtutes secundum circularem sanguinis motum explicans*, Leiden.

Lange, F. A.: 1865, *Die Grundlegung der mathematischen Psychologie. Ein Versuch zur Nachweisung des Fundamentalen Fehlers bei Herbart und Drobisch*, Duisburg.

Lange, F. A.: 1870, 'Seelenlehre', in *Encyklopädie des gesamten Erziehungs- und Unterrichtswesens*, K. A. Schmid (ed.), Vol. VIII, Gotha.

Lange, F. A.: 1914, *Geschichte des Materialismus und Kritik seiner Bedeutung in der Gegenwart*, Vol. I (1866), 9th edn, third enlarged revised edition of the biographical preface and introduction with a critical comment by Hermann Cohen, Leipzig.

Lange, F. A.: 1974, *Geschichte des Materialismus und Kritik seiner Bedeutung in der Gegenwart*, 2 Vols., A. Schmid (ed.), Frankfurt a.M.

Laplace, P. S.: 1829, *Essai philosophique sur les probabilités*, 5th edn, Bruxelles.

Lazarus, M.: 1869, 'Rede auf W. Griesinger, *Archiv für Psychiatrie und Nervenkrankheiten*, 1869, 775–782.

Leibbrand, W.: 1953–1954, 'K. von Rokitansky und Schopenhauer', *Schopenhauer Jahrbuch* **35**, 75–77.

Leibbrand, W.: 1956, *Die spekulative Medizin der Romantik*, Hamburg.

Leibbrand, W. and A. Wettley: 1961, *Der Wahnsinn. Geschichte der abendländischen Psychopathologie*, Freiburg i. Br. and München.

Leibniz, G. W.: 1961, *Die philosophischen Schriften*, Vol. 6, C. I. Gerhardt (ed.), Hildesheim (reprint of the Berlin 1885 edn).

Leibniz, G. W.: 1961a, 'Principes de la Nature et de la Grace, fondés en raison' (1714), in G. W. Leibniz 1961, 598–606.

Lesky, E.: 1965, *Die Wiener medizinische Schule im 19. Jahrhundert*, Graz and Köln.

Liebert, A.: 1915, 'Johannes Müller, der Physiologe, in seinem Verhältnis zur Philosophie und in seiner Bedeutung für dieselbe', *Kant-Studien* **20**, 357–375.

Liebig, J. von: 1841, *Die organische Chemie in ihrer Anwendung auf Physiologie und Pathologie*, Braunschweig.

Liebmann, O.: 1912, *Kant und die Epigonen*, Berlin (reprint of the 1865 edn).

Lindeboom, G. A.: 1979, *Descartes and Medicine*, Amsterdam.

Linden, M.: 1976, *Untersuchungen zum Anthropologiebegriff des 18. Jahrhunderts*, Bonn and Frankfurt a.M.

Lindner, G. A.: 1858, *Lehrbuch der empirischen Psychologie nach genetischer Methode*, Cilli.

Lotze, R. H.: 1842, 'Leben, Lebenskraft', in R. Wagner (ed.) 1842–1853, I, XIX–LVIII.

Lotze, R. H.: 1846, 'Seele und Seelenleben', in R. Wagner (ed.) 1842–1853, III, 142–164.

Lotze, R. H.: 1848, *Allgemeine Pathologie und Therapie als mechanische Naturwissenschaften*, 2nd edn, Leipzig.

Lotze, R. H.: 1852, *Medicinische Psychologie oder Physiologie der Seele*, Leipzig.

Lotze, R. H.: 1891, *Kleine Schriften*, Vol. 3, Leipzig.

Lotze, R. H.: 1896, *Mikrokosmus. Ideen zur Naturgeschichte und Geschichte der Menschheit. Versuch einer Anthropologie*, Vol. I, 5th edn, Leipzig.

Lotze, R. H.: 1912, *System der Philosophie* (1874/79), Vol. II, G. Misch (ed.), Leipzig.

Mahoney, E. P. (ed.): 1974, *Medieval Aspects of Renaissance Learning*, Three Essays by P. O. Kristeller, translated by Edward P. Mahoney (ed.), Durham.

Mandelbaum, M.: 1974, *History, Man and Reason. A Study of Nineteenth Century Thought*, 2nd edn, Baltimore and London.

Marquard, O.: 1965, 'Zur Geschichte des philosophischen Begriffs "Anthropologie" seit dem Ende des 18. Jahrhunderts', *Collegium Philosophicum* 1965, 209–239.

Marquard, O.: 1971, 'Anthropologie', in J. Ritter 1971, 362–374.

Marquard, O.: 1973, *Schwierigkeiten mit der Geschichtsphilosophie. Aufsätze*, Frankfurt.

Marquard, O.: 1973a, 'Ueber einige Beziehungen zwischen Aesthetik und Therapeutik in der Philosophie des neunzehnten Jahrhunderts', in O. Marquard 1973, 85–105.

Marx, O. M.: 1972, 'Wilhelm Griesinger and the history of psychiatry: a reassessment', *Bulletin of the History of Medicine* **XLVI**, 6, 519–544.

Mayer, J. R.: 1842–1845, See W. Preyer 1889.

Mayer, J. R.: 1845, *Die organische Bewegung in ihrem Zusammenhange mit dem Stoffwechsel*, Heilbronn.

Melanchthon, Ph.: 1843–1860, 'Opera' (1562/64), in *Corpus Reformatorum*, Vol. I–XXVIII, ed. by C. G. Bretschneider and H. E. Bindseil, Halle and Braunschweig.

Mette, A.: 1976, *Wilhelm Griesinger, der Begründer der wissenschaftlichen Psychiatrie in Deutschland*, Leipzig.

Miethe, T. L. and V. J. Bourke: 1980, *Thomistic Bibliography, 1940–1978*, Westport and London.

Mill, J. S.: 1843, *A System of Logic, Ratiocinative and Inductive, being a Connected View of the Principles of Evidence, and the Methods of Scientific Investigation*, 2 vols, London.

Moleschott, J.: 1850, 'Lehre der Nahrungsmittel für das Volk', *Blätter für literarische Unterhaltung* 269, Nov. 9th, 1850.

Moleschott, J.: 1852, *Der Kreislauf des Lebens, physiologische Antworten auf Liebigs chemische Briefe*, Mainz.

Moleschott, J.: 1894, *Für meine Freunde. Lebenserinnerungen*, Giessen.

Müller, Chr. (ed.): 1973, *Lexikon der Psychiatrie*, Berlin, Heidelberg, and New York.

Müller, J.: 1822, *De Phoronomia Animalium*, Bonn.

Müller, J.: 1824, 'Von dem Bedürfniss der Physiologie nach einer philosophischen Naturbetrachtung. Eine öffentliche Vorlesung, gehalten auf der Universität zu Bonn am 19ten October 1824', in J. Müller 1826a, 3–36.

Müller, J.: 1826, *Ueber die phantastischen Gesichtserscheinungen. Eine physiologische Untersuchung mit einer physiologischen Urkunde des Aristoteles über den Traum, den Philosophen und Aerzten gewidmet*, Koblenz (reprint München 1967).

Müller, J.: 1826a, *Zur vergleichenden Physiologie des Gesichtssinnes des Menschen und der Thiere nebst einem Versuch über die Bewegungen der Augen und über den menschlichen Blick*, Leipzig.

Müller, J.: 1844, *Handbuch der Physiologie des Menschen*, Vol. I, 4th edn, Coblenz.

Nasse, F.: 1818, 'Vorbericht', *Zeitschrift für psychische Aerzte* 1818, Heft 1, 1–16.

Nasse, F.: 1818a, 'Ueber die Abhängigkeit oder Unabhängigkeit des Irreseyns von einem vorausgegangenen körperlichen Krankheitszustande', *Zeitschrift für psychische Aerzte* 1818, Heft 1, 128–140; 3, 409–456.

Nasse, F.: 1820, 'Vereintseyn von Seele und Leib oder Einsseyn?', *Zeitschrift für psychische Aerzte* 1820, Heft 1, 6–22.

Neuburger, M.: 1934, 'Rokitansky als Vorkämpfer der mechanistischen Forschungsmethode und der idealistischen Weltanschauung', *Wiener Klinische Wochenschrift* 1934, 12, 358–360.

Noreña, C. G.: 1975, *Studies in Spanish Renaissance Thought*, The Hague.

Olmsted, J. M. D.: 1944, *François Magendie, Pioneer in Experimental Physiology and Scientific Medicine in XIX Century France*, preface by John F. Fulton, New York.

Pagel, W.: 1967, *William Harvey's Biological Ideas. Selected Aspects and Historical Background*, Basel and New York.

Passmore, J.: 1966, *A Hundred Years of Philosophy*, 2nd edn, Harmondsworth.

Peters, R. S. (ed.): 1962, *Brett's History of Psychology*, London and New York.

Petersen, P.: 1921, *Geschichte der Aristotelischen Philosophie im Protestantischen Deutschland*, Leipzig.

Petry, M. J. (ed.): 1978, *Hegel's Philosophy of Subjective Spirit, Vol. 1: Introductions*. Edited and translated with an introduction and explanatory notes, Dordrecht and Boston.

Picavet, Fr.: 1971, *Les idéologues. Essai sur l'histoire des idées et des théories scientifi-*

ques, philosophiques, réligieuses, etc. en France depuis 1789, New York (reprint of the 1891 edition).

Pico della Mirandola, G.: 1942, *Oratio de hominis dignitate* (1487), E. Garin (ed.), Firenze.

Platner, E.: 1772, *Anthropologie für Aerzte und Weltweise*, Leipzig.

Platner, E.: 1790, *Neue Anthropologie für Aerzte und Weltweise mit besonderer Rücksicht auf Physiologie, Pathologie, Moralphilosophie und Aesthetik*, Leipzig.

Plato: 1958, *Platonis opera*, J. Burnet (ed.), 5 vols, Oxford.

Plato: 1958a, *Platonis res publica*, J. Burnet (ed.), Oxford.

Plessner, H.: 1975, *Die Stufen des Organischen und der Mensch. Einleitung in die philosophische Anthropologie*, 3rd edn, Berlin and New York.

Pongratz, L. J.: 1967, *Problemgeschichte der Psychologie*, Bern and München.

Preyer, W. (ed.): 1889, *Robert von Mayer über die Erhaltung der Energie. Briefe an Wilhelm Griesinger nebst dessen Antwortschreiben aus den Jahren 1842–1845*, Berlin.

Rádl, E.: 1970a, *Geschichte der biologischen Theorien in der Neuzeit*, Vol. I, 2nd edn, Hildesheim and New York (reprint of the 1913 edn, Leipzig and Berlin).

Rádl, E.: 1970b, *Geschichte der biologischen Theorien in der Neuzeit*, Vol. II: *Geschichte der Entwicklungstheorien in der Biologie des XIX Jahrhunderts*, Hildesheim and New York (reprint of the Leipzig 1909 edition).

Randall, J. H.: 1961, *The School of Padua and the Emergence of Modern Science*, Padova.

Randall, J. H.: 1962, *The Career of Philosophy*, Vol. I: *From the Middle Ages to the Enlightenment*, New York.

Rawidowicz, S.: 1964, *Ludwig Feuerbachs Philosophie* (1931), 2nd edn, Berlin.

Reil, J.: 1799, *Über die Erkenntnis und Kur der Fieber*, Vol. IV, Halle.

Reil, J.: 1803, *Rhapsodien über die Anwendung der psychischen Curmethode auf Geisteszerrüttungen*, Halle.

Riese, W.: 1959, *A History of Neurology*, New York.

Risse, G. B.: 1972, 'Kant, Schelling, and the early search for a philosophical "science" of medicine in Germany', *Journal of the History of Medicine and allied Sciences* 27, 145–158.

Ritter, J. (ed.): 1971, *Historisches Wörterbuch der Philosophie*, Vol. I, Darmstadt.

Ritter, J. and K. Gründer (eds.): 1976, *Historisches Wörterbuch der Philosophie*, Vol. IV, Darmstadt.

Rokitansky, K. von: 1842, *Handbuch der pathologischen Anatomie*, Vol. III (Vol. II: 1844; I: 1846), Wien.

Rokitansky, K. von: 1859, 'Zur Orientirung über Medicin und deren Praxis', *Almanach der kaiserlichen Akademie der Wissenschaften* 9, 2nd section, 119–152, Wien.

Rokitansky, K. von: 1862, 'Freiheit der Naturforschung'. (Die feierliche Eröffnung des Pathologisch-anatomischen und chemischen Institutes am k.k Allgemeinen Krankenhause am 24. Mai 1862, 11 ff.), Wien.

Rokitansky, K. von: 1867, 'Der selbständige Werth des Wissens', *Almanach der kaiserlichen Akademie der Wissenschaften* 17, 103–140 (Wien).

Rosenthal, D. M. (ed.): 1971, *Materialism and the Mind–Body Problem*, Englewood Cliffs N. J..

Ross, D.: 1961, *Aristotle. De Anima*, Edited with an introduction and commentary, Oxford.

Rössler, D.: 1971, 'Anthropologie, medizinische', in J. Ritter 1971, 374–376.

Rothschuh, K. E.: 1953, *Geschichte der Physiologie*, Berlin, Göttingen, and Heidelberg.

Rothschuh, K. E.: 1968, *Physiologie. Der Wandel ihrer Konzepte, Probleme und Methoden vom 16. bis 19. Jahrhundert*, Freiburg and München.

Rothschuh, K. E. (ed.): 1975, *Was ist Krankheit? Erscheinung, Erklärung, Sinngebung*, Darmstadt.

Sanders, C.: 1982, 'G. Verwey, Psychiatrie tussen anthropologie en natuurwetenschap', *ANTW* 74, 2, 128–130.

Schelling, F. W. J.: 1958, *Werke. Vierter Hauptband: Schriften zur Philosophie der Freiheit (1804–1815)*, M. Schröter (ed.), München (reprint of the München 1927 edn).

Schelling, F. W. J.: 1958a, *Vorrede zu den Jahrbüchern der Medicin als Wissenschaft* (1806), in F. W. J. Schelling 1958, 65–73.

Schelling, F. W. J.: 1958b, 'Vorläufige Bezeichnung des Standpunktes der Medicin nach Grundsätzen der Naturphilosophie' (1806), in F. W. J. Schelling 1958, 194–222.

Schelling, F. W. J.: 1975, *Ausgewählte Werke. Schriften von 1794–1798*, Darmstadt (reprint of the Stuttgart and Augsburg 1856–1857 edn).

Schelling, F. W. J.: 1975a, *Ausgewählte Werke. Schriften von 1799–1801*, Darmstadt (reprint of the Stuttgart and Augsburg 1858–1859 edn).

Schiller, F.: 1962, *Schillers Werke. Nationalausgabe*, Vol. XX, part I, Weimar.

Schiller, F.: 1962a, 'Versuch über den Zusammenhang der thierischen Natur des Menschen mit seiner geistigen' (1708), in F. Schiller 1962, 37–39.

Schipperges, H.: 1959, 'Leitlinien und Grenzen der Psychosomatik bei Friedrich Nasse', *Confinia Psychiatrica* 2, 19–37.

Schipperges, H.: 1964, 'Die Assimilation der arabischen Medizin durch das lateinische Mittelalter', *Südhoffs Archiv für Geschichte der Medizin und der Naturwissenschaften*, Beiheft 3.

Schleiden, M. J.: 1849, *Die Botanik als inductive Wissenschaft*. Vol. I: *Methodologische Grundlage. Vegetabilische Stofflehre. Die Lehre von der Pflanzenzelle*, 3rd edn, Leipzig (1st edn Leipzig 1842, entitled: *Grundzüge der wissenschaftlichen Botanik nebst einer methodologischen Einleitung als Anleitung zum Studium der Pflanze*).

Schleiden, M. J.: 1863, *Ueber den Materialismus der neueren deutschen Naturwissenschaft, sein Wesen und Geschichte*, Leipzig.

Schmidt, A. (ed.): 1967, *Ludwig Feuerbach, Anthropologischer Materialismus. Ausgewählte Schriften*, Frankfurt a.M. and Wien.

Schmidt, A. (ed.): 1977, *Drei Studien über Materialismus. Schopenhauer–Horkheimer–Glücksproblem*, München and Wien.

Schmidt, A. (ed.): 1977a, 'Schopenhauer und der Materialismus', in A. Schmidt 1977, 21–79.

Schomerus, H. G.: 1966, 'Gesundheit und Krankheit der Person in der medizinischen Anthropologie Johann Christian August Heinroths', *Jahrbuch für Psychologie, Psychotherapie und medizinische Anthropologie* 14, 309–328.

Schönpflug, U.: 1976, 'Ich', in J. Ritter and K. Gründer 1976, 6 ff..

Schopenhauer, A.: 1961, *Sämtliche Werke*, Wolfgang Freiherr von Löhneysen (ed.), Vol. I, Darmstadt.

Schopenhauer, A.: 1961a, 'Die Welt als Wille und Vorstellung', Vol. I, (1844[2] (1819))
    in A. Schopenhauer 1961, 7–558 (715) (1st edn 1819, 2nd edn 1844).
Schopenhauer, A.: 1961b, *Sämtliche Werke*, Wolfgang Freiherr von Löhneysen (ed.),
    Vol. II, Darmstadt.
Schopenhauer, A.: 1961c, 'Die Welt als Wille und Vorstellung', Vol. II, (1858[3] (1844))
    in A. Schopenhauer 1961b, 11–829 (1st edn 1844, 3rd edn 1858).
Schopenhauer, A.: 1962, *Sämtliche Werke*, Wolfgang Freiherr von Löhneysen (ed.),
    Vol. III, Darmstadt.
Schopenhauer, A.: 1962a, 'Über die vierfache Wurzel des Satzes vom zureichenden
    Grunde. Eine philosophische Abhandlung', in A. Schopenhauer 1962, 7–89 (1st
    edn. 1813, 2nd edn 1847).
Schopenhauer, A.: 1962b, 'Über das Sehn und die Farben. Eine Abhandlung', in A.
    Schopenhauer 1962, 193–297 (1st edn 1816, 2nd edn 1854).
Schopenhauer, A.: 1962c, 'Über den Willen in der Natur. Eine Erörterung der Bestäti-
    gungen, welche die Philosophie des Verfassers seit ihrem Auftreten durch die em-
    pirischen Wissenschaften erhalten hat', in A. Schopenhauer 1962, 301–479 (1st
    edn 1836).
Schopenhauer, A.: 1965, *Sämtliche Werke*, Wolfgang Freiherr von Löhneysen (ed.),
    Vol. V, Darmstadt.
Schopenhauer, A.: 1965a, 'Parerga und Paralipomena. Kleine philosophische Schriften,
    II', in A. Schopenhauer 1965, 9–773.
Schüling, H.: 1967, *Bibliographie der psychologischen Literatur des 16. Jahrhunderts*,
    Hildesheim.
Schwann, T.: 1839, *Mikroskopische Untersuchungen über die Uebereinstimmung in der
    Struktur und dem Wachstum der Thiere und Pflanzen*, Berlin.
Seidel, H.: 1971, *Der Begriff des Intellekts (νοῦσ) bei Aristoteles im philosophischen
    Zusammenhang seiner Hauptschriften*, Meisenheim am Glan.
Siebenthal, W. von: 1950, *Krankheit als Folge der Sünde*, Hannover.
Singer, Ch. and E. Underwood: 1962, *A Short History of Medicine*, 2nd edn, Oxford.
Snelders, H. A. M.: 1973, *De invloed van Kant, de Romantiek en de "Naturphilosophie"
    op de anorganische natuurwetenschappen in Duitsland*, Thesis, Univ. of Utrecht.
Snell, L: 1852, 'Ueber die veränderte Sprechweise und die Bildung neuer Worte und
    Ausdrücke im Wahnsinn', *Allgemeine Zeitschr. für Psychiatrie* 9, 11–24.
Snell, L: 1860, 'Die Personenverwechselung, als Symptom der Geistesstörung', *All-
    gemeine Zeitschrift für Psychiatrie* 17, 545–554.
Snell, L.: 1865, 'Ueber Monomanie als primäre Form der Seelenstörung', *Allgemeine
    Zeitschrift für Psychiatrie* 22, 368–381.
Snell, L.: 1872, 'Über die verschiedenen Formen der Melancholie', *Allgemeine Zeitschrift
    für Psychiatrie* 28, 222–229.
Snell, L.: 1872a, 'Zur Erinnerung an Maximilian Jacobi', *Allgemeine Zeitschrift für
    Psychiatrie* 28, 415–424.
Sombart, W.: 1938, 'Beiträge zur Geschichte der wissenschaftlichen Anthropologie',
    Sitzungsberichte der Preussischen Akademie der Wissenschaften (phil.-hist. Kl.) XIII,
    96–130, Berlin.
Spiegelberg, H.: 1972, *Phenomenology in psychology and psychiatry. A historical
    introduction*, Evanston.
Sprengel, K.: 1818, 'Ueber Plato's Lehre von den Geisteszerrüttungen', *Zeitschrift für*

psychische Aerzte (ed. F. Nasse) 2, Heft 2, 159–173.

Steffens, K.: 1922, *Anthropologie* (1822), introduced by H. Poppelbaum (ed.), Stuttgart.

Sticker, G.: 1939, 'Wunderlich, Roser, Griesinger, die drei Schwäbischen Reformatoren der Medizin', *Südhoffs Archiv für Geschichte der Medizin und der Naturwissenschaften* 32, Heft 4/5, 207–274.

Temkin, O.: 1946, 'Materialism in French and German physiology of the early 19th century', *Bulletin of the History of Medicine* 20, 322–327.

Thiele, R.: 1970, 'Wilhelm Griesinger 1817–1868', in K. Kolle 1970, 115–127.

Thomas Aquinas St.: 1953, *Summa Theologiae* (1266–72), tomus primus complectens primam partem, editio altera emendata, ed. Commissio piana, Ottawa.

Thomas Aquinas St.: 1965, *Summa Theologiae*, Vol. 21, J. P. Reid (ed.), London and New York.

Thijssen-Schoute, C. L.: 1954, 'Nederlands Cartesianisme', *Verhandelingen der Kon. Akad. van Wetenschappen, afd. Letterkunde*, Nieuwe Reeks, Vol. LX, Amsterdam.

Ueberweg, F.: 1923, *Die deutsche Philosophie des XIX Jahrhunderts und der Gegenwart*, 12th edn, revised by T. K. Oesterreich, Berlin.

Vaihinger, H.: 1876, *Hartmann, Dühring und Lange. Zur Geschichte der deutschen Philosophie im XIX Jahrhundert. Ein kritischer Essay*, Iserlohn.

Vaihinger, H.: 1913, *Die Philosophie des Als Ob. System der theoretischen, praktischen und religiösen Fiktionen der Menschheit auf Grund eines idealistischen Positivismus. Mit einem Anhang über Kant und Nietzsche*, 2nd edn, Berlin.

Vliegen, P.: 1973, 'Einheitspsychose', in Chr. Müller 1973, 148–150.

Vogel, C. J. de: 1974, unpublished lecture, Nijmegen.

Vogt, C.: 1855, *Köhlerglaube und Wissenschaft. Eine Streitschrift gegen Hofrath Rudolph Wagner in Göttingen* (1854), 4th edn, Giessen 1855.

Vorländer, K.: 1903, *Geschichte der Philosophie*, Vol II: *Philosophie der Neuzeit*, Leipzig.

Wagner, R.: 1842–1853, *Handwörterbuch der Physiologie mit Rücksicht auf physiologische Pathologie*, 6 vols, Braunschweig.

Wagner, R.: 1857, *Der Kampf um die Seele vom Standpunkt der Wissenschaft*, Göttingen.

Walch, J.: 1733, *Philosophisches Lexicon*, 2nd edn, Leipzig.

Wallace, W. A.: 1972, *Causality and Scientific Explanation, Vol. 1: Medieval and Early Classical Science*, Ann Arbor.

Weber, E.: 1907, *Die philosophische Scholastik des deutschen Protestantismus im Zeitalter der Orthodoxie*, Leipzig.

Wehnelt, B.: 1943, *Die Pflanzenpathologie der deutschen Romantik als Lehre vom kranken Leben und Bilden der Pflanzen, ihre Ideenwelt und ihre Beziehungen zu Medizin, Biologie und Naturphilosophie historisch-romantischer Zeit*, Bonn.

Weinmann, R.: 1895, *Die Lehre von den spezifischen Sinnesenergien*, Hamburg and Leipzig.

Weiss, G.: 1928, *Herbart und seine Schule*, München.

Wenin, Chr. and J–P. Deschepper: 1966, 'Chronique générale', *Revue philosophique de Louvain*, LXIV, 690–733.

Werner, A.: 1909, *Schellings Verhältnis zur Medizin und Biologie*, Thesis, Univ. of Leipzig, Paderborn.

Westphal, E.: 1868–1869, 'Nekrolog' (Wilhelm Griesinger), *Archiv für Psychiatrie und Nervenkrankheiten* 1, 760–782.

Windelband, W. and H. Heimsoeth: 1958, *Lehrbuch der Geschichte der Philosophie*, 15th edn, Tübingen.

Windischmann, C. J. H.: 1975, 'Der Ursprung der Krankheit – Die Ohnmacht des Menschen und die göttliche Hülfe' (from C. J. H. Windischmann, *Ueber Etwas, das der Heilkunst Noth thut. Ein Versuch zur Vereinigung dieser Kunst mit der christlichen Philosophie*, Leipzig 1824, 70–88), in K. E. Rothschuh 1975, 30–44.

Winkler Prins: 1975, 'mechanicisme', *Grote Winkler Prins*, A. J. Wiggers *et al.* (ed.), Vol. 12, 778, 7th edn, Amsterdam and Brussels.

Wittich, D.: 1971, *Vogt, Moleschott, Büchner. Schriften zum kleinbürgerlichen Materialismus in Deutschland*, introduced by Dieter Wittich (ed.), 2 vols, Berlin (DDR).

Wunderlich, C. A.: 1869, 'Wilhelm Griesinger, Nekrolog', *Archiv der Heilkunde* 10, 113–150.

Wundt, M.: 1939, *Die deutsche Schulmetaphysik des 17. Jahrhunderts*, Tübingen.

Wundt, M.: 1964, *Die deutsche Schulphilosophie im Zeitalter der Aufklärung*, Hildesheim (reprint of the Tübingen 1945 edition).

Wundt, W.: 1880, *Grundzüge der physiologischen Psychologie*, Vol. II, 2nd edn.

Wundt, W.: 1891, 'Zur Frage der Localisation der Grosshirnfunctionen', *Philosophische Studien*, VI, 1–25.

Wundt, W.: 1897, *Vorlesungen ueber die Menschen- und Thierseele*, 3rd edn, Hamburg and Leipzig.

Wundt, W.: 1908, *Grundzüge der physiologischen Psychologie*, Vol. I, 6th edn, Leipzig.

Wundt, W.: 1911, *Grundzüge der physiologischen Psychologie*, Vol. III, 6th edn, Leipzig.

Wyrsch, J.: 1956, *Zur Geschichte und Deutung der endogenen Psychosen*, Stuttgart.

Wyrsch, J.: 1957, 'Über Geschichte der Psychiatrie', *Bibl. Psychiatr. Neur.* 100, 21–41.

Wyss, D.: 1972, *Die tiefenpsychologischen Schulen von den Anfängen bis zur Gegenwart. Entwicklung, Probleme, Krisen*, 4th edn, Göttingen.

Zeller, E.: 1963, *Die Philosophie der Griechen in ihrer geschichtlichen Entwicklung*, Vol. II, part 2, 5th edn, Darmstadt (reprint of the 4th edn, Leipzig, 1921).

Zeller, E. A.: 1838, 'Ueber einige Hauptpunkte in der Erforschung und Heilung der Seelenstörungen', *Zeitschrift für die Beurtheilung und Heilung der krankhaften Seelenzustände* 1, 515–569.

Zeller, E. A.: 1840, '2. Bericht über die Wirksamkeit der Heilanstalt Winnenthal', *Medicinisches Correspondenz-Blatt des württembergischen ärztlichen Vereins.*

Zeller, E. A.: 1844, 'Bericht über die Wirksamkeit der Heilanstalt Winnenthal, vom 1. März 1840 bis 18. Febr. 1843', *Allgemeine Zeitschrift für Psychiatrie* 1, 1–79.

Zeller, G.: 1961, *Die Geschichte der Einheitspsychose vor Kraepelin*, unpublished.

Zeller, G.: 1968, 'Welcher psychiatrischen Schule hat Wilhelm Griesinger angehört? Ein Beitrag zum Verständnis seines Lebenswerkes und seiner Biographie', *Deutsches Medizinisches Journal* 19, Heft 9, 328–334.

Zeller, H.: 1921, 'Ernst-Albert Zeller (1804–1877)', in Th. Kirchhoff 1921, 208–218.

mechanical 236 n233
term 234 n205 (Müller)
See also conservation of energy, law
    of; specific energies, law(s) or
    theory of
epistemology, see theory of knowledge
experience 90
    positivist ethos of 231 n164
    its role in Herbart 120, 129, 150; cf.
        254 n501
    the standpoint of
        Griesinger and 63, 111, 114,
            118, 120, 131, 136, 138, 150
        Lotze and 244 n374
    See also common sense standpoint
explanation 181
    causal 190–191
    'mechanical' scientific 115–116
    teleological 57, 248 n435
    vitalistic 88

formalism, logical 43
freedom
    in Du Bois Reymond 68–69
    in Heinroth 9, 15–16, 18–19, 224
        n 75, see also health (psychic),
        anthropological conception of
    in Griesinger 245–246 n402, 247
        n429
    in Jacobi 29, 146
    in Kant, see freedom, practical; —,
        psychological; —, transcen-
        dental
    in Lange 186
    naturalisation of 232 n175
    and nature 46
    practical (Kant) 44
    psychological (Kant) 44–46, 48,
        232 n175
    in Renaissance thought 208–209,
        268 n23
    in Schiller 186
    in Schleiden 262 n638
    in Schwann 57
    in Zeller 145, 148

freethinkers humanitarian ethos 75, see
    also religion, humanistic

Galenism 213–214
gegenständliches Denken, see method of
    objective thinking

health (psychic, mental)
    absolutely different from mental
        illness 27–28, 258 n553
    anthropological conception of 15
Herbartianism 118, 177, 182, 250 n452
history
    of anthropology 211, 217, 226
        n105, 266–267 n1, 269 n33,
        see also history of empirical
        studies/science of man; – of
        philosophy of man
    of biology 213
    of empirical studies/science of man
        201–202, 211–212, 266 n1,
        271 n64
    of medicine xi, 4, 8, 94, 156, 168–
        169, 229 n139, 245 n389, 266
        n1
    of natural science 37
    of philosophy xi, xv, xviii–xix,
        4–7, 37, 40–41, 76, 118, 128,
        167, 176, 201, 221 n27, 234
        n 214, 266 n718, 266 n1, 267
        n12, 270 n49
        of man (i.e., philosophical anthro-
            pology) 202, 204
    of materialism 185, 238 n268,
        264 n669
    of Thomistic thinking 272 n69
    of physiology xi, 53, 59, 70–71,
        88, 94, 233 n187, 266 n1
    of psychiatry xi, xvii, 99, 103, 117,
        122, 144, 226 n104, 248
        n438, 255 n518, 255 n521,
        266 n1
    of psychoanalysis 122–123
    of psychology xi, 4, 41, 50, 117,
        122, 128, 176, 221 n27, 249
        n444, 266 n1, 270 n49

183, 263 n663, 264 n664,
264 n665
psychosomatic theory or thesis
in Nasse 8, 228–229 n139, see also
somaticism, Nasse and
in Zeller 143–144, 255 n521
see also somaticism, 'psychosomatic'character of
psychotherapy 22, 226 n101

rationalism 41, 232 n175
realism 12
of everyday life 172, see also natural
attitude
or materialism, as view of life 172
metaphysical (Herbart) 118–119,
122, 177, 249 n450
naive 108–109, 191
so-called, of Virchow 262 n638
transcendental 233 n195
reason 149
active, see *intellectus agens*
dual, theory of 205
passive, see *intellectus passivus*
pure (theoretical) 47, 195
reductionism 65, 75
anti-phenomenological 149–150;
cf. 252 n492
ontological 83, 183
physicalistic 170, 180, 236 n232
physiological 182, 231 n170
reflex action 104, 106–107, 113, 195,
253 n497
concept of 105
model, in Griesinger 106–107,
253 n498
reflex theory 105, 115, 118, 121, 131
reformation
of medicine 100, 139, 168, 255 n508
see also medicine, the 'new'
of physiology and pathology 91, see
also physiology, the 'new'
religion 13, 29, 141, 187–190, 192,
196, 224 n57, 226 n101, 253
n496, 265 n684
humanistic 74, 238 n268, see also

freethinkers humanitarian
ethos
materialism vs. 75, 78, 185, 187,
226 n101, see also materialism
conflict; materialism vs. Christian view of man
and medicine (psychiatry, psychotherapy) 226 n101
'mixture' of 25, 29, 91, 144,
148, 226 n10
and metaphysics 150
naturalism vs., see naturalism vs.
Christian personalistic view of
man
science and 152, 184, 188–189,
191–192
science vs. 73–74, 187

scepticism 88, 110, 148, 150, 189–
192, 208, 258 n552
schizophrenia 35, 257 n552
scholasticism 207–208, 212
Protestant 215, 271 n62
Schulphilosophie, see philosophy of the
School
science(s)
of life 160–161
of man 159, 221 n26, 271 n63
empirical, see empirical studies of
man
natural, see natural science
vs. religion, see religion, vs. science
scientism 237 n257, see also mechanism
as *Weltanschauung*
second Viennese school of medicine
168
Selbstverständnis, see selfconcept, self-
conception
selfconcept, selfconception
of life sciences 175
of man 114, 150, 211
Christian 205–206, 209, see also
person, man as
of the 'materialists' from about 1840
xii, 76–77, 80–83, 156, 163,
167, 174

therapy   91, 95, 101, 259 n579
Thomism   272 n69
transcendental
   presuppositions of experience   193
   the term   62 (Helmholtz), 266 n720
      (Lange)
   question   199
   See also critical philosophy; idea,
      transcendental; naturalism,
      transcendental

unitarian psychosis, theory of   141–143,
      150, 246 n408, 258 n552
unitarian view of man   141, 150
unitarism
   in general pathology   143
   in psychiatric theory   141
   in psychopathology   143, 150, see
      also unitarian psychosis,
      theory of
Universitätspsychiatrie, see psychiatry,
      university

vital force, see life force
vitalism   57, 67, 236 n237, 239 n280,
      241 n323
   dogmatic   233 n195
   Du Bois-Reymond and   66–67
   Magendie and   88
   mechanism vs.   37
   in Müller   56–57, 169
   in physiology   54, see also physiol-
      ogy, vitalistic
   Rokitansky vs.   169
   Schleiden vs.   78
   Schwann vs.   58, 78
   See also life force; materialism,
      vitalistic

zoology   202, 259 n579